ELECTROMAGNETIC FIELDS IN CAVITIES

IEEE PRESS SERIES ON ELECTROMAGNETIC WAVE THEORY

The IEEE Press Series on Electromagnetic Wave Theory consists of new titles as well as reissues and revisions of recognized classics in electromagnetic waves and applications which maintain long-term archival significance.

BOOKS IN THE IEEE PRESS SERIES ON ELECTROMAGNETIC WAVE THEORY

Chew, W. C., *Waves and Fields in Inhomogeneous Media*
Christopoulos, C., *The Transmission-Line Modeling Methods; TLM*
Clemmow, P. C., *The Plane Wave Spectrum Representation of Electromagnetic Fields*
Collin, R. E., *Field Theory for Guided Waves, Second Edition*
Collin, R. E., *Foundations for Microwave Engineering, Second Edition*
Dudley, D. G., *Mathematical Foundations for Electromagnetic Theory*
Elliott, R. S., *Antenna Theory and Design. Revised Edition*
Elliott, R. S., *Electromagnetics: History, Theory, and Applications*
Felsen, L. B., and Marcuvitz, N., *Radiation and Scattering of Waves*
Harrington, R. F., *Field Computation by Moment Methods*
Harrington, R. F, *Time Harmonic Electromagnetic Fields*
Hansen, T. B., and Yaghjian, A. D., *Plane-Wave Theory of Time-Domain Fields*
Hill, D. A., *Electromagnetic Fields in Cavities: Deterministic and Statistical Theories*
Ishimaru, A., *Wave Propagation and Scattering in Random Media*
Jones, D. S., *Methods in Electromagnetic Wave Propagation, Second Edition*
Josefsson, L., and Persson, P., *Conformal Array Antenna Theory and Design*
Lindell I. V., *Methods for Electromagnetic Field Analysis*
Lindell, I. V., *Differential Forms in Electromagnetics*
Stratton, J. A., *Electromagnetic Theory, A Classic Reissue*
Tai, C. T., *Generalized Vector and Dyadic Analysis, Second Edition*
Van Bladel, J, G., *Electromagnetic Fields, Second Edition*
Van Bladel, J. G., *Singular Electromagnetic Fields and Sources*
Volakis, et al., *Finite Element Method for Electromagnetics*
Zhu, Y., and Cangellaris, A., *Multigrid Finite Element Methods for Electromagnetic Field Modeling*

ELECTROMAGNETIC FIELDS IN CAVITIES

DETERMINISTIC AND STATISTICAL THEORIES

David A. Hill
Electromagnetics Division
National Institute of Standards and Technology

The IEEE Press Series on Electromagnetic Wave Theory
Andreas Cangellaris, *Series Editor*

IEEE

Celebrating 125 Years
of Engineering the Future

WILEY

A JOHN WILEY & SONS, INC., PUBLICATION

IEEE Press
445 Hoes Lane
Piscataway, NJ 08854

Published by John Wiley & Sons, Inc., Hoboken, New Jersey.
Published simultaneously in Canada.

Library of Congress Cataloging-in-Publication Data is available

ISBN: 978-0-470-46590-5

10 9 8 7 6 5 4 3 2 1

To Elaine

CONTENTS

PREFACE

The subject of electromagnetic fields (or acoustics) in cavities has a long history and a well-developed literature. So my first obligation is to justify devoting an entire book to the subject of electromagnetic fields in cavities. I have two primary motivations. First, the classical deterministic cavity theories are scattered throughout many book chapters and journal articles. In Part I (Deterministic Theory) of this book, I have attempted to consolidate much of this material into one place for the convenience of the reader. Second, in recent years it has become clear that statistical methods are required to predict and interpret the behavior of electromagnetic fields in large, complex cavities. Since these methods are in a rapidly developing stage, I have devoted Part II (Statistical Theories for Electrically Large Cavities) to a detailed description of current statistical theories and applications. My interest in statistical fields in cavities began while analysizing reverberation (or mode-stirred) chambers, which are intentionally designed to generate statistical fields for electromagnetic compatibility (EMC) testing.

Consider now the deterministic material covered in Part I. Chapter 1 includes Maxwell's equations and their use in deriving the resonant empty-cavity modes for cavities of general shape. The asymptotic result (for electrically large cavities) for the mode density (the number of resonant modes divided by a small frequency bandwidth) turns out to be a robust quantity because it depends only on the cavity volume and the frequency. Hence, this later turns out to be useful in Part II. Chapter 1 also covers cavity Q (as determined by wall losses), cavity excitation (the source problem), and perturbation theories (for small inclusions or small wall deformation). These topics are important for the design of high-Q microwave resonators and for measurement of material properties.

Chapters 2 through 4 cover the three cavity shapes (rectangular, circular cylindrical, and spherical) where the vector wave equation is separable and the resonant-mode fields and resonant frequencies can be determined by separation of variables. For each cavity shape, the cavity Q as determined by wall losses is analyzed. For practical cavity applications, cavities need to be excited, and the most compact description of cavity excitation is given via the dyadic Green's function. The specific form of the dyadic Green's function, as derived by C. T. Tai (the master of dyadic Green's functions) is given for the three separable cavity shapes. The dyadic Green's functions for perfectly conducting walls have infinities at resonant frequencies, but the inclusion of wall losses (finite Q) eliminates these infinities.

The statistical material in Part II is really the novel part of this book. Chapter 5 describes the motivation for statistical approaches: lack of detailed cavity information (including boundaries and loading); sensitivity of fields to cavity geometry and excitation; and interpretation of theoretical or measured results. The general point is that the field at a single frequency and a single point in a large, complex cavity can vary drastically because of standing waves. However, some of the field statistics are found to be quite well behaved and fairly insensitive to cavity parameters. Chapter 6 includes probability concepts that are well known in textbooks, but are included here in an effort to make the book self-contained and to define the notation to be used in later chapters.

Chapter 7 presents an extensive treatment of the statistical theory of reverberation chambers. A plane-wave integral representation of the fields is found to be convenient because each plane wave satisfies source-free Maxwell's equations, and the statistical properties are incorporated in the plane-wave coefficients. This theory is used to derive the statistical properties of the electric and magnetic fields, including the probability density functions of the scalar components and the squared magnitudes. The theoretical results in this chapter and following chapters are compared with experimental statistical results that have been obtained using mechanical stirring (paddle wheel) in the National Institute of Standards and Technology (NIST) reverberation chamber. The plane-wave integral representation is shown to be useful in deriving spatial correlation functions of fields and energy density, antenna or test-object responses, and a composite chamber Q that is the result of four types of power loss (wall loss, absorptive loading, aperture leakage, and antenna loading). Since reverberation chambers are reciprocal devices, their use in EMC emissions (total radiated power) measurements is also analyzed and demonstrated with a test object. Although the initial plane-wave integral representation was developed for regions well separated from sources, test objects, and walls, multiple-image theory has been used to derive boundary fields that satisfy the required wall boundary conditions and evolve uniformly to the expected results at large distances from walls.

Chapter 8 uses the fundamentals of Chapter 7 to treat aperture excitation of electrically large cavities, an important problem in EMC applications. Power balance is used to derive a statistical solution for the field strength within the cavity, and experimental results are used to check the theoretical results.

Chapter 9 examines cases that deviate from the statistically spatial uniformity environment of Chapter 7. In place of mechanical stirring, frequency stirring (expanding the bandwidth from the usual continuous-wave (cw) case) is analyzed for its ability to generate a spatially uniform field. The effect of direct-path coupling from the transmitting antenna (unstirred energy) is analyzed and measured, and the usual Rayleigh probability density function (PDF) is replaced by the Rice PDF.

Chapter 10 covers several applications of reverberation chambers to practical issues. Nested reverberation chambers connected by an aperture with a shielding material are used to evaluate the shielding effectiveness of thin materials. The shielding effectiveness (SE) of shielded enclosures is evaluated by several methods for both large and small enclosures. The measurement of chamber Q is used to infer the efficiency of a test antenna or the absorption cross section of a lossy material.

Chapter 11 represents a departure from the rest of Part II and discusses various models for indoor wireless propagation. This subject is important to the very large wireless communication industry when either the receiver or transmitter (or both) is located inside a building. With the exception of some metal-wall factories, buildings and rooms have fairly low Q values and are typically treated with empirical propagation models. Some of the models for path loss, temporal characteristics (including RMS delay spread), and angle of arrival are discussed, along with measured data. The possibility of simulating an indoor wireless communication system by loading a reverberation chamber or by varying the ratio of stirred to unstirred energy is also investigated.

This book has ten appendices. Appendices A, B, and C cover standard material on vector analysis and special functions and are included primarily to keep the book self-contained. Appendix D on the role of chaos in cavity fields is included because a large literature is developing on this subject, and some inconsistencies have appeared. A brief discussion of ray chaos and wave chaos is included in an effort to clarify the subject. Appendices E and F are included because they treat the response of two simple antennas (short electric dipole and small loop) where we can readily show that their responses reduce to the general result for an antenna in a reverberation chamber. Appendix G uses ray theory to illustrate that mode stirrers must be both electrically large and large compared to chamber dimensions to stir the fields effectively. Appendix H treats the canonical spherical absorber as a good test case for theoretical and measured absorption in a reverberation chamber. Appendix I utilizes Bethe hole theory to derive the transmission cross section of a small circular aperture (another canonical geometry) averaged over incidence angle and polarization for reverberation chamber application. Appendix J on scaling is included because many laboratory scale models must be scaled in size and frequency to compare with real-world objects (such as aircraft cavities), and material scaling presents some difficulties.

Some of the material in this book is new, but much of it is a restatement of results already available in the literature. Because of the large literature on fields in cavities and the rapid development of statistical methods, is it unavoidable that some important references have been omitted. For such omissions, I offer my apologies to the authors.

This book is intended for use by researchers, practicing engineers, and graduate students. In particular, the material is applicable to microwave resonators (Part I), electromagnetic compatibility (Part II), and indoor wireless communications (Chapter 11), but the theory is sufficiently general to cover other applications. Most of the material in this book could be covered in a one-semester graduate course. Problems are included at the ends of the chapters for use by students or readers who would like to dig deeper into selected topics.

I express my sincere appreciation to everyone who in any way contributed to the creation of this book. I thank my colleagues in NIST and researchers outside NIST for many illuminating discussions. Also, I thank Drs. Perry Wilson, Robert Johnk, Claude

Weil, and David Smith for reviewing the manuscript. Most of all, my NIST colleagues who performed many hours of measurements and data processing, particularly Galen Koepke and John Ladbury, are to be thanked for providing experimental results for comparisons with theory and for injecting a dose of reality to the complex subject of statistical fields in cavities.

DAVID A. HILL

PART I

DETERMINISTIC THEORY

CHAPTER 1

Introduction

The cavities discussed in Part I consist of a region of finite extent bounded by conducting walls and filled with a uniform dielectric (usually free space). After a brief discussion of fundamentals of electromagnetic theory, the general properties of cavity modes and their excitation will be given in this chapter. The remaining three chapters of Part I give detailed expressions for the modal resonant frequencies and field structures, quality (Q) factor [1], and Dyadic Green's Functions [2] for commonly used cavities of separable geometries (rectangular cavity in Chapter 2, circular cylindrical cavity in Chapter 3, and spherical cavity in Chapter 4). The International System of Units (SI) is used throughout.

1.1 MAXWELL'S EQUATIONS

Since this book deals almost exclusively with time-harmonic fields, the field and source quantities have a time variation of $\exp(-i\omega t)$, where the angular frequency ω is given by $\omega = 2\pi f$. The time dependence is suppressed throughout. The differential forms of Maxwell's equations are most useful in modal analysis of cavity fields. If we follow Tai [2], the three independent Maxwell equations are:

$$\nabla \times \vec{E} = i\omega\vec{B}, \tag{1.1}$$

$$\nabla \times \vec{H} = \vec{J} - i\omega\vec{D}, \tag{1.2}$$

$$\nabla \bullet \vec{J} = i\omega\rho, \tag{1.3}$$

where $\vec{E}$ is the electric field strength (volts/meter), $\vec{B}$ is the magnetic flux density (teslas), $\vec{H}$ is the magnetic field strength (amperes/meter), $\vec{D}$ is the electric flux density (coulombs/meter^2), $\vec{J}$ is the electric current density (amperes/meter^2), and ρ is the electric charge density (coulombs/meter^3). Equation (1.1) is the differential form of Faraday's law, (1.2) is the differential form of the Ampere-Maxwell law, and (1.3) is the equation of continuity.

Electromagnetic Fields in Cavities: Deterministic and Statistical Theories, by David A. Hill

Two dependent Maxwell equations can be obtained from (1.1)–(1.3). Taking the divergence of (1.1) yields:

$$\nabla \bullet \vec{B} = 0 \tag{1.4}$$

Taking the divergence of (1.2) and substituting (1.3) into that result yields

$$\nabla \bullet \vec{D} = \rho \tag{1.5}$$

Equation (1.4) is the differential form of Gauss's magnetic law, and (1.5) is the differential form of Gauss's electric law. An alternative point of view is to consider (1.1), (1.2), and (1.5) as independent and (1.3) and (1.4) as dependent, but this does not change any of the equations. Sometimes a magnetic current is added to the right side of (1.1) and a magnetic charge is added to the right side of (1.4) in order to introduce duality [3] into Maxwell's equations. However, we choose not to do so.

The integral or time dependent forms of (1.1)–(1.5) can be found in numerous textbooks, such as [4]. The vector phasors, for example $\vec{E}$, in (1.1)–(1.5) are complex quantities that are functions of position $\vec{r}$ and angular frequency ω, but this dependence will be omitted except where required for clarity. The time and space dependence of the real field quantities, for example electric field $\vec{\mathcal{E}}$, can be obtained from the vector phasor quantity by the following operation:

$$\vec{\mathcal{E}}(\vec{r}, t) = \sqrt{2}\,\mathrm{Re}[\vec{E}(\vec{r}, \omega)\exp(-i\omega t)], \tag{1.6}$$

where Re represents the real part. The introduction of the $\sqrt{2}$ factor in (1.6) follows Harrington's notation [3] and eliminates a 1/2 factor in quadratic quantities, such as power density and energy density. It also means that the vector phasor quantities represent root-mean-square (RMS) values rather than peak values.

In order to solve Maxwell's equations, we need more information in the form of the constitutive relations. For isotropic media, the constitutive relations are written:

$$\vec{D} = \varepsilon \vec{E}, \tag{1.7}$$

$$\vec{B} = \mu \vec{H}, \tag{1.8}$$

$$\vec{J} = \sigma \vec{E}, \tag{1.9}$$

where ε is the permittivity (farads per meter), μ is the permeability (henrys/meter), and σ is the conductivity (siemens/meter). In general, ε, μ, and σ are frequency dependent and complex. Actually, there are more general constitutive relations [5] than those shown in (1.7)–(1.9), but we will not require them.

In many problems, $\vec{J}$ is treated as a source current density rather than an induced current density, and the problem is to determine $\vec{E}$ and $\vec{H}$ subject to specified boundary conditions. In this case (1.1) and (1.2) can be written:

$$\nabla \times \vec{E} = i\omega\mu\vec{H}, \tag{1.10}$$

$$\nabla \times \vec{H} = \vec{J} - i\omega\varepsilon\vec{E} \tag{1.11}$$

Equations (1.10) and (1.11) are two vector equations in two vector unknowns ($\vec{E}$ and $\vec{H}$) or equivalently six scalar equations in six scalar unknowns. By eliminating either $\vec{H}$ in (1.10) or $\vec{E}$ in (1.11), we can obtain inhomogeneous vector wave equations:

$$\nabla \times \nabla \times \vec{E} - k^2\vec{E} = i\omega\mu\vec{J}, \tag{1.12}$$

$$\nabla \times \nabla \times \vec{H} - k^2\vec{H} = \nabla \times \vec{J}, \tag{1.13}$$

where $k = \omega\sqrt{\mu\varepsilon}$. Chapters 2 through 4 will contain sections where dyadic Green's functions provide compact solutions to (1.12) and (1.13) and satisfy the boundary conditions at the cavity walls.

1.2 EMPTY CAVITY MODES

Consider a simply connected cavity of arbitrary shape with perfectly conducting electric walls as shown in Figure 1.1. The interior of the cavity is filled with a homogeneous dielectric of permittivity ε and permeability μ. The cavity has volume V and surface area S. Because the walls have perfect electric conductivity, the tangential electric field at the wall surface is zero:

$$\hat{n} \times \vec{E} = 0, \tag{1.14}$$

where $\hat{n}$ is the unit normal directed outward from the cavity. Because the cavity is source free and the permittivity is independent of position, the divergence of the electric field is zero:

$$\nabla \bullet \vec{E} = 0 \tag{1.15}$$

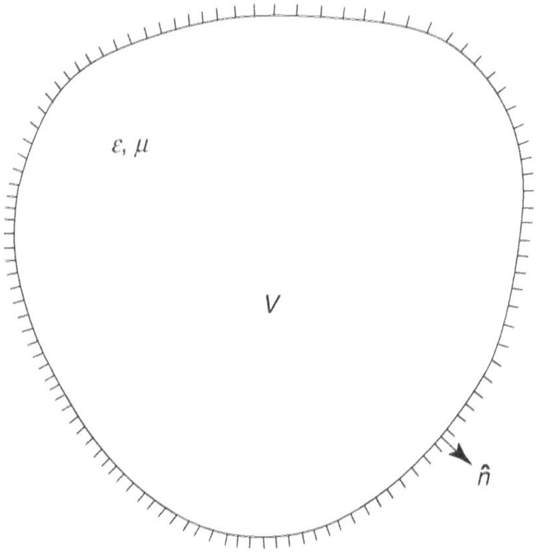

FIGURE 1.1 Empty cavity of volume V with perfectly conducting walls.

If we set the current $\vec{J}$ equal to zero in (1.12), we obtain the homogeneous vector wave equation:

$$\nabla \times \nabla \times \vec{E} - k^2 \vec{E} = 0 \tag{1.16}$$

We can work directly with (1.16) in determining the cavity modes, but it is simpler and more common [6, 7] to replace the double curl operation by use of the following vector identity (see Appendix A):

$$\nabla \times \nabla \times \vec{E} = \nabla(\nabla \bullet \vec{E}) - \nabla^2 \vec{E} \tag{1.17}$$

Since the divergence of $\vec{E}$ is zero, (1.17) can be used to reduce (1.16) to the vector Helmholtz equation:

$$(\nabla^2 + k^2)\vec{E} = 0. \tag{1.18}$$

The simplest form of the Laplacian operator ∇^2occurs in rectangular coordinates, where $\nabla^2 \vec{E}$ reduces to:

$$\nabla^2 \vec{E} = \hat{x}\nabla^2 E_x + \hat{y}\nabla^2 E_y + \hat{z}\nabla^2 E_z, \tag{1.19}$$

where $\hat{x}$, $\hat{y}$, and $\hat{z}$ are unit vectors.

We assume that the permittivity ε and the permeability μ of the cavity are real. Then nontrivial (nonzero) solutions of (1.14), (1.15), and (1.18) occur when k is equal to one of an infinite number of discrete, real eigenvalues k_p (where $p = 1, 2, 3, \ldots$). For each eigenvalue k_p, there exists an electric field eigenvector $\vec{E}_p$. (There can be degenerate cases where two or more eigenvectors have the same eigenvalue.) The pth eigenvector satisfies:

$$(-\nabla \times \nabla \times + k_p^2)\vec{E}_p = (\nabla^2 + k_p^2)\vec{E}_p = 0 \quad (\text{in } V), \tag{1.20}$$

$$\nabla \bullet \vec{E}_p = 0 \quad (\text{in } V), \tag{1.21}$$

$$\hat{n} \times \vec{E}_p = 0 \quad (\text{on } S). \tag{1.22}$$

For convenience (and without loss of generality), each electric field eigenvector can be chosen to be real ($\vec{E}_p = \vec{E}_p^*$, where * indicates complex conjugate).

The corresponding magnetic field eigenvector $\vec{H}_p$ can be determined from (1.1) and (1.8):

$$\vec{H}_p = \frac{1}{i\omega_p \mu} \nabla \times \vec{E}_p, \tag{1.23}$$

where the angular frequency ω_p is given by:

$$\omega_p = \frac{k_p}{\sqrt{\mu\varepsilon}} \tag{1.24}$$

Hence, the pth normal mode of the resonant cavity has electric and magnetic fields, $\vec{E}_p$ and $\vec{H}_p$, and a resonant frequency f_p $(= \omega_p/2\pi)$. The magnetic field is then pure imaginary $(\vec{H}_p = -\vec{H}_p^*)$ and has the same phase throughout the cavity (as does $\vec{E}_p$).

For the pth mode, the time-averaged values of the electric stored energy $\overline{W}_{ep}$ and the magnetic stored energy $\overline{W}_{mp}$ are given by the following integrals over the cavity volume [3]:

$$\overline{W}_{ep} = \frac{\varepsilon}{2} \iiint_V \vec{E}_p \bullet \vec{E}_p^* \mathrm{d}V, \tag{1.25}$$

$$\overline{W}_{mp} = \frac{\mu}{2} \iiint_V \vec{H}_p \bullet \vec{H}_p^* \mathrm{d}V \tag{1.26}$$

(The complex conjugate in (1.25) is not actually necessary when $\vec{E}_p$ is real, but it increases the generality to cases where $\vec{E}_p$ is not chosen to be real.) In general, the complex Poynting vector $\vec{S}$ is given by [3]:

$$\vec{S} = \vec{E} \times \vec{H}^* \tag{1.27}$$

If we apply Poynting's theorem to the pth mode, we obtain [6]:

$$\oiint_S (\vec{E}_p \times \vec{H}_p^*) \bullet \hat{n} \mathrm{d}S = 2i\omega_p(\overline{W}_{ep} - \overline{W}_{mp}) \tag{1.28}$$

Since $\hat{n} \times \vec{E}_p = 0$ on S, the left side of (1.28) equals zero, and for each mode we have:

$$\overline{W}_{ep} = \overline{W}_{mp} = \overline{W}_p/2 \tag{1.29}$$

Thus, the time-averaged electric and magnetic stored energies are equal to each other and are equal to one half the total time-averaged stored energy $\overline{W}_p$ at resonance. However, since (1.23) shows that the electric and magnetic fields are 90 degrees out of phase, the total energy in the cavity oscillates between electric and magnetic energy.

Up to now we have discussed only the properties of the fields and the energy of an individual cavity mode. It is also important to know what the distribution of the resonant frequencies is. In general, this depends on cavity shape, but the problem has been examined from an asymptotic point of view for electrically large cavities. Weyl [8] has studied this problem for general cavities, and Liu et al. [9] have studied the problem in great detail for rectangular cavities. For a given value of wavenumber k, the asymptotic expression (for large $kV^{1/3}$) for the number of modes N_s with eigenvalues less than or equal to k is [8, 9]:

$$N_s(k) \cong \frac{k^3 V}{3\pi^2} \tag{1.30}$$

The subscript s on N indicates that (1.30) is a smoothed approximation, whereas N determined by mode counting has step discontinuities at each mode. It is usually more useful to know the number of modes as a function of frequency. In that case, (1.30) can be written:

$$N_s(f) \cong \frac{8\pi f^3 V}{3c^3} \tag{1.31}$$

where $c\ (= 1/\sqrt{\mu\varepsilon})$ is the speed of light in the medium (usually free space). The f^3 dependence in (1.31) indicates that the number of modes increases rapidly at high frequencies.

The mode density D_s is also an important quantity because it is an indicator of the separation between the modes. By differentiating (1.30), we obtain:

$$D_s(k) = \frac{\mathrm{d}N_s(k)}{\mathrm{d}k} \cong \frac{k^2 V}{\pi^2} \tag{1.32}$$

The mode density as a function of frequency is obtained by differentiating (1.31):

$$D_s(f) = \frac{\mathrm{d}N_s(f)}{\mathrm{d}f} \cong \frac{8\pi f^2 V}{c^3} \tag{1.33}$$

The f^2 dependence in (1.33) indicates that the mode density also increases rapidly for high frequencies. The approximate frequency separation (in Hertz) between modes is given by the reciprocal of (1.33).

1.3 WALL LOSSES

For cavities with real metal walls, the wall conductivity σ_w is large, but finite. In this case, the eigenvalues and resonant frequencies become complex. An exact calculation of the cavity eigenvalues and eigenvectors is very difficult, but an adequate approximate treatment is possible for highly conducting walls. This allows us to obtain an approximate expression for the cavity quality factor Q_p [1].

The exact expression for the time-average power $\bar{P}_p$ dissipated in the walls can be obtained by integrating the normal component of the real part of the Poynting vector (defined in 1.27) over the cavity walls:

$$\bar{P}_p = \oiint_S \mathrm{Re}(\vec{E}_p \times \vec{H}_p^*) \bullet \hat{n} \mathrm{d}S \tag{1.34}$$

For simplicity and to compare with earlier work [6], we assume that the cavity medium and the cavity walls have free-space permeability μ_0, as shown in Figure 1.2. Using a vector identity, we can rewrite (1.34) as:

$$\bar{P}_p = \oiint_S \mathrm{Re}[(\hat{n} \times \vec{E}_p) \bullet \vec{H}_p^*] \mathrm{d}S \tag{1.35}$$

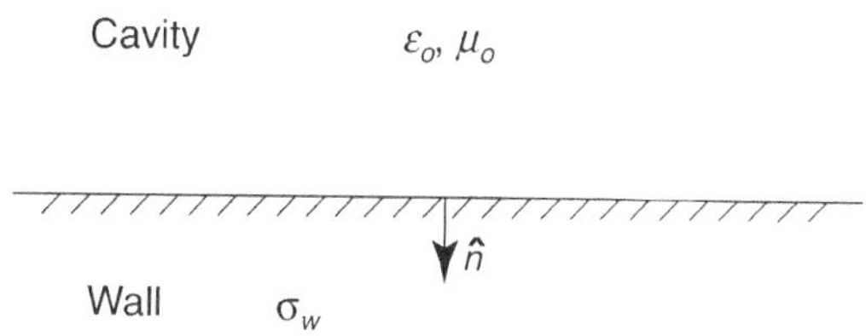

FIGURE 1.2 Cavity wall with conductivity σ_W.

In (1.35), we can approximate $\vec{H}_p$ by its value for the case of the lossless cavity. For $\hat{n} \times \vec{E}_p$, we can use the surface impedance boundary condition [10]:

$$\hat{n} \times \vec{E}_p \cong \eta \vec{H}_p \quad \text{on } S \tag{1.36}$$

where:

$$\eta \cong \sqrt{\frac{\omega_p \mu_0}{i\sigma_w}} \tag{1.37}$$

By substituting (1.36) and (1.37) into (1.35), we obtain:

$$\bar{P}_p \cong R_s \oint\!\!\oint_S \vec{H}_p \bullet \vec{H}_p^* \mathrm{d}S \tag{1.38}$$

where the surface resistance R_s is the real part of η:

$$R_s \cong \mathrm{Re}(\eta) \cong \sqrt{\frac{\omega_p \mu_0}{2\sigma_w}} \tag{1.39}$$

The quality factor Q_p for the pth mode is given by [1, 6]:

$$Q_p = \omega_p \frac{\overline{W}_p}{\bar{P}_p} \tag{1.40}$$

where $\overline{W}_p$ $(= 2\overline{W}_{mp} = 2\overline{W}_{ep})$ is the time-averaged total stored energy. Substituting (1.26) and (1.38) into (1.40), we obtain:

$$Q_p \cong \omega_p \frac{\mu_0 \iiint_V \vec{H}_p \bullet \vec{H}_p^* \mathrm{d}V}{R_s \oint\!\!\oint_S \vec{H}_p \bullet \vec{H}_p^* \mathrm{d}S} \tag{1.41}$$

where $\vec{H}_p$ is the magnetic field of the pth cavity mode without losses. An alternative to (1.41) can be obtained by introducing the skin depth δ [3]:

$$Q_p \cong \frac{2 \iiint_V \vec{H}_p \bullet \vec{H}_p^* \mathrm{d}V}{\delta \oint\!\!\oint_S \vec{H}_p \bullet \vec{H}_p^* \mathrm{d}S} \tag{1.42}$$

where $\delta = \sqrt{2/(\omega_p \mu_0 \sigma_w)}$. In order to accurately evaluate (1.41) or (1.42), we need to know the magnetic field distribution of the pth mode, and in general this depends on the cavity shape and resonant frequency ω_p. This will be pursued in the next three chapters.

A rough approximation for (1.42) has been obtained by Borgnis and Papas [6]:

$$Q_p \cong \frac{2 \iiint_V \mathrm{d}V}{\delta \oiint_S \mathrm{d}S} = \frac{2V}{\delta S} \tag{1.43}$$

For highly conducting metals, such as copper, δ is very small compared to the cavity dimensions. Hence, the quality factor Q_p is very large. This is why metal cavities make very effective resonators. Even though (1.43) is a very crude approximation to (1.42)—it essentially assumes that $\vec{H}_p$ is independent of position—it is actually close to another approximation that has been obtained by two unrelated methods. Either by taking a modal average about the resonant frequency for rectangular cavities [9] or by using a plane-wave integral representation for stochastic fields in a multimode cavity of arbitrary shape (see either Section 8.1 or [11]), the following expression for Q has been obtained:

$$Q \cong \frac{3V}{2\delta S} \tag{1.44}$$

Hence, (1.43) exceeds (1.44) by a factor of only $\frac{4}{3}$. It is actually possible to improve the approximation in (1.43) and bring it into agreement with (1.44) by imposing the boundary conditions for $\vec{H}_p$ on S. If we take the z axis normal to S at a given point, then the normal component H_{pz} is zero on S. However, the x component is at a maximum because it is a tangential component:

$$H_{px} = H_{pm} \quad \text{on } S \tag{1.45}$$

We can make a similar argument for H_{py}. Hence, we can approximate the surface integral in (1.42) as:

$$\oiint_S \vec{H}_p \bullet \vec{H}_p^* \mathrm{d}S \cong 2|H_{pm}|^2 S \tag{1.46}$$

For the volume integral, we can assume that all three components of $\vec{H}_p$ contribute equally if the cavity is electrically large. However, since each rectangular component is a standing wave with approximately a sine or cosine spatial dependence, then a factor of $\frac{1}{2}$ occurs from integrating a sine-squared or cosine-squared dependence over an integer number of half cycles in V. Hence, the volume integral in (1.42) can be written:

$$\iiint_V \vec{H}_p \bullet \vec{H}_p^* \mathrm{d}V \cong \frac{3}{2}|H_{pm}|^2 V \tag{1.47}$$

If we substitute (1.46) and (1.47) into (1.42), then we obtain:

$$Q_p \cong \frac{2}{\delta}\frac{(3/2)|H_{pm}|^2 V}{2|H_{pm}|^2 S} = \frac{3V}{2\delta S} \tag{1.48}$$

which is in agreement with (1.44). Hence, the single-mode approximation, the modal average for rectangular cavities [9], and the plane-wave integral representation for stochastic fields in a multimode cavity [11] all yield the same approximate value for Q.

When cavities have no loss, the fields of a resonant mode oscillate forever in time with no attenuation. However, with wall loss present, the fields and stored energy decay with time after any excitation ceases. For example, the incremental change in the time-averaged total stored energy in a time increment $\mathrm{d}t$ can be written:

$$\mathrm{d}\overline{W}_p = -\overline{P}_p \mathrm{d}t \tag{1.49}$$

By substituting (1.40) into (1.49), we can derive the following first-order differential equation:

$$\frac{\mathrm{d}\overline{W}_p}{\mathrm{d}t} = -\frac{\omega_p}{Q_p}\overline{W}_p \tag{1.50}$$

For the initial condition, $\overline{W}_p|_{t=0} = \overline{W}_{p0}$, the solution to (1.50) is:

$$\overline{W}_p = \overline{W}_{p0}\exp(-t/\tau_p), \quad \text{for } t \geq 0 \tag{1.51}$$

where $\tau_p = Q_p/\omega_p$. Hence, the energy decay time τ_p of the pth mode is the time required for the time-average energy to decay to $1/e$ of its initial value. Equations (1.49)–(1.51) assume that the decay time τ_p is large compared to the averaging period $1/f_p$. This is assured if Q_p is large.

By a similar analysis when the energy is switched off at $t = 0$, we find that the fields of the pth mode, $\vec{E}_p$ and $\vec{H}_p$, also have an exponential decay, but that the decay time is $2\tau_p$. This is equivalent to replacing the resonant frequency ω_p for a lossless cavity by the complex frequency $\omega_p\left(1-\frac{i}{2Q_p}\right)$ corresponding to a lossy cavity [6]. We can use this result to determine the bandwidth of the pth mode [6]. If E_{pm} is any scalar component of the electric field of the pth mode, then its time dependence $\widetilde{E}_{pm}(t)$ when the mode is suddenly excited at $t = 0$ can be written:

$$\widetilde{E}_{pm}(t) = E_{pm0}\exp\left(-i\omega_p t - \frac{\omega_p t}{2Q_p}\right)\mathrm{U}(t), \tag{1.52}$$

where U is the unit step function and E_{pm0} is independent of t. The Fourier transform of (1.52) is:

$$E_{pm}(\omega) = \frac{E_{pm0}}{2\pi}\int_0^{\infty}\exp\left[-i\omega_p t - \frac{\omega_p t}{2Q_p} + i\omega t\right]\mathrm{d}t, \tag{1.53}$$

which can be evaluated to yield:

$$E_{pm0}(\omega) = \frac{E_{pm0}}{2\pi} \frac{1}{i(\omega_p - \omega) + \dfrac{\omega_p}{2Q_p}} \tag{1.54}$$

The absolute value of (1.54) is:

$$|E_{pm}(\omega)| = \frac{|E_{pm0}|Q_p}{\pi\omega_p} \frac{1}{\sqrt{1 + \left[\dfrac{2Q_p(\omega - \omega_p)}{\omega_p}\right]^2}} \tag{1.55}$$

The maximum of (1.55) occurs at $\omega = \omega_p$:

$$|E_{pm}(\omega_p)| = \frac{|E_{pm0}|Q_p}{\pi\omega_p} \tag{1.56}$$

This maximum value is seen to be proportional to Q_p. The frequencies at which (1.55) drops to $\frac{1}{\sqrt{2}}$ times its maximum value are called the half-power frequencies, and their separation $\Delta\omega$ (or Δf in Hertz) is related to Q_p by:

$$\frac{\Delta\omega}{\omega_p} = \frac{\Delta f}{f_p} = \frac{1}{Q_p} \tag{1.57}$$

Hence Q_p is a very important property of a cavity mode because it controls both the maximum field amplitude and the mode bandwidth.

1.4 CAVITY EXCITATION

Cavities are typically excited by short monopoles, small loops, or apertures. Complete theories for the excitation of modes in a cavity have been given by Kurokawa [12] and Collin [13]. According to Helmholtz's theorem, the electric field in the interior of a volume V bounded by a closed surface S can be written as the sum of a gradient and a curl as follows [13]:

$$\begin{aligned} \vec{E}(\vec{r}) = &-\nabla\left[\iiint\limits_V \frac{\nabla_0 \bullet \vec{E}(\vec{r}_0)}{4\pi R} \mathrm{d}V_0 - \oiint\limits_S \frac{\hat{n} \bullet \vec{E}(\vec{r}_0)}{4\pi R} \mathrm{d}S_0\right] \\ &+ \nabla \times \left[\iiint\limits_V \frac{\nabla_0 \times \vec{E}(\vec{r}_0)}{4\pi R} \mathrm{d}V_0 - \oiint\limits_S \frac{\hat{n} \times \vec{E}(\vec{r}_0)}{4\pi R} \mathrm{d}S_0\right], \end{aligned} \tag{1.58}$$

where $R = |\vec{r} - \vec{r}_0|$ and $\hat{n}$ is the outward unit normal to the surface S. Equation (1.58) gives the conditions for which the electric field $\vec{E}(\vec{r})$ can be either a purely solenoidal or a purely irrotational field. A purely solenoidal (zero divergence) field must satisfy the conditions $\nabla \bullet \vec{E} = 0$ in V and $\hat{n} \bullet \vec{E} = 0$ on S. In this case, there is no volume or surface charge associated with the field. In the following chapters, we will see that some modes are purely solenoidal in the volume V, but are not purely solenoidal because the mode has surface charge ($\hat{n} \bullet \vec{E} \neq 0$ on S). A purely irrotational or lamellar field (zero curl) must satisfy the conditions $\nabla \times E = 0$ in V and $\hat{n} \times E = 0$ on S. For a cavity with perfectly conducting walls, $\hat{n} \times E = 0$ on S. However, for a time varying field, $\nabla \times E \neq 0$ in V. Hence, in general the electric field is not purely solenoidal or irrotational.

For the modal expansion of the electric field, we follow Collin [13]. The solenoidal modes $\vec{E}_p$ satisfy (1.20)–(1.22). The irrotational modes $\vec{F}_p$ are solutions of:

$$(\nabla^2 + l_p^2)\vec{F}_p = 0 \quad (\text{in } V), \tag{1.59}$$

$$\nabla \times \vec{F}_p = 0 \quad (\text{in } V), \tag{1.60}$$

$$\hat{n} \times \vec{F}_p = 0 \quad (\text{on } S) \tag{1.61}$$

These irrotational modes are generated from scalar functions Φ_p that are solutions of:

$$(\nabla^2 + l_p^2)\Phi_p = 0 \quad (\text{in } V), \tag{1.62}$$

$$\Phi_p = 0 \quad (\text{on } S), \tag{1.63}$$

$$l_p\vec{F}_p = \nabla\Phi_p \tag{1.64}$$

The factor l_p in (1.64) yields the desired normalization for $\vec{F}_p$ when Φ_p is normalized.

The $\vec{E}_p$ modes are normalized so that:

$$\iiint_V \vec{E}_p \bullet \vec{E}_p \mathrm{d}V = 1 \tag{1.65}$$

(The normalization in (1.65) can be made consistent with the energy relationship in (1.25) if we set $\overline{W} = \varepsilon$.) The scalar functions Φ_p are similarly normalized:

$$\iiint_V \Phi_p^2 \mathrm{d}V = 1 \tag{1.66}$$

From (1.64), the normalization for the $\vec{F}_p$ modes can be written:

$$\iiint_V \vec{F}_p \bullet \vec{F}_p \mathrm{d}V = \iiint_V l_p^{-2} \nabla\Phi_p \bullet \nabla\Phi_p \mathrm{d}V \tag{1.67}$$

To evaluate the right side of (1.67), we use the vector identity for the divergence of a scalar times a vector:

$$\nabla \bullet (\Phi_p \nabla\Phi_p) = \Phi_p \nabla^2\Phi_p + \nabla\Phi_p \bullet \nabla\Phi_p \tag{1.68}$$

From (1.62), (1.63), (1.68), and the divergence theorem, we can evaluate the right side of (1.67):

$$\iiint_V l_p^{-2}\nabla\Phi_p \bullet \nabla\Phi_p \mathrm{d}V = \iiint_V \Phi_p^2 \mathrm{d}V + l_p^{-2} \oint\!\!\!\oint_S \Phi_p \frac{\partial \Phi_p}{\partial n} \mathrm{d}S = 1, \tag{1.69}$$

since the second integral on the right side is zero. Thus the $\vec{F}_p$ modes are also normalized:

$$\iiint_V \vec{F}_p \bullet \vec{F}_p \mathrm{d}V = 1 \tag{1.70}$$

We now turn to mode orthogonality. To show that the $\vec{E}_p$ and $\vec{F}_p$ modes are orthogonal, we begin with the following vector identity:

$$\nabla \bullet (\vec{F}_q \times \nabla \times \vec{E}_p) = \nabla \times \vec{F}_q \bullet \nabla \times \vec{E}_p - \vec{F}_q \bullet \nabla \times \nabla \times \vec{E}_p \tag{1.71}$$

Substituting (1.20) and (1.60) into the right side of (1.71), we obtain:

$$\nabla \bullet (\vec{F}_q \times \nabla \times \vec{E}_p) = -k_p^2 \vec{F}_q \bullet \vec{E}_p \tag{1.72}$$

Using the divergence theorem and the vector identity, $\vec{A} \bullet \vec{B} \times \vec{C} = \vec{C} \bullet \vec{A} \times \vec{B}$, in (1.72), we can obtain:

$$k_p^2 \iiint_V \vec{F}_q \bullet \vec{E}_p \mathrm{d}V = -\oint\!\!\!\oint_S \hat{n} \times \vec{F}_q \bullet \nabla \times \vec{E}_p \mathrm{d}S \tag{1.73}$$

Substituting (1.61) into (1.73), we obtain the desired orthogonality result:

$$k_p^2 \iiint_V \vec{F}_q \bullet \vec{E}_p \mathrm{d}V = 0 \tag{1.74}$$

The modes $\vec{E}_p$ are also mutually orthogonal. By dotting $\vec{E}_q$ into (1.20), reversing the subscripts, subtracting the results, and integrating over V, we obtain:

$$(k_q^2 - k_p^2) \iiint_V \vec{E}_p \bullet \vec{E}_q = \iiint_V (\vec{E}_p \bullet \nabla \times \nabla \times \vec{E}_q - \vec{E}_q \bullet \nabla \times \nabla \times \vec{E}_p) \mathrm{d}V \tag{1.75}$$

By using the vector identity, $\nabla \bullet \vec{A} \times \vec{B} = \vec{B} \bullet \nabla \times \vec{A} - \vec{A} \bullet \nabla \times \vec{B}$, the right side of (1.75) can be rewritten:

$$(k_q^2 - k_p^2) \iiint_V \vec{E}_p \bullet \vec{E}_q = \iiint_V \nabla \bullet (\vec{E}_q \times \nabla \times \vec{E}_p - \vec{E}_p \times \nabla \times \vec{E}_q) \mathrm{d}V \tag{1.76}$$

By using the divergence theorem and (1.22), we obtain the desired result:

$$(k_q^2-k_p^2)\iiint\limits_V \vec{E}_p \bullet \vec{E}_q = -\oint\limits_S (\hat{n}\times\vec{E}_p \bullet \nabla\times\vec{E}_q - \hat{n}\times\vec{E}_q \bullet \nabla\times\vec{E}_p)\mathrm{d}S = 0 \qquad (1.77)$$

When $k_q^2 \neq k_p^2$, the modes $\vec{E}_p$ and $\vec{E}_q$ are orthogonal. For degenerate modes that have the same eigenvalue ($k_p = k_q$), we can use the Gram-Schmidt orthogonalization procedure to construct a new subset of orthogonal modes [13].

We now consider cavity excitation by an electric current $\vec{J}$. The electric field $\vec{E}$ satisfies (1.12). We can expand the electric field in terms of the $\vec{E}_p$ and $\vec{F}_p$ modes:

$$\vec{E} = \sum_p (A_p\vec{E}_p + B_p\vec{F}_p), \qquad (1.78)$$

where A_p and B_p are constants to be determined. Substitution of (1.78) into (1.12) yields

$$\sum_p [(k_p^2-k^2)A_p\vec{E}_p - k^2 B_p\vec{F}_p] = i\omega\mu\vec{J} \qquad (1.79)$$

If we scalar multiply (1.79) by $\vec{E}_p$ and $\vec{F}_p$ and integrate over the volume V, we obtain:

$$(k_p^2-k^2)A_p = i\omega\mu\iiint\limits_V \vec{E}_p(\vec{r}')\bullet\vec{J}(\vec{r}')\mathrm{d}V', \qquad (1.80)$$

$$-k^2 B_p = i\omega\mu\iiint\limits_V \vec{F}_p(\vec{r}')\bullet\vec{J}(\vec{r}')\mathrm{d}V' \qquad (1.81)$$

Substitution of (1.80) and (1.81) into (1.78) gives the solution for $\vec{E}$:

$$\vec{E}(\vec{r}) = i\omega\mu\iiint\limits_V \sum_p \left[\frac{\vec{E}_p(\vec{r})\vec{E}_p(\vec{r}')}{k_p^2-k^2} - \frac{\vec{F}_p(\vec{r})\vec{F}_p(\vec{r}')}{k^2}\right]\bullet\vec{J}(\vec{r}')\mathrm{d}V' \qquad (1.82)$$

The summation quantity is the dyadic Green's function $\overleftrightarrow{G}_e$ for the electric field in the cavity [2, 13]:

$$\overleftrightarrow{G}_e(\vec{r},\vec{r}') = \sum_p \left[\frac{\vec{E}_p(\vec{r})\vec{E}_p(\vec{r}')}{k_p^2-k^2} - \frac{\vec{F}_p(\vec{r})\vec{F}_p(\vec{r}')}{k^2}\right] \qquad (1.83)$$

The summation over integer p actually represents a triple sum over a triple set of integers. The specific details will be given in the next three chapters.

Equations (1.82) and (1.83) have singularities at $k^2 = k_p^2$. However, if we include wall loss as in Section 1.3, then we can replace k_p by $k_p(1-\frac{i}{2Q_p})$. Then there are no singularities for real k (except at the source point, $r = r'$, which will be discussed later).

1.5 PERTURBATION THEORIES

When a cavity shape is deformed or the dielectric is inhomogeneous, the analysis is generally difficult, and numerical methods are required. However, if the shape deformation or the dielectric inhomogeneity is small, then perturbation techniques [14] are applicable.

1.5.1 Small-Sample Perturbation of a Cavity

If a small sample of dielectric or magnetic material of volume V_s is introduced into a cavity (as in Figure 1.3), the resonant frequency ω_p of the cavity is changed by a small amount $\delta\omega$. If the sample has loss, then $\delta\omega$ becomes complex and a damping factor occurs (the cavity Q is changed). If the sample is properly positioned, the measurement of the complex frequency change $\delta\omega$ can be used to infer the complex permittivity or permeablility of the sample [15].

If $\vec{E}_p$ and $\vec{H}_p$ are the unperturbed fields of the pth cavity mode and $\vec{E}_1$ and $\vec{H}_1$ are the perturbation fields due to the introduced sample, then the total perturbed fields $\vec{E}'$ and $\vec{H}'$ are:

$$\vec{E}' = \vec{E}_p + \vec{E}_1, \tag{1.84}$$

$$\vec{H}' = \vec{H}_p + \vec{H}_1 \tag{1.85}$$

The (complex) frequency of oscillation is $\omega_p + \delta\omega$. Outside the sample, the magnetic and electric flux densities, $\vec{B}'$ and $\vec{D}'$, are given by:

$$\vec{B}' = \vec{B}_p + \vec{B}_1 = \mu(\vec{H}_p + \vec{H}_1), \tag{1.86}$$

$$\vec{D}' = \vec{D}_p + \vec{D}_1 = \varepsilon(\vec{E}_p + \vec{E}_1) \tag{1.87}$$

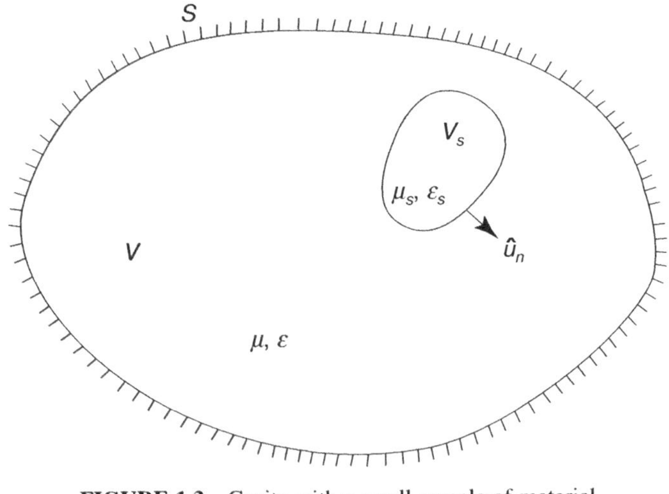

FIGURE 1.3 Cavity with a small sample of material.

Inside the sample, we have:

$$\vec{B}' = \mu_s \vec{H}' = \vec{B}_p + \vec{B}_1 = \mu \vec{H}_p + \mu[\kappa_{sm}(\vec{H}_p + \vec{H}_1) - \vec{H}_p], \tag{1.88}$$

$$\vec{D}' = \varepsilon_s \vec{E}' = \vec{D}_p + \vec{D}_1 = \varepsilon \vec{E}_p + \varepsilon[\kappa_{se}(\vec{E}_p + \vec{E}_1) - \vec{E}_p], \tag{1.89}$$

where μ_s and ε_s are the permeability and permittivity of the sample and κ_{sm} and κ_{se} are the relative permeability and permittivity of the sample. Here we assume that the sample is isotropic, but for anisotropic materials these quantities become tensors.

Throughout the cavity, the total fields satisfy Maxwell's curl equations:

$$\nabla \times (\vec{E}_p + \vec{E}_1) = i(\omega_p + \delta\omega)(\vec{B}_p + \vec{B}_1), \tag{1.90}$$

$$\nabla \times (\vec{H}_p + \vec{H}_1) = -i(\omega_p + \delta\omega)(\vec{D}_p + \vec{D}_1) \tag{1.91}$$

The unperturbed fields satisfy:

$$\nabla \times \vec{E}_p = i\omega_p \vec{B}_p, \tag{1.92}$$

$$\nabla \times \vec{H}_p = -i\omega_p \vec{D}_p \tag{1.93}$$

Subtracting (1.92) from (1.90) and (1.93) from (1.91), we obtain:

$$\nabla \times \vec{E}_1 = i[\omega_p + \delta\omega(\vec{B}_p + \vec{B}_1)], \tag{1.94}$$

$$\nabla \times \vec{H}_1 = -i[\omega_p \vec{D}_1 + \delta\omega(\vec{D}_p + \vec{D}_1)] \tag{1.95}$$

If we scalar multiply (1.94) by $\vec{H}_p$ and (1.95) by $\vec{E}_p$ and add the results, we obtain:

$$\begin{aligned}&\vec{H}_p \bullet \nabla \times \vec{E}_1 + \vec{E}_p \bullet \nabla \times \vec{H}_1 \\ &= -i\omega_p(\vec{E}_p \bullet \vec{D}_1 - \vec{B}_1 \bullet \vec{H}_p) - i\delta\omega(\vec{E}_p \bullet \vec{D}_p + \vec{E}_p \bullet \vec{D}_1 - \vec{H}_p \bullet \vec{B}_p - \vec{H}_p \bullet \vec{B}_1)\end{aligned} \tag{1.96}$$

Using (1.92)–(1.95) and vector identities, we can write the right side of (1.96) in the two following forms:

$$\begin{aligned}&\vec{H}_p \bullet \nabla \times \vec{E}_1 + \vec{E}_p \bullet \nabla \times \vec{H}_1 \\ &= \vec{E}_1 \bullet \nabla \times \vec{H}_p + \vec{H}_1 \bullet \nabla \times \vec{E}_p - \nabla \bullet (\vec{H}_p \times \vec{E}_1 + \vec{E}_p \times \vec{H}_1) \\ &= -i\omega_p(\vec{D}_p \bullet \vec{E}_1 - \vec{B}_p \bullet \vec{H}_1) - \nabla \bullet (\vec{H}_p \times \vec{E}_1 + \vec{E}_p \times \vec{H}_1)\end{aligned} \tag{1.97}$$

If we substitute (1.94) and (1.95) into (1.97) and evaluate the result outside the sample, we obtain:

$$i\delta\omega(\varepsilon \vec{E}_p \bullet \vec{E}_p + \varepsilon \vec{E}_p \bullet \vec{E}_1 - \mu \vec{H}_p \bullet \vec{H}_p - \mu \vec{H}_p \bullet \vec{H}_1) = \nabla \bullet (\vec{H}_p \times \vec{E}_1 + \vec{E}_0 \times \vec{H}_1) \tag{1.98}$$

The perturbation fields $\vec{E}_1$ and $\vec{H}_1$ are not necessarily small everywhere in the cavity. However, if (1.98) is integrated over the volume $V - V_s$, it is possible to neglect contributions of terms involving $\vec{E}_1$ and $\vec{H}_1$ when the sample volume V_s is small. Taking into account that $\vec{E}_p$ and $\vec{E}_1$ are normal to S, and using the divergence theorem and vector identities, we obtain:

$$-i\delta\omega \int\limits_{V-V_s} (\vec{B}_p \bullet \vec{H}_p - \vec{D}_p \bullet \vec{E}_p)\mathrm{d}V = \int\limits_{\Sigma} [(\hat{u}_n \times \vec{E}_1) \bullet \vec{H}_p + (\hat{u}_n \times \vec{H}_1) \bullet \vec{E}_p]\mathrm{d}\Sigma, \quad (1.99)$$

where $\hat{u}_n$ is the outward unit normal from the sample and Σ is the surface of the sample.

Comparing the right sides of (1.96) and (1.97), we obtain:

$$\begin{aligned} &i\omega_p(\vec{E}_1 \bullet \vec{D}_p - \vec{B}_p \bullet \vec{H}_1) + i(\omega_p + \delta\omega)(\vec{B}_1 \bullet \vec{H}_p - \vec{E}_p \bullet \vec{D}_1) \\ &+ i\delta\omega(\vec{H}_p \bullet \vec{B}_p - \vec{E}_p \bullet \vec{D}_p) = \nabla \bullet (\vec{E}_1 \times \vec{H}_p + \vec{H}_1 \times \vec{E}_p) \end{aligned} \quad (1.100)$$

If we neglect $\delta\omega$ in the factor $(\omega_p + \delta\omega)$, integration of (1.100) over the sample volume yields:

$$\begin{aligned} &i\delta\omega \int\limits_{V_s} (\vec{B}_p \bullet \vec{H}_p - \vec{D}_p \bullet \vec{E}_p)\mathrm{d}V_s + i\omega_p \int\limits_{V_s} (\vec{E}_1 \bullet \vec{D}_p - \vec{E}_p \bullet \vec{D}_1 - \vec{B}_p \bullet \vec{H}_1 + \vec{B}_1 \bullet \vec{H}_p)\mathrm{d}V_s \\ &= \int\limits_{\Sigma} [(\hat{u}_n \times \vec{E}_1) \bullet \vec{H}_p + (\hat{u}_n \times \vec{H}_1) \bullet E_p]\mathrm{d}\Sigma \end{aligned} \quad (1.101)$$

The surface integrals in (1.99) and (1.101) are equal. Thus we can equate the left sides of (1.99) and (1.101) to obtain:

$$\frac{\delta\omega}{\omega_p} = \frac{\int\limits_{V_s} [(\vec{E}_1 \bullet \vec{D}_p - \vec{E}_p \bullet \vec{D}_1) - (\vec{H}_1 \bullet \vec{B}_p - \vec{H}_p \bullet \vec{B}_1)]\mathrm{d}V_s}{\int\limits_{V} (\vec{E}_p \bullet \vec{D}_p - \vec{H}_p \bullet \vec{B}_p)\mathrm{d}V} \quad (1.102)$$

Inside the sample, we can write the constitutive relations, (1.7) and (1.8), in more convenient forms:

$$\vec{D}_1 = \varepsilon_0 \vec{E} + \vec{P} \quad \text{and} \quad \vec{B}_1 = \mu_0 \vec{H}_1 + \mu_0 \vec{M}, \quad (1.103)$$

where ε_0 and μ_0 are the permittivity and permeability of free space, $\vec{P}$ is the electric polarization, and $\vec{M}$ is the magnetic polarization. For convenience, we will assume in the rest of this section that the cavity permittivity $\varepsilon = \varepsilon_0$ and the cavity permeability $\mu = \mu_0$. If we substitute (1.103) into (1.102), we obtain:

$$\frac{\delta\omega}{\omega_p} = \frac{\mu_0 \int\limits_{V_s} \vec{H}_p \bullet \vec{M}\mathrm{d}V_s - \int\limits_{V_s} \vec{E}_p \bullet \vec{P}\mathrm{d}V_s}{\int\limits_{V} (\vec{E}_p \bullet \vec{D}_p - \vec{H}_p \bullet \vec{B}_p)\mathrm{d}V} \quad (1.104)$$

If the sample volume V_s is very small, $\vec{E}_p$ and $\vec{H}_p$ are nearly constant throughout the sample volume, and (1.104) can be approximated as:

$$\frac{\delta\omega}{\omega_p} = \frac{\mu_0 \vec{H}_p \bullet \vec{P}_m - \vec{E}_p \bullet \vec{P}_e}{\int\limits_{V} (\vec{E}_p \bullet \vec{D}_p - \vec{H}_p \bullet \vec{B}_p)\mathrm{d}V}, \quad (1.105)$$

where $\vec{P}_e$ and $\vec{P}_m$ are the quasi-static electric and magnetic dipole moments induced in the sample by the cavity modal fields $(\vec{E}_p, \vec{H}_p)$.

For a spherical sample of radius a, the induced dipole moments are [15, 16]:

$$\vec{P}_e = 4\pi a^3 \varepsilon_0 \frac{\kappa_{se}-1}{\kappa_{se}+2} \vec{E}_p(P), \tag{1.106}$$

$$\vec{P}_m = 4\pi a^3 \frac{\kappa_{sm}-1}{\kappa_{sm}+2} \vec{H}_p(P), \tag{1.107}$$

where P is the location of the center of the sphere. If we substitute (1.25), (1.26), (1.29), (1.106), (1.107) into (1.05), we obtain the following resonant frequency shift:

$$\frac{\delta\omega}{\omega_p} = -\frac{2\pi a^3}{W}\left[\mu_0 \frac{\kappa_{sm}-1}{\kappa_{sm}+2}|\vec{H}_p(P)|^2 + \varepsilon_0 \frac{\kappa_{se}-1}{\kappa_{se}+2}|\vec{E}_p(P)|^2\right] \tag{1.108}$$

Equation (1.108) is the desired mathematical result, which can be applied to a number of measurements. Consider first the case where the spherical sample is located at a point where the electric field $\vec{E}_p(P)$ is zero. If the relative permeability κ_{sm} of the sample is known, then (1.108) can be used to determine the square of the magnetic field at P:

$$|\vec{H}_p(P)|^2 = -\frac{\delta\omega}{\omega_p}\frac{W}{2\pi a^3 \mu_0}\frac{\kappa_{sm}+2}{\kappa_{sm}-1} \tag{1.109}$$

If the magnitude of the square of the magnetic field at P is known (measured), then (1.108) can be used to determine κ_{sm}:

$$\kappa_{sm} = 2\frac{\frac{\pi a^3}{W}\mu_0|\vec{H}_p(P)|^2 - \frac{\delta\omega}{\omega_p}}{\frac{2\pi a^3}{W}\mu_0|\vec{H}_p(P)|^2 + \frac{\delta\omega}{\omega_p}} \tag{1.110}$$

If $\delta\omega$ is real, then κ_{sm} is real and the sample has no magnetic loss. However, if $\delta\omega$ is complex, then κ_{sm} is complex and the sample does have magnetic loss. The imaginary part of the resonant frequency is related to the cavity Q from the expression for a complex resonant frequency $\omega_p(1-\frac{i}{Q})$. Hence the change in the imaginary part of the resonant frequency is determined from the change in Q. This is typically determined by measuring the half-power bandwidth, which is given by (1.57).

In the analogous case, the spherical sample is located at a point where the magnetic field $\vec{H}_p(P)$ is zero. If the relative permittivity κ_{se} of the sample is known, then (1.108) can be used to determine the square of the electric field at P:

$$|\vec{E}_p(P)|^2 = -\frac{\delta\omega}{\omega_p}\frac{W}{2\pi a^3 \varepsilon_0}\frac{\kappa_{se}+2}{\kappa_{se}-1} \tag{1.111}$$

This method has been used to map the electric field along the axis of a linear accelerator [15]. If the magnitude of the square of the electric field at P is known (measured), then (1.108) can be used to determine κ_{se}:

$$\kappa_{se} = 2\frac{\dfrac{\pi a^3}{\overline{W}}\varepsilon_0|\vec{E}_p(P)|^2 - \dfrac{\delta\omega}{\omega_p}}{\dfrac{2\pi a^3}{\overline{W}}\varepsilon_0|\vec{E}_p(P)|^2 + \dfrac{\delta\omega}{\omega_p}} \tag{1.112}$$

Similar to (1.110), $\delta\omega$ can be either real (lossless dielectric sample) or complex (lossy dielectric sample).

1.5.2 Small Deformation of Cavity Wall

Here we consider the change in the resonant frequency of a cavity mode due to a small deformation in the cavity wall. This case is useful in determining the effects of small accidental deformations or intentional displacements of pistons or membranes on the resonant frequencies.

Our derivation is similar to that of Argence and Kahan [7], but with somewhat different notation. We first write Maxwell's equation for the curl of $\vec{E}_p$ and the complex conjugate for Maxwell's equation for the curl of $\vec{H}_p$ for the pth mode of the unperturbed cavity:

$$\nabla \times \vec{E}_p = i\omega_p\mu\vec{H}_p, \tag{1.113}$$

$$\nabla \times \vec{H}_p^* = -i\omega_p\varepsilon\vec{E}_p^*, \tag{1.114}$$

where the electric current term is omitted in (1.114) for this source-free case. If we scalar multiply (1.113) by $\vec{H}_p^*$ and (1.114) by $\vec{E}_p$ and take the difference, we obtain:

$$\vec{H}_p^* \bullet \nabla \times \vec{E}_p - \vec{E}_p \bullet \nabla \times \vec{H}_p^* = -i\omega_p(\mu\vec{H}_p \bullet \vec{H}_p^* - \varepsilon\vec{E}_p \bullet \vec{E}_p^*) \tag{1.115}$$

If we integrate (1.115) over the volume V, the two terms on the right side can be written in terms of the time-averaged magnetic and electric energies from (1.25) and (1.26). The left side of (1.115) can be converted to a divergence via a vector identity and converted to a surface integral over S by use of the divergence theorem. The result is:

$$-\oiint_S (\vec{E}_p \times \vec{H}_p^*) \bullet \hat{n}\,dS = 2i\omega(\overline{W}_{mp} - \overline{W}_{ep}) \tag{1.116}$$

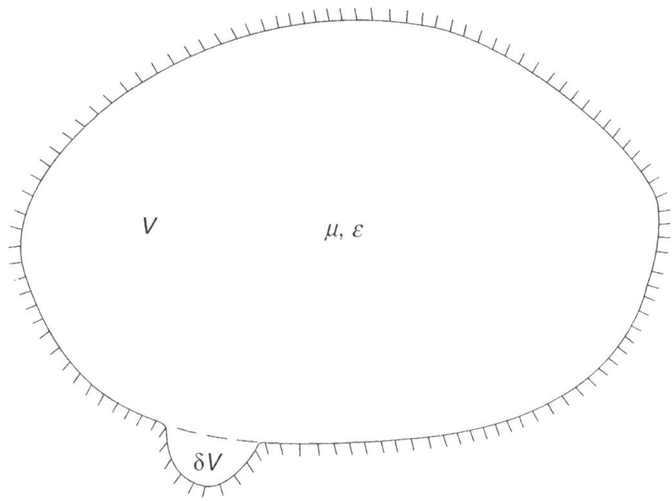

FIGURE 1.4 Cavity with a small deformation δV in the cavity wall.

Equation (1.116) can be written in the form:

$$\Phi_p = -\oiint_S (\vec{E}_p \times \vec{H}_p^*) \bullet \hat{n} dS = 2i\omega \iiint_V \tau_p dV, \tag{1.117}$$

where:

$$\tau_p = \frac{\mu}{2} \vec{H}_p \bullet \vec{H}_p^* - \frac{\varepsilon}{2} \vec{E}_p \bullet \vec{E}_p^*, \tag{1.118}$$

which is the difference between the time-average magnetic and electric energy densities.

We consider now a small deformation in the cavity wall, as shown in Figure 1.4. We write the perturbed electric field $\vec{E}'$ and magnetic field $\vec{H}'$as in (1.84) and (1.85). The resonant frequency of the deformed cavity is $\omega_p + \delta\omega$. The analogy to (1.117) for the perturbed cavity is:

$$\Phi' = \Phi_p + \delta\Phi = 2i(\omega_p + \delta\omega) \iiint_{V+\delta V} (\tau_p + \delta\tau) dV \tag{1.119}$$

Subtracting (1.117) from (1.119) and neglecting second-order terms, we obtain:

$$\delta\Phi = 2i\omega_p \iiint_V \delta\tau dV + 2i\delta\omega \iiint_V \tau dV + 2i\omega_p \iiint_{\delta V} \tau dV \tag{1.120}$$

The perturbed fields satisfy the following Maxwell curl equations, which are equivalent to (1.90) and (1.91):

$$\nabla \times (\vec{H}_p + \vec{H}_1) = i\varepsilon(\omega_p + \delta\omega)(\vec{E}_p + \vec{E}_1), \tag{1.121}$$

$$\nabla \times (\vec{E}_p + \vec{E}_1) = -i\mu(\omega_p + \delta\omega)(\vec{H}_p + \vec{H}_1) \tag{1.122}$$

By subtracting the complex conjugate of (1.114) from (1.121) and (1.113) from (1.122), we obtain:

$$\nabla \times \vec{H}_1 = i\varepsilon(\omega_p \vec{E}_1 + \vec{E}_p \delta\omega), \tag{1.123}$$

$$\nabla \times \vec{E}_1 = -i\mu(\omega_p \vec{H}_1 + \vec{H}_p \delta\omega) \tag{1.124}$$

We can write τ' in a manner analogous to (1.118):

$$\tau' = \frac{\mu}{2}(\vec{H}_p + \vec{H}_1) \bullet (\vec{H}_p^* + \vec{H}_1^*) - \frac{\mu}{2}(\vec{E}_p + \vec{E}_1) \bullet (\vec{E}_p^* + \vec{E}_1^*) \tag{1.125}$$

If we subtract (1.118) from (1.125) and ignore second order terms (such as $\vec{H}_1 \bullet \vec{H}_1^*$), we obtain:

$$\delta\tau = \tau' - \tau_p = \frac{\mu}{2}(\vec{H}_p \bullet \vec{H}_1^* + \vec{H}_p^* \bullet \vec{H}_1) - \frac{\varepsilon}{2}(\vec{E}_p \bullet \vec{E}_1^* + \vec{E}_p^* \bullet \vec{E}_1) \tag{1.126}$$

By substituting the curl equations from this section into (1.126) and using a vector identity, we can multiply the result by $2i\omega_p$to obtain:

$$2i\omega_p \delta\tau = i\nabla \bullet \mathrm{Im}(\vec{E}_p \times \vec{H}_1) + i\varepsilon\delta\omega \vec{E}_p \bullet \vec{E}_p^* \tag{1.127}$$

If we substitute (1.127) into (1.120), we obtain:

$$\delta\Phi = 2i \oiint_S [\mathrm{Im}(\vec{E}_p^* \times \delta\vec{H}_1)] \bullet \hat{n} \mathrm{d}S + i\delta\omega \iiint_V (\mu\vec{H}_p \bullet \vec{H}_p^* + \varepsilon\vec{E}_p \bullet \vec{E}_p^*) \mathrm{d}V + i\omega_p \iiint_{\delta V} (\mu\vec{H}_p \bullet \vec{H}_p^* - \varepsilon\vec{E}_p \bullet \vec{E}_p^*) \mathrm{d}V \tag{1.128}$$

Because the cavity walls are assumed to be perfectly conducting, the tangential component of the electric field is zero and $\delta\Phi = 0$. Similarly:

$$\oiint_S [\mathrm{Im}(\vec{E}_p^* \times \vec{H}_1)] \bullet \hat{n} \mathrm{d}V = 0 \tag{1.129}$$

By using $\delta\Phi = 0$ and (1.129) in (1.128), we obtain the desired result for the relative shift in the resonant frequency of the deformed cavity:

$$\frac{\delta\omega}{\omega_p} = -\frac{\iiint\limits_{\delta V}(\mu\vec{H}_p \bullet \vec{H}_p^* - \varepsilon\vec{E}_p \bullet \vec{E}_p^*)dV}{\iiint\limits_{V}(\mu\vec{H}_p \bullet \vec{H}_p^* + \varepsilon\vec{E}_p \bullet \vec{E}_p^*)dV} \tag{1.130}$$

Equation (1.130) can be written in a simpler form if we define time-average electric and magnetic energy densities for the pth mode:

$$\bar{w}_{pe} = \frac{\varepsilon}{2}\vec{E}_p \bullet \vec{E}_p^* \quad \text{and} \quad \bar{w}_{pm} = \frac{\mu}{2}\vec{H}_p \bullet \vec{H}_p^* \tag{1.131}$$

If we substitute (1.131) into (1.130), we can simplify the result to:

$$\begin{aligned}\frac{\delta\omega}{\omega_p} &= \frac{-1}{\overline{W}_p}\iiint\limits_{\delta V}(\bar{w}_{pm} - \bar{w}_{pe})dV \\ &\approx \frac{(\bar{w}_{pe} - \bar{w}_{pm})\delta V}{\overline{W}_p}\end{aligned} \tag{1.132}$$

In the second result in (1.132), $\bar{w}_{pe}$ and $\bar{w}_{pm}$ are the time-averaged electric and magnetic energies at the volume deformation. Equation (1.132) shows that if the cavity is compressed ($\delta V < 0$) in a region where $\bar{w}_{pm} > \bar{w}_{pe}$, then $\delta\omega > 0$ and the resonant frequency is increased. However, if $\delta V < 0$ and $\bar{w}_{pm} < \bar{w}_{pe}$, then $\delta\omega < 0$ and the resonant frequency is decreased. For δV positive, the results are reversed. The result in (1.132) is identical to that given by Borgnis and Papas [6].

PROBLEMS

1-1 Derive (1.3) from (1.2) and (1.5). This shows that the continuity equation can be derived from two of Maxwell's equations.

1-2 Show that (1.17) is satisfied in rectangular coordinates where $\vec{E} = \hat{x}E_x + \hat{y}E_y + \hat{z}E_z$. Combine that result with (1.15) and (1.16) to derive the vector Helmholtz equation in (1.18).

1-3 Apply the boundary condition, $\hat{n} \times \vec{E}_p = 0$ on S, to (1.28) to show that $\overline{W}_{ep} = \overline{W}_{mp}$ as in (1.29). Hint: use the vector identity (A19). Is the boundary condition, $\hat{n} \bullet \vec{H} = 0$ on S, sufficient to derive the same result?

1-4 Using the smoothed approximations in (1.31) and (1.33), determine the mode number and mode density for an empty cavity of volume 1 m^3 at a frequency of 1 GHz. What is the mode separation?

1-5 Show that the $1/e$ decay time of the fields of the pth mode is $2Q_p/\omega_p$.

1-6 In (1.82), show that the coupling of the current source $\vec{J}$ to $\vec{F}_p$ is zero if $\nabla \bullet \vec{J} = 0$ and the normal component of $\vec{J}$ is zero at the boundary of the source region. Hint: use the divergence theorem.

1-7 Does a small loop current, $\vec{J} = \hat{\phi}\frac{I_0}{\rho_0}\delta(\rho-\rho_0)$, satisfy the current conditions for problem 1-6?

1-8 Does a short dipole current, $\vec{J} = I_0\delta(x)\delta(y)U(\frac{l}{2}-|z|)$, satisfy the current conditions for problem 1-6?

1-9 Consider a small lossless dielectric sphere, $\text{Re}(\kappa_{se}) > 1, \text{Im}(\kappa_{se}) = 0$, and $\kappa_{sm} = 0$, inserted in a lossless cavity. From (108), what is the sign of the resonant frequency shift $\delta\omega$? What is the physical explanation for this sign?

1-10 Consider a small lossy dielectric sphere, $\text{Re}(\kappa_{se}) > 1, \text{Im}(\kappa_{se}) > 0$, and $\kappa_{sm} = 0$, inserted in a lossless cavity. From (108), what is the sign of the imaginary part of the frequency shift $\text{Im}(\delta\omega)$? What is the physical explanation for this sign?

CHAPTER 2

Rectangular Cavity

The rectangular cavity is the first of three separable geometries we will consider. (See Chapters 3 and 4 for the circular cylindrical cavity and the spherical cavity.) The geometry for a general rectangular cavity with sides of length a, b, and c is shown in Figure 2.1. Rectangular cavities are used as single-mode resonators [13], for making dielectric or permeability measurements [17], or as reverberation (mode-stirred) chambers [9, 18], where a stirrer is added to yield a multi-mode cavity.

2.1 RESONANT MODES

The simplest method for constructing the resonant modes for a rectangular cavity is to derive modes that are transverse electric (TE) or transverse magnetic (TM) to one of the three axes. In keeping with standard waveguide notation [13], we choose the z axis. The TE modes can also be called magnetic modes because the E_z component is zero. Similarly, the TM modes can be called electric modes because the H_z component is zero.

From (1.18) and (1.19), we see that the z component of the electric field E^{TM}_{zmnp} of a TM mode satisfies the scalar Helmholtz equation:

$$(\nabla^2 + k^2_{mnp})E^{TM}_{zmnp} = 0, \tag{2.1}$$

where k_{mnp} is an eigenvalue to be determined. (The triple subscript mnp takes the place of p in Chapter 1.) From the electric field boundary condition in (1.22), the solution to (2.1) is:

$$E^{TM}_{zmnp} = E_0 \sin\frac{m\pi x}{a} \sin\frac{n\pi y}{b} \cos\frac{p\pi z}{c}, \tag{2.2}$$

Electromagnetic Fields in Cavities: Deterministic and Statistical Theories, by David A. Hill

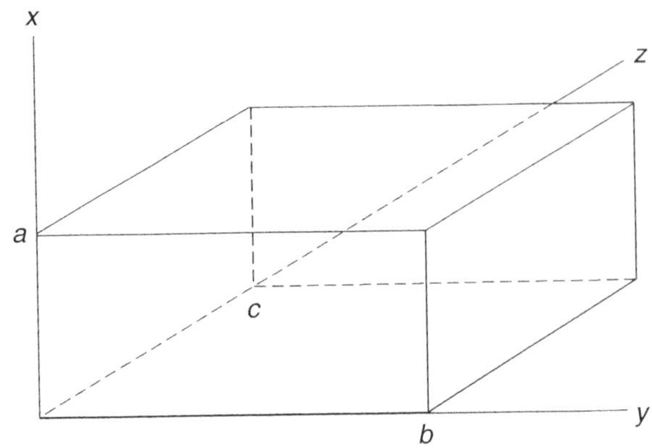

FIGURE 2.1 Rectangular cavity.

where E_0 is an arbitrary constant with units of V/m and m, n, and p are integers. The eigenvalues k_{mnp} satisfy:

$$k_{mnp}^2 = \left(\frac{m\pi}{a}\right)^2 + \left(\frac{n\pi}{b}\right)^2 + \left(\frac{p\pi}{c}\right)^2 \tag{2.3}$$

For convenience, we can also write (2.3) as:

$$k_{mnp}^2 = k_x^2 + k_y^2 + k_z^2,$$

where

$$k_x = \frac{m\pi}{a}, \quad k_y = \frac{n\pi}{b}, \quad k_z = \frac{p\pi}{c}. \tag{2.4}$$

The electric and magnetic fields can be obtained from an electric Hertz vector [13] which has only a z component Π_e:

$$\vec{\Pi}_e = \hat{z}\Pi_e \tag{2.5}$$

Curl operations on $\vec{\Pi}_e$ yield [13]:

$$\vec{E} = \nabla \times \nabla \times \vec{\Pi}_e \quad \text{and} \quad \vec{H} = -i\omega\varepsilon\nabla \times \vec{\Pi}_e \tag{2.6}$$

From (2.2) and (2.6), we can determine that the z component of the electric Hertz vector for the mnp mode must take the form:

$$\Pi_{emnp} = \frac{E_{zmnp}^{TM}}{k_{mnp}^2 - k_z^2} = \frac{E_0}{k_{mnp}^2 - k_z^2} \sin\frac{m\pi x}{a} \sin\frac{n\pi y}{b} \cos\frac{p\pi z}{c} \tag{2.7}$$

The z component of the electric field is given in (2.2), and the transverse components are determined from (2.6) and (2.7):

$$
\begin{aligned}
E^{TM}_{xmnp} &= -\frac{k_x k_z E_0}{k^2_{mnp} - k^2_z} \cos\frac{m\pi x}{a} \sin\frac{n\pi y}{b} \sin\frac{p\pi z}{c}, \\
E^{TM}_{ymnp} &= \frac{k_y k_z E_0}{k^2_{mnp} - k^2_z} \sin\frac{m\pi x}{a} \cos\frac{n\pi y}{b} \sin\frac{p\pi z}{c}
\end{aligned}
\tag{2.8}
$$

The z component of the magnetic field is zero (by definition for a TM mode), and the transverse components of the magnetic field are determined from (2.6) and (2.7):

$$
\begin{aligned}
H^{TM}_{xmnp} &= -\frac{i\omega_{mnp}\varepsilon k_y E_0}{k^2_{mnp} - k^2_z} \sin\frac{m\pi x}{a} \cos\frac{n\pi y}{b} \cos\frac{p\pi z}{c}, \\
H^{TM}_{ymnp} &= \frac{i\omega_{mnp}\varepsilon k_x E_0}{k^2_{mnp} - k^2_z} \cos\frac{m\pi x}{a} \sin\frac{n\pi y}{b} \cos\frac{p\pi z}{c}
\end{aligned}
\tag{2.9}
$$

By requiring that E^{TM}_{zmnp} be nonzero, the allowable values of the mode numbers are $m = 1, 2, 3, \ldots$; $n = 1, 2, 3, \ldots$; and $p = 0, 1, 2, \ldots$.

The TE (or magnetic) modes are derived in an analogous manner. The z component of the magnetic field satisfies the scalar Helmholtz equation, and the boundary conditions require that it takes the form:

$$
H^{TE}_{zmnp} = H_0 \cos\frac{m\pi x}{a} \cos\frac{n\pi y}{b} \sin\frac{p\pi z}{c}, \tag{2.10}
$$

where H_0 where is an arbitrary constant with units of A/m. The eigenvalues and axial wave numbers are the same as those of the TM modes in (2.3) and (2.4).

The electric and magnetic fields can be obtained from a magnetic Hertz vector [13] that has only a z component Π_h:

$$
\vec{\Pi}_h = \hat{z}\Pi_h \tag{2.11}
$$

Curl operations on $\vec{\Pi}_h$ yield [13]:

$$
\vec{H} = \nabla \times \nabla \times \vec{\Pi}_h \quad \text{and} \quad \vec{E} = i\omega\mu\nabla \times \Pi_h \tag{2.12}
$$

From (2.10) and (2.12), we can determine that the z component of the magnetic Hertz vector for the mnp mode must take the form:

$$
\Pi_{hmnp} = \frac{H^{TE}_{zmnp}}{k^2_{mnp} - k^2_z} = \frac{H_0}{k^2_{mnp} - k^2_z} \cos\frac{m\pi z}{a} \cos\frac{n\pi y}{b} \sin\frac{p\pi z}{c} \tag{2.13}
$$

The z component of the magnetic field is given in (2.10), and the transverse components of the magnetic field are determined from (2.13) and (2.17):

$$\begin{aligned} H_{xmnp}^{TE} &= -\frac{H_0 k_x k_y}{k_{mnp}^2 - k_z^2} \sin\frac{m\pi x}{a} \cos\frac{n\pi y}{b} \cos\frac{p\pi z}{c}, \\ H_{ymnp}^{TE} &= \frac{H_0 k_y k_z}{k_{mnp}^2 - k_z^2} \cos\frac{m\pi x}{a} \sin\frac{n\pi y}{b} \sin\frac{p\pi z}{c} \end{aligned} \tag{2.14}$$

The z component of the electric field is zero (by definition for a TE mode), and the transverse components of the electric field are determined from (2.12) and (2.13):

$$\begin{aligned} E_{xmnp}^{TE} &= -\frac{i\omega_{mnp}\mu k_y H_0}{k_{mnp}^2 - k_z^2} \cos\frac{m\pi x}{a} \sin\frac{n\pi y}{b} \sin\frac{p\pi z}{c}, \\ E_{ymnp}^{TE} &= \frac{i\omega_{mnp}\mu k_x H_0}{k_{mnp}^2 - k_z^2} \sin\frac{m\pi x}{a} \cos\frac{n\pi y}{b} \sin\frac{p\pi z}{c} \end{aligned} \tag{2.15}$$

The allowable values of the mode numbers are $m = 0, 1, 2, \ldots$; $n = 0, 1, 2, \ldots$; and $p = 1, 2, 3, \ldots$ with the exception that $m = n = 0$ is not allowed.

The resonant frequencies f_{mnp} can be determined from (2.3):

$$f_{mnp} = \frac{1}{2\sqrt{\mu\varepsilon}} \sqrt{\left(\frac{m}{a}\right)^2 + \left(\frac{n}{b}\right)^2 + \left(\frac{p}{c}\right)^2} \tag{2.16}$$

If m, n, and p are all nonzero, then two modes are degenerate (the TE_{mnp} and TM_{mnp} modes have the same resonant frequency). For $a < b < c$, the lowest resonant frequency occurs for the TE_{011} mode. An example of the instantaneous electric and magnetic field patterns for the TE_{011} mode are shown in Figure 2.2 [3]. Table 2.1 [3] shows the ratio f_{mnp}/f_{011} for the case $a \leq b \leq c$.

For use as single-mode resonators (filters or electromagnetic material property measurements), the goal is to excite only a single mode at its resonant frequency or at its perturbed resonant frequency for material measurements [17]. However, for use of a rectangular cavity as a reverberation chamber (mode-stirred chamber) [18, 19], a large metal stirrer is used to vary both the resonant frequencies and the excitation of multiple modes. In this case, it is useful to know the locations of resonant frequencies over a large bandwidth. Liu, Chang, and Ma [9] have thoroughly studied the resonant frequencies of rectangular cavities with application to reverberation chambers. They determined the total number N of modes with eigenvalues k_{mnp} less than or equal to k by computer counting using (2.3). N as a function of k or f is discontinuous, but they have also derived a smooth approximation N_s given by [9]:

$$N_s(k) = \frac{abc}{3\pi^2} k^3 - \frac{a + b + c}{2\pi} k + \frac{1}{2} \tag{2.17}$$

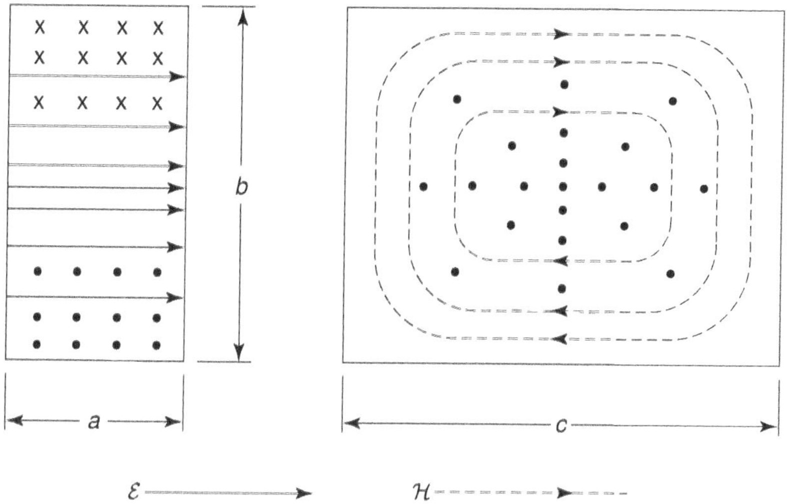

FIGURE 2.2 Instantaneous electric $\mathcal{E}$ and magnetic $\mathcal{H}$ field lines for the TE_{011} cavity mode [3].

The first term on the right side of (2.17) is Weyl's classical approximation N_W [9], which is valid for cavities of general shape and can be written in terms of the cavity volume V:

$$N_W(k) = \frac{Vk^3}{3\pi^2} \tag{2.18}$$

The extra terms in (2.17) are specific to the rectangular shape. The mode numbers in (2.17) and (2.18) can also be written as functions of frequency f:

$$N_s(f) = \frac{8\pi}{3}abc\frac{f^3}{v^3} - (a+b+c)\frac{f}{v} + \frac{1}{2} \tag{2.19}$$

TABLE 2.1 $\frac{f_{mnp}}{f_{011}}$ **for a Rectangular Cavity,** $a \le b \le c$ **[3].**

$\frac{b}{a}$	$\frac{c}{a}$	TE_{011}	TE_{101}	TM_{110}	TM_{111} TE_{111}	TE_{012}	TE_{021}	TE_{201}	TE_{102}	TM_{120}	TM_{210}	TM_{112} TE_{112}
1	1	1	1	1	1.22	1.58	1.58	1.58	1.58	1.58	1.58	1.73
1	2	1	1	1.26	1.34	1.26	1.84	1.84	1.26	2.00	2.00	1.55
2	2	1	1.58	1.58	1.73	1.58	1.58	2.91	2.00	2.00	2.91	2.12
2	4	1	1.84	2.00	2.05	1.26	1.84	3.60	2.00	2.53	3.68	2.19
4	4	1	2.91	2.91	3.00	1.58	1.58	5.71	3.16	3.16	5.71	3.24
4	8	1	3.62	3.65	3.66	1.26	1.84	7.20	3.65	4.03	7.25	3.82
4	16	1	3.88	4.00	4.01	1.08	1.96	7.76	3.91	4.35	7.83	4.13

and:

$$N_W(f) = \frac{8\pi V f^3}{3 \; v^3}, \tag{2.20}$$

where $v = 1/\sqrt{\mu\varepsilon}$ is the speed of light in the medium (usually free space). Equations (2.17)–(2.20) are asymptotic high-frequency approximations that are valid when the cavity dimensions are somewhat greater than a half wavelength.

Numerical results for N (by computer counting), N_s, and N_W are shown in Figure 2.3 for the NIST reverberation chamber ($a = 2.74$ m, $b = 3.05$ m, and $c = 4.57$ m). The extra terms in N_s improve the agreement obtained with Weyl's formula. The smooth mode density $D_s(f)$ is also shown in Figure 2.3. It is obtained by differentiating (2.19):

$$D_s(f) = \frac{dN_s(f)}{df} = 8\pi abc\frac{f^2}{v^3} - \frac{a+b+c}{v} \tag{2.21}$$

The Weyl approximation again equals the first term:

$$D_W(f) = \frac{dN_W(f)}{df} = 8\pi V\frac{f^2}{v^3} \tag{2.22}$$

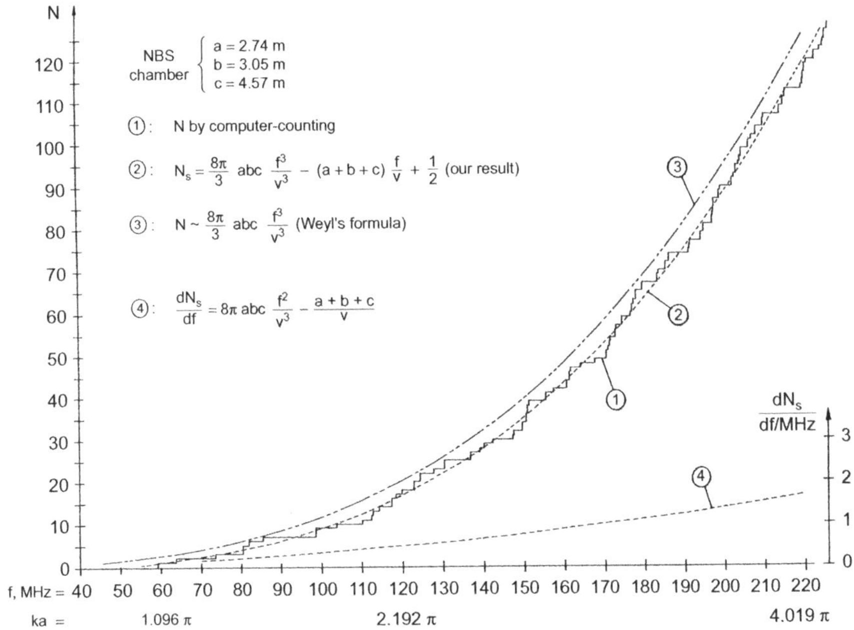

FIGURE 2.3 Mode number and mode density as a function of frequency for the NBS (NIST) chamber [9].

The mode density is an important reverberation chamber design parameter because it determines how many modes are present in a small bandwidth about a given frequency. For example, Figure 2.3 shows that the NIST reverberation chamber [19] has a mode density somewhat greater than one mode per megahertz at a frequency of 200 MHz. Experience has shown that the NIST chamber provides adequate performance at frequencies above 200 MHz, but not below 200 MHz, where the mode density is too low to obtain spatial field uniformity [19].

2.2 WALL LOSSES AND CAVITY *Q*

An expression for cavity Q due to wall losses of cavities of arbitrary shape was given in (1.41). For rectangular cavities, the expressions for the magnetic field are known, and the integrals can be evaluated to determine Q for the various mode types and numbers. Harrington [3, p. 190] has given expressions for the Q values of TE and TM modes of arbitrary order.

To illustrate the details of the evaluation of Q, we will derive Q for the specific case of a TM mode where none of the indices is equal to zero. We write (1.41) in the following form:

$$Q_{mnp}^{TM} = \omega_{mnp} \frac{\mu \iiint\limits_V \vec{H}_{mnp}^{TM} \bullet \vec{H}_{mnp}^{TM*} \, dV}{R_s \oiint\limits_S \vec{H}_{mnp}^{TM} \bullet \vec{H}_{mnp}^{TM*} \, dS}, \tag{2.23}$$

where we have replaced free space μ_0 by μ for greater generality, and the magnetic field expression is given by (2.9). The dot product in (2.23) can be written:

$$\begin{aligned} \vec{H}_{mnp}^{TM} \bullet \vec{H}_{mnp}^{TM*} = & \frac{\omega_{mnp}^2 \varepsilon^2 |E_0|^2}{(k_{mnp}^2 - k_z^2)^2} \left(k_y^2 \sin^2 \frac{m\pi x}{a} \cos^2 \frac{n\pi y}{b} \right. \\ & \left. + k_x^2 \cos^2 \frac{m\pi x}{a} \sin^2 \frac{n\pi y}{b} \right) \cos^2 \frac{p\pi z}{c} \end{aligned} \tag{2.24}$$

The volume integral in the numerator of (2.23) involves integrals of trigonometric functions over x, y, and z, and the result using (2.24) is:

$$\iiint\limits_V \vec{H}_{mnp}^{TM} \bullet \vec{H}_{mnp}^{TM*} \, dV = \frac{\omega_{mnp}^2 \varepsilon^2 |E_0|^2 abc}{8(k_{mnp}^2 - k_z^2)^2} (k_x^2 + k_y^2) \tag{2.25}$$

The closed surface integral in the denominator of (2.23) involves integrals of trigonometric functions over two of the three rectangular coordinates on six rectangular surfaces, and the result using (2.24) is:

$$\oiint_S \vec{H}^{TM}_{mnp} \bullet \vec{H}^{TM*}_{mnp}\, dS = \frac{\omega^2_{mnp}\varepsilon^2|E_0|^2}{2(k^2_{mnp}-k^2_z)^2[k^2_x b(a+c)+k^2_y a(b+c)]} \tag{2.26}$$

From (2.23), (2.25), and (2.26), we can write Harrington's result for Q^{TM}_{mnp}:

$$Q^{TM}_{mnp} = \frac{\eta abck^2_{xy}k_{mnp}}{4R_s[k^2_x b(a+c)+k^2_y a(b+c)]}, \tag{2.27}$$

where $\eta = \sqrt{\mu/\varepsilon}$ and $k^2_{xy} = k^2_x + k^2_y$. The Q expressions for the other modes can be derived by the same method and are given by [3, p. 190]:

$$Q^{TM}_{mn0} = \frac{\eta abck^3_{mn0}}{2R_s(abk^2_{mn0}+2bck^2_x+2ack^2_y)}, \tag{2.28}$$

$$Q^{TE}_{mnp} = \frac{\eta abck^2_{xy}k^3_{mnp}}{4R_s[bc(k^4_{xy}+k^2_y k^2_z)+ac(k^4_{xy}+k^2_z)+abk^2_{xy}k^2_z]}, \tag{2.29}$$

$$Q^{TE}_{0np} = \frac{\eta abck^3_{0np}}{2R_s(bck^2_{0np}+2ack^2_y+2abk^2_z)}, \tag{2.30}$$

$$Q^{TE}_{mop} = \frac{\eta abck^3_{mop}}{2R_s(ack^2_{mop}+2bck^2_x+2abk^2_z)} \tag{2.31}$$

The expressions (2.27)–(2.31) for quality factor are fairly complex, but it is possible to obtain a composite $\widetilde{Q}$ by averaging $\frac{1}{Q}$ values over the resonant modes [9]. This has been done by taking into account that each combination of m, n, and p (taking on values of positive integers) corresponds to two modes (TE and TM). For large values of ka, kb, and kc, the average over a small range of k gives the following result [9]:

$$\widetilde{Q} \equiv \frac{1}{\langle 1/Q \rangle} = \frac{3\eta kabc}{4R_s S}\frac{1}{1+\dfrac{3\pi}{8k}\left(\dfrac{1}{a}+\dfrac{1}{b}+\dfrac{1}{c}\right)}, \tag{2.32}$$

where $S = 2(ab+bc+ac)$ is the surface area. We can modify (2.32) by recognizing that abc is the volume V of the the cavity. We can also extend (2.32) to the case where the walls are of magnetic permeability μ_w (as for example steel walls). Then (2.32) can be written:

$$\widetilde{Q} = \frac{3V}{2\mu_r S\delta_s}\frac{1}{1+\dfrac{3\pi}{8k}\left(\dfrac{1}{a}+\dfrac{1}{b}+\dfrac{1}{c}\right)}, \tag{2.33}$$

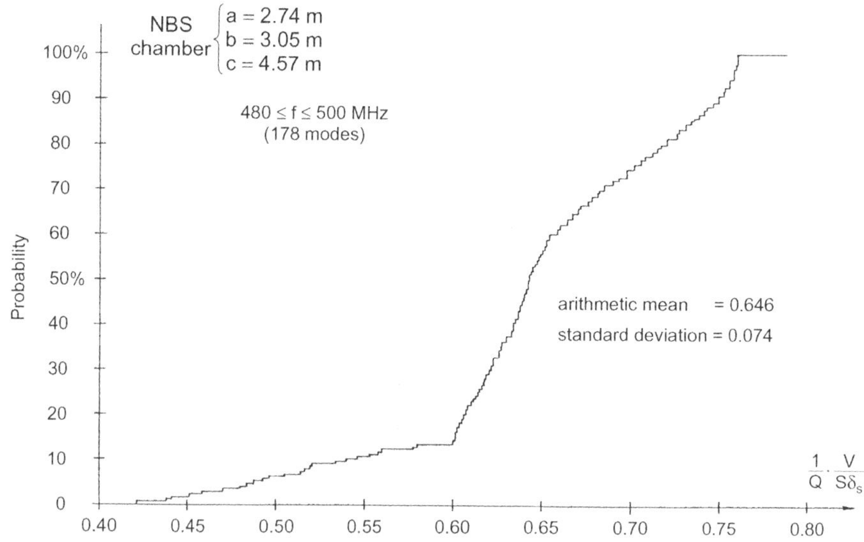

FIGURE 2.4 Cumulative distribution of the normalized $1/Q$ values in the 480 MHz to 500 MHz frequency band for the NBS (NIST) chamber [9].

where $\mu_r = \mu_w/\mu_0$ and $\delta_s = \sqrt{2/(\omega\mu_w\sigma_w}$. If ka, kb, and kc are sufficiently large and $\mu_r = 1$, then (2.33) reduces to (1.44), which applies to general cavity shapes. As a numerical check on (2.32) or (2.33), a numerical average of $\frac{1}{Q}$ was taken for a frequency range of 480 to 500 MHz for the dimensions of the NIST reverberation chamber. This 20 MHz bandwidth included 178 modes, and the spread of the inverse Q values is shown in Figure 2.4 [9]. The mean value of $\frac{V}{(QS\delta_s)}$ is 0.646, which is close to the expected analytical result of $\frac{2}{3}$ (for $\mu_r = 1$), and the standard deviation (0.074) is fairly small. Further numerical results are given in [9].

2.3 DYADIC GREEN'S FUNCTIONS

Dyadic Green's functions [2] provide a compact notation for determining the electric and magnetic fields due to current sources. For example, the excitation of a rectangular cavity by a dipole, monopole, or loop antenna can be treated by use of Dyadic Green's functions. (The electric field in the source region requires special treatment [20], but the electric dyadic Green's function is still useful there.) The electric $\overleftrightarrow{G}_e$ and magnetic $\overleftrightarrow{G}_m$ dyadic green's functions satisfy the following differential equations:

$$\nabla \times \nabla \times \overleftrightarrow{G}_e(\vec{r},\vec{r}\,') - k^2\overleftrightarrow{G}_e(\vec{r},\vec{r}\,') = \overleftrightarrow{I}\delta(\vec{r}-\vec{r}\,'), \tag{2.34}$$

$$\nabla \times \nabla \times \overleftrightarrow{G}_m(\vec{r},\vec{r}\,') - k^2\overleftrightarrow{G}_m(\vec{r},\vec{r}\,') = \nabla \times [\overleftrightarrow{I}\delta(\vec{r}-\vec{r}\,')], \tag{2.35}$$

where $\overleftrightarrow{I}$ is the unit dyadic:

$$\overleftrightarrow{I} = \hat{x}\hat{x} + \hat{y}\hat{y} + \hat{z}\hat{z}, \tag{2.36}$$

and $\delta(\vec{r}-\vec{r}\prime)$ is the three-dimensional delta function:

$$\delta(\vec{r}-\vec{r}\prime) = \delta(x-x')\delta(y-y')\delta(z-z') \tag{2.37}$$

The double arrow above the Green's functions indicates a three- by -three dyadic.

In addition to the differential equations, (2.34) and (2.35), we need to specify boundary conditions to make the dyadic Green's functions unique. For the electric dyadic Green's function, the boundary condition is analogous to that for the electric field in (1.22):

$$\hat{n} \times \overleftrightarrow{G}_e(\vec{r},\vec{r}\prime) = 0 \tag{2.38}$$

at $x = 0$ and a, $y = 0$ and b, and $z = 0$ and c. For the magnetic dyadic Green's function, the boundary condition is similar to (2.38) except that it involves the curl [2]:

$$\hat{n} \times \nabla \times \overleftrightarrow{G}_m(\vec{r},\vec{r}\prime) = 0 \tag{2.39}$$

at $x = 0$ and a, $y = 0$ and b, and $z = 0$ and c.

The solution to (2.34) and (2.38) for the electric dyadic Green's function is [2]:

$$\overleftrightarrow{G}_e(\vec{r},\vec{r}\prime) = -\frac{\hat{z}\hat{z}}{k^2}\delta(\vec{r}-\vec{r}\prime) + \frac{2}{ab}\sum_{m=0}^{\infty}\sum_{n=0}^{\infty}\frac{(2-\delta_0)}{k_c^2 k_g \sin k_g c}\begin{bmatrix} \vec{M}_{eo}(c-z)\vec{M}'_{eo}(z') - \vec{N}_{oe}(c-z)\vec{N}'_{oe}(z') \\ \vec{M}_{eo}(z)\vec{M}'_{eo}(c-z) - \vec{N}_{oe}(z)\vec{N}'_{oe}(c-z') \end{bmatrix} \begin{matrix} z > z' \\ z < z' \end{matrix}, \tag{2.40}$$

where:

$$\vec{M}_{eo}(z) = \nabla \times (\hat{z}C_x C_y \sin k_g z), \tag{2.41}$$

$$\vec{N}_{oe}(z) = \frac{1}{k}\nabla \times \nabla \times (\hat{z}S_x S_y \cos k_g z), \tag{2.42}$$

$C_x = \cos k_x x$, $C_y = \cos k_y y$, $S_x = \sin k_x x$, $S_y = \sin k_y y$, $k_x = \frac{m\pi}{a}$, $k_y = \frac{n\pi}{b}$, $k_c^2 = k_x^2 + k_y^2$, $k_g^2 = k^2 - k_c^2$, and $\delta_0 = \begin{cases} 1, m & \text{or} \quad n = 0 \\ 0, m & \text{and} \quad n \neq 0 \end{cases}$. The $\vec{M}_{eo}(z)$ vectors give the electric field of the TE modes as given previously in (2.15), and the $\vec{N}_{oe}(z)$ vectors give the electric field of the TM modes as given previously in (2.2) and (2.8). The primed quantities, $\vec{M}'_{eo}$ and

$\vec{N}'_{eo}$, relate to the effect of the location and polarization of the electric dipole source:

$$\vec{M}'_{eo}(z') = \nabla' \times [C'_x C'_y \sin k_g z' \hat{z}], \tag{2.43}$$

$$\vec{N}'_{oe}(z') = \frac{1}{k} \nabla' \times \nabla' \times \left[S'_x S'_y \cos k_g z' \hat{z} \right], \tag{2.44}$$

where $C'_x = \cos k_x x'$, $C'_y = \cos k_y y'$, $S'_x = \sin k_x x'$ and $S'_y = \sin k_y y'$.

When the excitation frequency corresponds to that of a resonant mode such that

$$k_g = \frac{p\pi}{c}, \quad p = 0, 1, 2, \ldots$$

or

$$\sqrt{\left(\frac{2\pi}{\lambda}\right)^2 - \left(\frac{m\pi}{a}\right)^2 - \left(\frac{n\pi}{b}\right)^2} = \frac{p\pi}{c}, \tag{2.45}$$

then $\vec{G}_e$ in (2.40) is singular. However, if we include wall loss as in Section 1.3, then we can replace k_g by k_g^l, where:

$$k_g^l \approx \sqrt{k^2 - \left[\left(\frac{m\pi}{a}\right)^2 + \left(\frac{n\pi}{b}\right)^2\right]\left(1 - \frac{2i}{Q_{mnp}}\right)} \tag{2.46}$$

We have neglected the Q_{mnp}^{-2} term in (2.45) because Q_{mnp} is large. The introduction of the $\frac{2i}{Q_{mnp}}$ term in (2.46) means that k_g^l cannot be real for real k. (We cannot have both m and n equal to zero.) Consequently, the sine term in the denominator of (2.40) cannot be zero:

$$\sin k_g^l c \neq 0, \tag{2.47}$$

and the singularities of (2.40) at the resonant frequencies no longer occur.

The solution to (2.35) and (2.39) for the magnetic dyadic Green's function can be derived from the curl of the electric dyadic Green's function [2]:

$$\overleftrightarrow{G}_m(\vec{r}, \vec{r}') = \nabla \times \overleftrightarrow{G}_e(\vec{r}, \vec{r}') \tag{2.48}$$

In order to apply (2.48), we need the expressions for the curls [2] of the relevant vector terms in (2.40):

$$\nabla \times \vec{M}_{eo}(z) = k\vec{N}_{eo}(z), \tag{2.49}$$

$$\nabla \times \vec{N}_{oe}(z) = k\vec{M}_{oe}(z) \tag{2.50}$$

If we substitute (2.40), (2.49), and (2.50) into (2.48), we can obtain the desired expression for $\overleftrightarrow{G}_m$:

$$\overleftrightarrow{G}_m(\vec{r},\vec{r}\prime) = \frac{2k}{ab}\sum_{m=0}^{\infty}\sum_{n=0}^{\infty}\frac{(2-\delta_0)}{k_c^2 k_g \sin k_g c}\begin{bmatrix}\vec{N}_{eo}(c-z)\vec{M}'_{eo}(z')-\vec{M}_{oe}(c-z)\vec{N}'_{eo}(z') \\ \vec{N}_{eo}(z)\vec{M}'_{eo}(c-z')-\vec{M}_{oe}(z)\vec{N}'_{oe}(c-z')\end{bmatrix}\begin{matrix} z > z' \\ z < z'\end{matrix} \tag{2.51}$$

In contrast to (2.40), (2.51) does not include a delta function because it is cancelled by the derivative of the discontinuity in (2.40) at $z = z'$.

2.3.1 Fields in the Source-Free Region

Consider a volume current density $\vec{J}(\vec{r}\prime)$ confined to a volume V' in a rectangular cavity, as shown in Figure 2.5. The observation point $\vec{r}$ is located within the cavity but outside the volume V'. The electric field can be written as an integral over the source volume [2]:

$$\vec{E}(r) = i\omega\mu_0 \iiint_{V'} \overleftrightarrow{G}_e(\vec{r},\vec{r}\prime)\bullet\vec{J}(\vec{r}\prime)\mathrm{d}V', \tag{2.52}$$

where $\overleftrightarrow{G}_e$ is given by (2.40). Similarly, the magnetic field can be written as an integral over the source volume [2]:

$$\vec{H}(\vec{r}) = \iiint_{V\prime} \overleftrightarrow{G}_m(\vec{r},\vec{r}\prime)\bullet\vec{J}(\vec{r}\prime)\mathrm{d}V', \tag{2.53}$$

where $\overleftrightarrow{G}_m$ is given by (2.51). The volume integrals in (2.52) and (2.53) are well behaved because $\overleftrightarrow{G}_e(\vec{r},\vec{r}\prime)$ and $\overleftrightarrow{G}_m(\vec{r},\vec{r}\prime)$ are well behaved for $\vec{r} \neq \vec{r}\prime$.

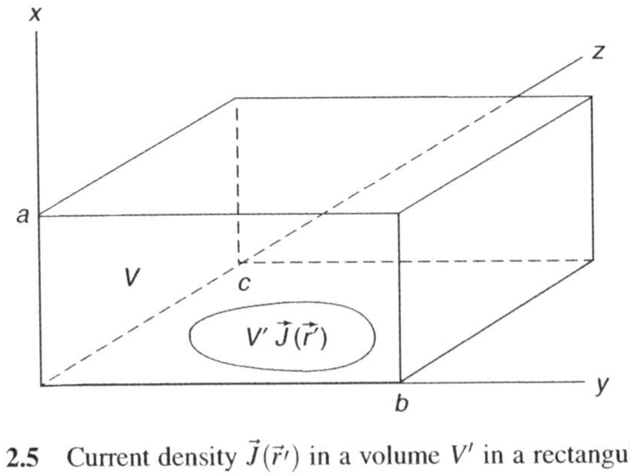

FIGURE 2.5 Current density $\vec{J}(\vec{r}\prime)$ in a volume V' in a rectangular cavity.

2.3.2 Fields in the Source Region

In the source region, we must deal with singularities in the Green's functions at $\vec{r} = \vec{r}'$. In evaluating the magnetic field, the singularity in $\overleftrightarrow{G}_m(\vec{r}, \vec{r}')$ at $\vec{r} = \vec{r}'$ is integrable, and (2.53) can still be used to calculate $\vec{H}$.

The evaluation of the electric field in the source region has been the subject of much discussion [20, 21]. The outcome of this discussion is that (2.52) cannot be used in the source region. It is necessary to replace (2.52) with a principle volume integral that excludes a small volume V_δ about $\vec{r} = \vec{r}'$ in the integration and adds a term proportional to the electric current. The details of the derivation are given in [20] and [21]; here we give only the final result:

$$\vec{E}(r) = i\omega\mu_0 \lim_{\delta \to 0} \iiint_{V'-V_\delta} \overleftrightarrow{G}_e(\vec{r}, \vec{r}') \bullet \vec{J}(\vec{r}')\mathrm{d}V' + \frac{\overleftrightarrow{L} \bullet \vec{J}(\vec{r})}{i\omega\varepsilon_0}, \tag{2.54}$$

where the source dyad $\overleftrightarrow{L}$ is given by [20]:

$$\overleftrightarrow{L} = \frac{1}{4\pi} \iint_{S_\delta} \frac{\hat{n}'\hat{e}_{R'}}{R'^2} \mathrm{d}R' \tag{2.55}$$

The geometry for determining $\overleftrightarrow{L}$ is shown in Figure 2.6. Numerically, the analytical limit in (2.54) is achieved if the maximum chord length δ satisfies [20]:

$$\delta \le \frac{\lambda}{2\pi}, \tag{2.56}$$

where λ is the free-space wavelength. Thus the maximum chord length of the principle volume needs to be small compared to a free-space wavelength, provided that the

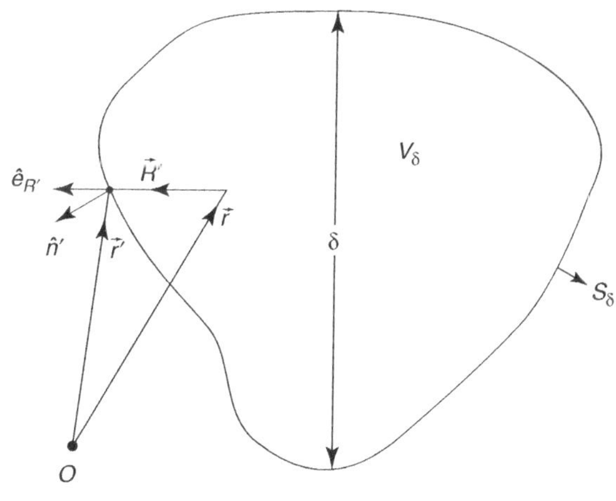

FIGURE 2.6 Principle volume V_δ in the current source region.

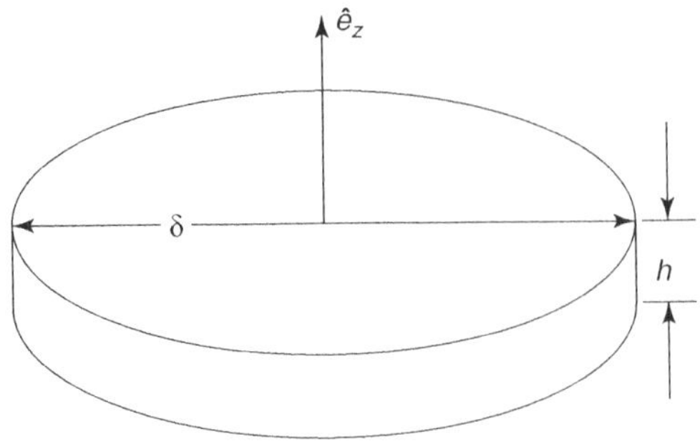

FIGURE 2.7 Principle volume in the shape of a pill box.

source current $\vec{J}$ does not vary appreciably over the same principle volume. The shape of principle volume is arbitrary, but for the rectangular cavity geometry of Figure 2.1, the most logical shape is a thin pill box, as shown in Figure 2.7, where $h/\delta \to 0$. In this case, $\overleftrightarrow{L}$ is given by [20]:

$$\overleftrightarrow{L} = \hat{e}_z \hat{e}_z \tag{2.57}$$

Note that the coefficient of the delta function in (2.40) also contains $\hat{e}_z \hat{e}_z$. Further discussion of this term is contained in [22].

PROBLEMS

2-1 Although it is conventional to derive modes that are TM and TE to the z axis, consider modes that are TM to the x axis in Figure 2.1. Start with the x components of the electric and magnetic fields written as:

$$E_{xmnp}^{TM_x} = E_{0x} \cos\frac{m\pi x}{a} \sin\frac{n\pi y}{b} \sin\frac{p\pi z}{c} \quad \text{and} \quad H_{xmnp}^{TM_x} = 0$$

Derive the expressions for the other four field components, $E_{ymnp}^{TM_x}$, $E_{zmnp}^{TM_x}$, $H_{ymnp}^{TM_x}$, and $H_{zmnp}^{TM_x}$, of the $\mathrm{TM_x}$ mode.

2-2 Consider now the $\mathrm{TE_x}$ mode. Start with the x components of the electric and magnetic fields written as:

$$E_{xmnp}^{TE_x} = 0 \quad \text{and} \quad H_{xmnp}^{TE_x} = H_{0x} \sin\frac{m\pi x}{a} \cos\frac{n\pi y}{b} \cos\frac{p\pi z}{c}$$

Derive the expressions for the other four field components, $E_{ymnp}^{TE_x}$, $E_{zmnp}^{TE_x}$, $H_{ymnp}^{TE_x}$, and $H_{zmnp}^{TE_x}$.

2-3 If we write the mode fields in vector form, show that the TM_x mode field can be written as a linear combination of the TM and TE mode fields:

$$\vec{E}_{mnp}^{TM_x} = A\vec{E}_{mnp}^{TM} + B\vec{E}_{mnp}^{TE}$$

Derive the expressions for A and B.

2-4 Show that the TE_x mode field can also be written as a linear combination of the TM and TE mode fields:

$$\vec{E}_{mnp}^{TE_x} = C\vec{E}_{mnp}^{TM} + D\vec{E}_{mnp}^{TE}$$

Derive the expressions for C and D.

2-5 Derive (2.18) from (2.3). Hint: construct a k_x, k_y, k_z lattice with appropriate spacings from (2.4). Then determine the number of resonant frequencies in one eighth of a sphere of radius k. Take account of the TM, TE mode degeneracy.

2-6 Derive (2.25) from (2.24).

2-7 Derive (2.26) from (2.24).

2-8 Derive (2.29) using the same method as that for (2.27).

2-9 Show that (2.40) satisfies (2.34).

2-10 Show that (2.40) satisfies (2.38).

2-11 Show that (2.51) satisfies (2.35).

2-12 Show that (2.51) satisfies (2.39).

2-13 Show that (2.53) is integrable in the source region V'. Are there requirements on the source current $\vec{J}(r')$ for this to hold?

CHAPTER 3

Circular Cylindrical Cavity

The circular cylindrical cavity is the second of three separable geometries that we will consider. The geometry for a cylindrical cavity of radius a and length d is shown in Figure 3.1. Circular cylindrical cavities are used as single-mode resonators [13] or for making dielectric or permeability measurements [23,24].

3.1 RESONANT MODES

The standard method for constructing the resonant modes for a circular cylindrical cavity is to derive modes that are TE or TM to the z axis. The TE modes can also be called magnetic modes because the E_z component is zero. Similarly, the TM modes can be called electric modes because the H_z component is zero.

From (1.18) and (1.19), we see that the z component of the electric field E_{znpq}^{TM} of a TM mode satisfies the scalar Helmholtz equation:

$$(\nabla^2 + k_{npq}^2) E_{znpq}^{TM} = 0, \tag{3.1}$$

where k_{npq} is an eigenvalue to be determined. The triple subscript will be explained as we proceed with the solution of (3.1). In cylindrical coordinates (ρ,ϕ,z), the first term in (3.1) can be written (see Appendix A):

$$\nabla^2 E_{znpq}^{TM} = \frac{1}{\rho}\frac{\partial}{\partial \rho}\left(\rho \frac{\partial E_{znpq}^{TM}}{\partial \rho}\right) + \frac{1}{\rho^2}\frac{\partial^2 E_{znpq}^{TM}}{\partial \phi^2} + \frac{\partial^2 E_{znpq}^{TM}}{\partial z^2} \tag{3.2}$$

If we use separation of variables, we can write E_{znpq}^{TM} as [3]:

$$E_{znpq}^{TM} = R(\rho)\Phi(\phi)Z(z) \tag{3.3}$$

If we substitute (3.2) and (3.3) into (3.1) and divide by E_{znpq}^{TM}, we obtain:

$$\frac{1}{\rho R}\frac{\mathrm{d}}{\mathrm{d}\rho}\left(\rho \frac{\mathrm{d}R}{\mathrm{d}\rho}\right) + \frac{1}{\rho^2 \Phi}\frac{\mathrm{d}^2\Phi}{\mathrm{d}\phi^2} + \frac{1}{Z}\frac{\mathrm{d}^2 Z}{\mathrm{d}z^2} + k_{npq}^2 = 0 \tag{3.4}$$

Electromagnetic Fields in Cavities: Deterministic and Statistical Theories, by David A. Hill

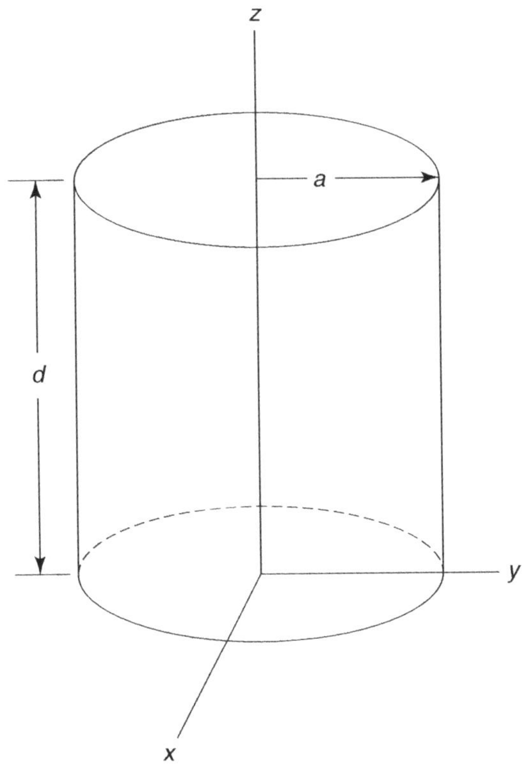

FIGURE 3.1 Circular cylindrical cavity.

Since the third term in (3.4) depends only on z, we can write it as:

$$\frac{1}{Z}\frac{d^2Z}{dz^2} = -k_z^2, \tag{3.5}$$

where k_z is a separation constant to be determined later. If we substitute (3.5) into (3.4) and multiply by ρ^2, we obtain:

$$\frac{\rho}{R}\frac{d}{d\rho}\left(\rho\frac{dR}{d\rho}\right) + \frac{1}{\Phi}\frac{d^2\Phi}{d\phi^2} + (k_{npq}^2 - k_z^2)\rho^2 = 0 \tag{3.6}$$

The second term in (3.6) depends only on ϕ; so we can write it as:

$$\frac{1}{\Phi}\frac{d^2\Phi}{d\phi^2} = -n^2 \tag{3.7}$$

If we substitute (3.7) into (3.6), replace $k_{npq}^2 - k_z^2$ by k_ρ^2, and multiply by R, we obtain:

$$\rho\frac{d}{d\rho}\left(\rho\frac{dR}{d\rho}\right) + \left[(k_\rho\rho)^2 - n^2\right]R = 0 \tag{3.8}$$

This is Bessel's equation [25] of order n. For convenience, we can rewrite (3.5) and (3.7) as:

$$\frac{d^2Z}{dz^2} + k_z^2 Z = 0, \tag{3.9}$$

$$\frac{d^2\Phi}{d\phi^2} + n^2\Phi = 0 \tag{3.10}$$

With (3.8) – (3.10) we have separated (3.1) into three ordinary differential equations with known solutions. Since the normal derivative of E_{znpq}^{TM} is zero at $z = 0$ and d, the solution to (3.9) is:

$$Z(z) = \cos\left(\frac{q\pi}{d}z\right), \quad q = 0, 1, 2, \ldots \tag{3.11}$$

Since Φ must be periodic in 2π, the solutions to (3.10) are:

$$\Phi(\phi) = \begin{Bmatrix} \sin n\phi \\ \cos n\phi \end{Bmatrix}, \quad n = 0, 1, 2, \ldots \tag{3.12}$$

From the electric field boundary condition in (1.22), the Bessel function [25] solution to (3.8) that is finite at $\rho = 0$ can be written:

$$R(\rho) = \mathrm{J}_n(k_\rho \rho), \tag{3.13}$$

where $k_\rho = x_{np}/a$, and x_{np} is the pth zero of the nth order Bessel function:

$$\mathrm{J}_n(x_{np}) = 0, \quad \text{where } n = 0, 1, 2, \ldots \quad \text{and } p = 1, 2, 3, \ldots \tag{3.14}$$

Some of the lower-order zeros of J_n are shown in Table 3.1 [13].

The z component of the electric field of a TM mode can be written:

$$E_{znpq}^{TM} = E_0 \mathrm{J}_n\left(\frac{x_{np}}{a}\rho\right)\begin{Bmatrix} \sin n\phi \\ \cos n\phi \end{Bmatrix}\cos\left(\frac{q\pi}{d}z\right), \tag{3.15}$$

where $k_{npq}^2 = \left(\frac{q\pi}{d}\right)^2 + \left(\frac{x_{np}}{a}\right)^2$ and E_0 is an arbitrary constant with units of V/m.

TABLE 3.1 Roots of $\mathrm{J}_n(p_{nm}) = 0$ [13].

n	p_{n1}	p_{n2}	p_{n3}	p_{n4}
0	2.405	5.520	8.654	11.792
1	3.832	7.016	10.174	13.324
2	5.135	8.417	11.620	14.796

As with the rectangular cavity, the electric and magnetic fields can be obtained from an electric Hertz vector [13] that has only a z component Π_e:

$$\vec{\Pi}_e = \hat{z}\Pi_e \tag{3.16}$$

Curl operations on $\vec{\Pi}_e$ yield [13]:

$$\vec{E} = \nabla \times \nabla \times \vec{\Pi}_e \quad \text{and} \quad \vec{H} = -i\omega\varepsilon\nabla \times \vec{\Pi}_e \tag{3.17}$$

From (3.15) and (3.17), we can determine that the z component of the electric Hertz vector for the npq mode must take the form:

$$\Pi_{enpq} = \frac{E_{znpq}^{TM}}{k_{npq}^2 - (q\pi/d)^2} = \frac{E_0}{k_{npq}^2 - (q\pi/d)^2} \mathrm{J}_n\left(\frac{x_{np}}{a}\rho\right)\left\{\begin{array}{c}\sin n\phi \\ \cos n\phi\end{array}\right\}\cos\left(\frac{q\pi}{d}z\right) \tag{3.18}$$

The z component of the electric field is given in (3.15), and the transverse components are determined from (3.17) and (3.18):

$$E_{\rho npq}^{TM} = \frac{-E_0}{k_{npq}^2 - (q\pi/d)^2} \frac{q\pi}{d}\frac{x_{np}}{a} \mathrm{J}_n'\left(\frac{x_{np}}{a}\rho\right)\left\{\begin{array}{c}\sin n\phi \\ \cos n\phi\end{array}\right\}\sin\left(\frac{q\pi}{d}z\right), \tag{3.19}$$

$$E_{\phi npq}^{TM} = \frac{-E_0}{k_{npq}^2 - (q\pi/d)^2} \frac{1}{\rho}\frac{nq\pi}{d} \mathrm{J}_n\left(\frac{x_{np}}{a}\rho\right)\left\{\begin{array}{c}\cos n\phi \\ -\sin n\phi\end{array}\right\}\sin\left(\frac{q\pi}{d}z\right), \tag{3.20}$$

where J'_n is the derivative of J_n with respect to the argument. The z component of the magnetic field is zero (by definition for a TM mode), and the transverse components of the magnetic field are determined from (3.17) and (3.18):

$$H_{\rho npq}^{TM} = \frac{-i\omega_{npq}\varepsilon E_0}{k_{npq}^2 - (q\pi/d)^2} \frac{n}{\rho} \mathrm{J}_n\left(\frac{x_{np}}{a}\rho\right)\left\{\begin{array}{c}\cos n\phi \\ -\sin n\phi\end{array}\right\}\cos\left(\frac{q\pi}{d}z\right), \tag{3.21}$$

$$H_{\phi npq}^{TM} = \frac{i\omega_{npq}\varepsilon E_0}{k_{npq}^2 - (q\pi/d)^2} \frac{x_{np}}{a} \mathrm{J}_n'\left(\frac{x_{np}}{a}\rho\right)\left\{\begin{array}{c}\sin n\phi \\ \cos n\phi\end{array}\right\}\cos\left(\frac{q\pi}{d}z\right) \tag{3.22}$$

The allowable values for n, p, and q are $n = 0, 1, 2, \ldots$; $p = 1, 2, 3, \ldots$; and $q = 0, 1, 2, \ldots$.

The TE (or magnetic) modes are derived in an analogous manner. The z component of the magnetic field satisfies the scalar Helmholtz equation, and the boundary conditions require that it takes the form:

$$H_{znpq}^{TE} = H_0\mathrm{J}_n\left(\frac{x'_{np}}{a}\rho\right)\left\{\begin{array}{c}\sin n\phi \\ \cos n\phi\end{array}\right\}\sin\left(\frac{q\pi}{d}z\right), \tag{3.23}$$

where H_0 is an arbitrary constant with units of A/m, n and p are integers, and x'_{np} is the pth zero of the derivative of J_n: $\mathrm{J}'_n(x'_{np}) = 0$. Some of the lower-order zeros of J'_n are shown in Table 3.2 [13].

TABLE 3.2 Roots of $J'_n(p'_{nm}) = 0$ [13].

n	p'_{n1}	p'_{n2}	p'_{n3}	p'_{n4}
0	3.832	7.016	10.174	13.324
1	1.841	5.331	8.536	11.706
2	3.054	6.706	9.970	13.170

The electric and magnetic fields can be determined from a magnetic Hertz vector [13] that has only a z component Π_h:

$$\vec{\Pi}_h = \hat{z}\Pi_h \tag{3.24}$$

Curl operations on (3.24) yield [13]:

$$\vec{H} = \nabla \times \nabla \times \vec{\Pi}_h \quad \text{and} \quad \vec{E} = i\omega\mu\nabla \times \vec{\Pi}_h \tag{3.25}$$

From (3.23) and (3.25), we can determine that the z component of the magnetic Hertz vector for the npq mode must take the form:

$$\Pi_{hnpq} = \frac{H^{TE}_{znpq}}{k^2_{npq} - (q\pi/d)^2} = \frac{H_0}{k^2_{npq} - (q\pi/d)^2} J_n\left(\frac{x'_{np}}{a}\rho\right)\begin{Bmatrix} \sin n\phi \\ \cos n\phi \end{Bmatrix}\sin\left(\frac{q\pi}{d}z\right) \tag{3.26}$$

The z component of the magnetic field is given in (3.23), and the transverse components are determined from (3.25) and (3.26):

$$H^{TE}_{\rho npq} = \frac{H_0}{k^2_{npq} - (q\pi/d)^2}\frac{q\pi}{d}\frac{x'_{np}}{a} J'_n\left(\frac{x'_{np}}{a}\rho\right)\begin{Bmatrix} \sin n\phi \\ \cos n\phi \end{Bmatrix}\cos\left(\frac{q\pi}{d}z\right), \tag{3.27}$$

$$H^{TE}_{\phi npq} = \frac{H_0}{k^2_{npq} - (q\pi/d)^2}\frac{q\pi}{d}\frac{n}{\rho} J_n\left(\frac{x'_{np}}{a}\rho\right)\begin{Bmatrix} \cos n\phi \\ -\sin n\phi \end{Bmatrix}\cos\left(\frac{q\pi}{d}z\right) \tag{3.28}$$

The z component of the electric field is zero (by definition for a TE mode), and the transverse components of the electric field are determined from (3.25) and (3.26):

$$E^{TE}_{\rho npq} = \frac{i\omega\mu H_0}{k^2_{npq} - (q\pi/d)^2}\frac{n}{\rho} J_n\left(\frac{x'_{np}}{a}\rho\right)\begin{Bmatrix} \cos n\phi \\ -\sin n\phi \end{Bmatrix}\sin\left(\frac{q\pi}{d}z\right), \tag{3.29}$$

$$E^{TE}_{\phi npq} = \frac{-i\omega\mu H_0}{k^2_{npq} - (q\pi/d)^2}\frac{x'_{np}}{a} J'_n\left(\frac{x'_{np}}{a}\rho\right)\begin{Bmatrix} \sin n\phi \\ \cos n\phi \end{Bmatrix}\sin\left(\frac{q\pi}{d}z\right) \tag{3.30}$$

The allowable values for the mode numbers are $n = 0, 1, 2, \ldots$; $p = 1, 2, 3, \ldots$; and $q = 1, 2, 3, \ldots$.

The resonant wavenumbers for the TM and TE modes are:

$$k_{npq}^{TM} = \sqrt{\left(\frac{\chi_{np}}{a}\right)^2 + \left(\frac{q\pi}{d}\right)^2}, \tag{3.31}$$

$$k_{npq}^{TE} = \sqrt{\left(\frac{\chi'_{np}}{a}\right)^2 + \left(\frac{q\pi}{d}\right)^2} \tag{3.32}$$

By setting $f = k/(2\pi\sqrt{\mu\varepsilon})$, we can determine the resonant frequencies of the TM and TE modes:

$$f_{npq}^{TM} = \frac{1}{2\pi\sqrt{\mu\varepsilon}}\sqrt{\left(\frac{\chi_{np}}{a}\right)^2 + \left(\frac{q\pi}{d}\right)^2}, \tag{3.33}$$

$$f_{npq}^{TE} = \frac{1}{2\pi\sqrt{\mu\varepsilon}}\sqrt{\left(\frac{\chi'_{np}}{a}\right)^2 + \left(\frac{q\pi}{d}\right)^2} \tag{3.34}$$

For $n > 0$, each n represents represents a pair of degenerate TM and TE modes (cos $n\phi$ or sin $n\phi$ variation).

Table 3.3 shows the normalized resonant frequencies for various values of d/a [3]. For $d/a < 2$, the TM_{010} mode is dominant (has the lowest resonant frequency). The field distribution for the TM_{010} mode is shown in Figure 3.2 [3]. For $d/a \geq 2$, the TE_{111} mode is the dominant mode.

For use as single-mode resonators (filters or electromagnetic property measurements), the goal is to excite only a single mode at its resonant frequency or at its perturbed resonant frequency for material measurements [23,24]. However, for use of a cylindrical cavity as a reverberation chamber (mode-stirred chamber) [18,19], it is useful to know the number of modes available for stirring over a large bandwidth. The number of modes with eigenvalues k_{npq} less than k can be approximated by (2.18) because that expression applies to cavities of arbitrary shape. The volume V of a cylindrical cavity is given by:

$$V = \pi a^2 d \tag{3.35}$$

TABLE 3.3 $\frac{f_{npq}}{f_{\text{dominant}}}$ **for a Circular Cavity of Radius a and Length d [3].**

$\frac{d}{a}$	TM_{010}	TE_{111}	TM_{110}	TM_{011}	TE_{211}	TM_{111} TE_{011}	TE_{112}	TM_{210}	TM_{020}
0	1.0	∞	1.59	∞	∞	∞	∞	2.13	2.29
0.5	1.0	2.72	1.59	2.80	2.90	3.06	5.27	2.13	2.29
1.0	1.0	1.50	1.59	1.63	1.80	2.05	2.72	2.13	2.29
2.0	1.0	1.0	1.59	1.19	1.42	1.72	1.50	2.13	2.29
3.0	1.13	1.0	1.80	1.24	1.52	1.87	1.32	2.41	2.60
4.0	1.20	1.0	1.91	1.27	1.57	1.96	1.30	2.56	3.00
∞	1.30	1.0	2.08	1.31	1.66	2.08	1.0	2.78	3.00

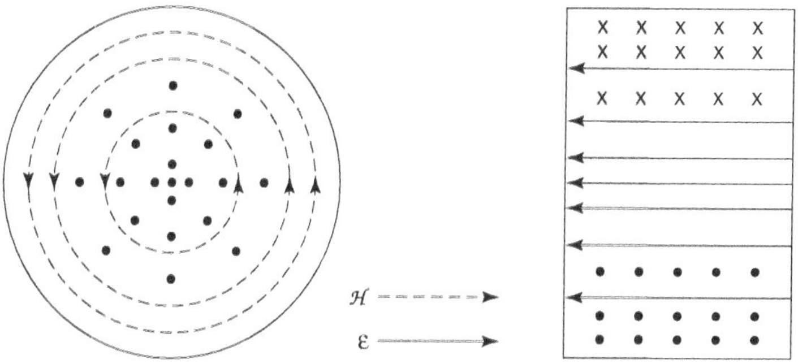

FIGURE 3.2 Instantaneous electric $\mathcal{E}$ and magnetic $\mathcal{H}$ field lines for the TM_{010} cavity mode [3].

If we substitute (3.35) into (2.18), the Weyl approximation for the number of modes is:

$$N_W(k) = \frac{a^2 dk^3}{2\pi} \tag{3.36}$$

If we wish to write the number of modes in terms of frequency f, we can replace k by $2\pi f/v$ in (3.37) to obtain:

$$N_W(f) = 4\pi^2 a^2 d(f/v)^3 \tag{3.37}$$

The mode density (modes/Hz) can be obtained by differentiating (3.37) with respect to f:

$$D_W(f) = \frac{dN_W(f)}{df} = 12\pi^2 a^2 d \frac{f^2}{v^3} \tag{3.38}$$

3.2 WALL LOSSES AND CAVITY *Q*

An expression for cavity Q due to wall losses of cavities of arbitrary shape was given in (1.41). For cylindrical cavities, the expressions for the magnetic field are known, and the integrals can be evaluated to determine Q for the various mode types and numbers. Harrington [3, p. 257] has given the Q expressions for the TE and TM modes of arbitrary order.

To illustrate the details of the evaluation of Q, we will derive Q for the specific case of the TM_{010} mode, which is the dominant mode (lowest resonant frequency) for $d/a < 2$. We first write (1.41) in the following form:

$$Q_{010}^{TM} = \frac{\omega_{010}\mu}{R_s} \frac{\iiint\limits_V \vec{H}_{010}^{TM} \bullet \vec{H}_{010}^{TM*} \, dV}{\oiint\limits_S \vec{H}_{010}^{TM} \bullet \vec{H}_{010}^{TM*} \, dS}, \tag{3.39}$$

where the magnetic field (which includes only a ϕ component) is given by (3.22). The dot product in (3.39) can be written:

$$\vec{H}_{010}^{TM} \bullet \vec{H}_{010}^{TM*} = \frac{\omega_{010}^2 \varepsilon^2 |E_0|^2}{k_{010}^4} \frac{x_{01}^2}{a^2} J_0'^2 \left(\frac{x_{01}}{a} \rho \right) \cos^2 \phi \tag{3.40}$$

The volume integral in the numerator of (3.39) can be written:

$$\iiint_V \vec{H}_{010}^{TM} \bullet \vec{H}_{010}^{TM*} \mathrm{d}V = \int_0^d \int_0^{2\pi} \int_0^a \vec{H}_{010}^{TM} \bullet \vec{H}_{010}^{TM*} \rho \mathrm{d}\rho \mathrm{d}\phi \mathrm{d}z \tag{3.41}$$

The ϕ and z integrations in (3.41) are easily performed. The ρ integration can be done by use of the following known integral [26, p. 634]:

$$\int_0^a J_0'^2 \left(\frac{x_{01}}{a} \rho \right) \rho \mathrm{d}\rho = \int_0^a J_1^2 \left(\frac{x_{01}}{a} \rho \right) \rho \mathrm{d}\rho = \frac{a^2}{2} J_1^2(x_{01}) \tag{3.42}$$

The expressions in (3.40) and (3.42) can be used to obtain the following result for the volume integral in (3.41):

$$\iiint_V \vec{H}_{010}^{TM} \bullet \vec{H}_{010}^{TM*} \mathrm{d}V = \frac{\pi d |E_0|^2 \eta x_{01}^2 J_1^2(x_{01})}{2k_{010}^2} \tag{3.43}$$

The surface integral in the denominator of (3.39) can be written:

$$\oiint_S \vec{H}_{010}^{TM} \bullet \vec{H}_{010}^{TM*} \mathrm{d}S = 2 \int_0^{2\pi} \int_0^a \vec{H}_{010}^{TM} \bullet \vec{H}_{010}^{TM*} \rho \mathrm{d}\rho \mathrm{d}\phi + da \int_0^{2\pi} \vec{H}_{010}^{TM} \bullet \vec{H}_{010}^{TM*} |_{\rho=a} \mathrm{d}\phi \tag{3.44}$$

With the aid of the ρ integral result in (3.42), we can evaluate (3.44):

$$\oiint_S \vec{H}_{010}^{TM} \bullet \vec{H}_{010}^{TM*} \mathrm{d}S = \frac{\pi(a+d)\eta |E_0|^2 x_{01}^2 J_1^2(x_{01})}{k_{010}^2 a} \tag{3.45}$$

If we substitute (3.43) and (3.45) into (3.39) and use the relationship $k_{010}a = x_{01}$, we obtain the desired result for Q_{010}^{TM}:

$$Q_{010}^{TM} = \frac{\eta x_{01} d}{2R_s(a+d)} \tag{3.46}$$

The Q expressions for general TM and TE modes can also be determined from (1.41), but the algebra is more complex. The resultant expressions are [3, p. 257]:

$$Q_{npq}^{TM} = \frac{\eta \sqrt{x_{np}^2 + (q\pi a/d)^2}}{2R_s(1+a/d)}, \tag{3.47}$$

$$Q_{npq}^{TE} = \frac{\eta[x'^2_{np} + (q\pi a/d)^2]^{3/2}(x'^2_{np} - n^2)}{2R_s\left[\left(\frac{nq\pi a}{d}\right)^2 + x'^4_{np} + \frac{2a}{d}\left(\frac{q\pi a}{d}\right)^2 (x'^2_{np} - n^2)\right]} \tag{3.48}$$

As a consistency check, it is easy to show that (3.47) reduces to (3.46) for $n = q = 0$ and $p = 1$.

3.3 DYADIC GREEN'S FUNCTIONS

Dyadic Green's functions for a circular cylindrical cavity have been derived by Tai [2] in a similar manner as for the rectangular cavity. They are again useful in providing a compact notation for determining the electric and magnetic fields due to current sources. Circular cylindrical cavities are typically excited by a dipole, monopole, or loop antenna, and dyadic Green's functions are useful for the analysis of such sources. (The electric field inside the source region requires special treatment [20], but the electric dyadic Green's function is still useful there.)

The electric $\overleftrightarrow{G}_e$ and magnetic $\overleftrightarrow{G}_m$ dyadic Green's functions satisfy the differential equations given in (2.34) and (2.35). In addition to the differential equations, we need to specify boundary conditions to make the dyadic Green's functions unique. The electric dyadic Green's function needs to satisfy (2.38) at $\rho = a$ and $z = 0$ and d. The magnetic dyadic Green's function needs to satisfy (2.39) at $\rho = a$ and $z = 0$ and d.

The solution for the electric dyadic Green's function is [2]:

$$\overleftrightarrow{G}_e(\vec{r},\vec{r}') = -\frac{\hat{z}\hat{z}}{k^2}\delta(\vec{r}-\vec{r}') + \sum_{n=0}^{\infty}\sum_{p=1}^{\infty}\frac{2-\delta_0}{2\pi}\left\{\frac{1}{\left(\frac{x_{np}}{a}\right)^2 I_\mu k_\mu \sin k_\mu d}\begin{matrix}\vec{M}_{npo}(d-z)\vec{M}'_{npo}(z') \\ \vec{M}_{npo}(z) \rightharpoonup M'_{npo}(d-z')\end{matrix}\right.$$

$$\left. - \frac{1}{\left(\frac{x_{np}}{a}\right)^2 I_\lambda k_\lambda \sin k_\lambda d}\begin{matrix}\vec{N}_{npe}(d-z)\vec{N}'_{npe}(z') \\ \vec{N}_{npe}(z)\vec{N}'_{npe}(d-z')\end{matrix}\right\}, \begin{matrix} z > z' \\ z < z' \end{matrix}, \tag{3.49}$$

where:

$$\vec{M}_{npo}(z) = \nabla \times \left[\hat{z}J_n\left(\frac{x_{np}}{a}\rho\right)\begin{matrix}\cos n\phi \\ \sin n\phi\end{matrix}\sin k_\mu z\right], \tag{3.50}$$

$$\vec{N}_{npe}(z) = \frac{1}{k}\nabla \times \nabla \times \left[\hat{z}J_n\left(\frac{x'_{np}}{a}\rho\right)\begin{matrix}\cos n\phi \\ \sin n\phi\end{matrix}\cos k_\lambda z\right], \tag{3.51}$$

$k_\mu = \sqrt{k^2 - (x_{nm}/a)^2}$, $k_\lambda = \sqrt{k^2 - (x'_{np}/a)^2}$, $I_\mu = \frac{a^2}{2x'^2_{np}}(x'^2_{np} - n^2)\mathrm{J}_n^2(x'_{np})$, $I_\lambda = \frac{a^2}{2}\mathrm{J}'^2_n(x_{np})$, and $\delta_0 = \begin{cases} 1, & n = 0 \\ 0, & n \neq 0 \end{cases}$. The $\vec{M}_{npo}$ vectors give the electric field of the TE modes as given previously in (3.29) and (3.30), and the $\vec{N}_{npe}$ vectors give the electric field of the TM modes as given previously in (3.15), (3.19), and (3.20). The primed vectors, $\vec{M}'_{npo}$ and $\vec{N}'_{npe}$, relate to the effect of the location and polarization of the electric dipole source:

$$\vec{M}'_{npo}(z') = \nabla' \times \left[\mathrm{J}_n\left(\frac{x_{np}}{a}\rho'\right) \begin{matrix} \cos n\phi' \\ \sin n\phi' \end{matrix} \sin k_\mu z'\right], \tag{3.52}$$

$$\vec{N}'_{npe}(z') = \frac{1}{k}\nabla' \times \nabla' \times \left[\mathrm{J}_n\left(\frac{x'_{np}}{a}\rho'\right) \begin{matrix} \cos n\phi' \\ \sin n\phi' \end{matrix} \cos k_\lambda z'\right] \tag{3.53}$$

When the excitation frequency corresponds to that of a resonant mode such that:

$$k_\mu = \frac{q\pi}{d}, \quad q = 0, 1, 2, \ldots \tag{3.54}$$

or

$$\sqrt{k^2 - \left(\frac{x_{np}}{a}\right)^2} = \frac{q\pi}{d},$$

then $\vec{G}_e$ in (3.49) is singular because $\sin k_\mu d = 0$. However, if we include wall loss as in Section 1.3, we can replace k_μ by k^l_μ, where:

$$k^l_\mu \approx \sqrt{k^2 - \left(\frac{x_{np}}{a}\right)^2\left(1 - \frac{2i}{Q_{npq}}\right)} \tag{3.55}$$

We have neglected the Q^{-2}_{npq} term in (3.55) because Q_{npq} is large. The introduction of the $\frac{2i}{Q_{npq}}$ term in (3.55) means that k_μ cannot be real for real k. Consequently, the $\sin k_\mu$ factor in the denominator of (3.49) cannot be zero. The same considerations apply to the case where:

$$k_\lambda = \frac{q'\pi}{d}, \quad q' = 0, 1, 2, \ldots$$

or

$$\sqrt{k^2 - \left(\frac{x'_{np}}{z}\right)^2} = \frac{q'\pi}{d} \tag{3.56}$$

If we include wall loss, we can replace k_λ by k^l_λ, where:

$$k^l_\lambda \approx \sqrt{k^2 - \left(\frac{x'_{np}}{a}\right)^2\left(1 - \frac{2i}{Q_{npq'}}\right)} \tag{3.57}$$

As with (3.55), we have neglected the $Q_{npq'}^{-2}$ term in (3.77) because $Q_{npq'}$ is large. Since k_λ cannot be real for real k, the $\sin k_\lambda$ factor in the denominator of (3.49) cannot be zero.

The solution to (2.35) and (2.39) for the magnetic dyadic Green's function can be obtained from the curl of the electric dyadic Green's function [2] as in (2.48). In order to apply (2.48), we need the expressions for the curls [2] of the relevant vector terms in (3.49):

$$\nabla \times \overleftrightarrow{M}_{npo}(z) = k\overleftrightarrow{N}_{npo}(z), \tag{3.58}$$

$$\nabla \times \overleftrightarrow{N}_{npe}(z) = k\overleftrightarrow{M}_{npe}(z) \tag{3.59}$$

If we substitute (3.49), (3.58), and (3.59) into (2.48), we obtain the desired expression for $\overleftrightarrow{G}_m$:

$$\overleftrightarrow{G}_m(\vec{r},\vec{r}') = \sum_{n=0}^{\infty}\sum_{p=1}^{\infty}\frac{k(2-\delta_0)}{2\pi}\left\{\frac{1}{\left(\frac{x'_{np}}{a}\right)^2 I_\mu k_\mu \sin k_\mu d}\begin{matrix}\vec{N}_{npo}(d-z)\vec{M}'_{npo}(z') \\ \vec{N}_{npo}(z)\vec{M}'_{npo}(d-z)\end{matrix}\right.$$

$$\left.-\frac{1}{\left(\frac{x_{np}}{a}\right)^2 I_\lambda k_\lambda \sin k_\lambda d}\begin{matrix}\vec{M}_{npe}(d-z)\vec{N}'_{npe}(z') \\ \vec{M}_{npe}(z)\vec{N}'_{npe}(d-z')\end{matrix}\right\},\quad \begin{matrix} z > z' \\ z < z' \end{matrix} \tag{3.60}$$

In contrast to (3.49), (3.60) does not include a delta function because it is cancelled by the derivative of the discontinuity in (3.49) at $z = z'$.

3.3.1 Fields in the Source-Free Region

Consider a volume current density $\vec{J}(\vec{r}')$ confined to a volume V' in a circular cylindrical cavity, as shown in Figure 3.3. The observation point r is located within the cavity, but outside the volume V'. The electric field can be written as an integral over the source volume [2]:

$$\vec{E}(r) = i\omega\mu \iiint_{V'} \overleftrightarrow{G}_e(\vec{r}, r') \bullet \vec{J}(r')\mathrm{d}V', \tag{3.61}$$

where $\overleftrightarrow{G}_e$ is given by (3.49). Similarly, the magnetic field can be written as an integral over the source volume [2]:

$$\vec{H}(\vec{r}) = \iiint_{V'} \overleftrightarrow{G}_m(\vec{r}, \vec{r}') \bullet \vec{J}(\vec{r}')\mathrm{d}V', \tag{3.62}$$

where $\overleftrightarrow{G}_m$ is given by (3.60). The volume integrals in (3.61) and (3.62) are well-behaved because $\overleftrightarrow{G}_e$ and $\overleftrightarrow{G}_m$ are well-behaved for $\vec{r} \neq r'$.

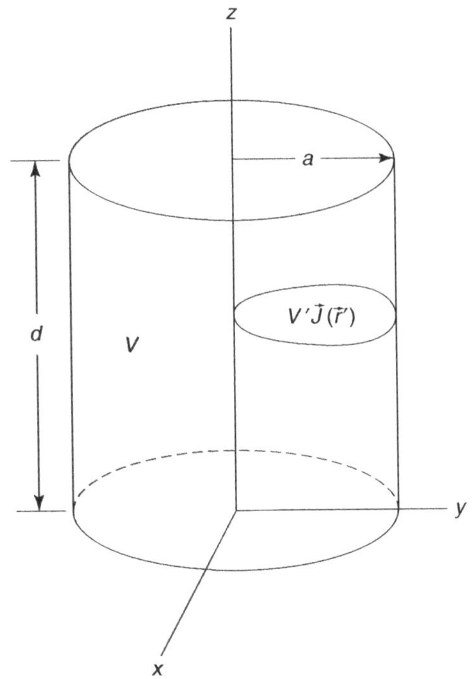

FIGURE 3.3 Current density $\vec{J}(\vec{r}')$ in a volume V' in a circular cylindrical cavity.

3.3.2 Fields in the Source Region

In the source region, we must deal with the singularities in the Green's functions at $\vec{r} = \vec{r}'$. The formal results are the same as those for the rectangular cavity in Section 2.3. In evaluating the magnetic field, the singularity in $\overleftrightarrow{G}_m(\vec{r}, \vec{r}')$ at $\vec{r} = \vec{r}'$ is integrable, and (3.62) can still be used to calculate $\vec{H}$.

The evaluation of the electric field has been discussed in Section 2.3, and (3.61) needs to be modified to (2.54)–(2.57). The only difference is that $\overleftrightarrow{G}_e$ for the rectangular cavity is replaced by $\overleftrightarrow{G}_e$ for the cylindrical cavity as given by (3.49).

PROBLEMS

3-1 Consider a vacuum-filled cylindrical cavity as in Figure 3.1 with $d = 2$ cm and $a = 1$ cm. Determine the resonant frequencies of the TM_{010} and TE_{111} modes. Are they equal as indicated in Table 3.3?

3-2 For copper walls ($\sigma_W = 5.7 \times 10^7$ S/m), what are the Q values for the two modes in Problem 3-1?

3-3 Derive (3.47) from (1.41).

3-4 Derive (3.48) from (1.41).

3-5 Show that (3.49) satisfies (2.34).

3-6 Show that (3.49) satisfies (2.38) at $\rho = a$ and $z = 0$ and d.

3-7 Show that (3.60) satisfies (2.35).

3-8 Show that (3.60) satisfies (2.39) at $\rho = a$ and $z = 0$ and d.

3-9 Show that (3.62) is integrable in the source region V'. Are there requirements on the source current $\vec{J}(r')$ for this to hold?

CHAPTER 4

Spherical Cavity

The spherical cavity is the third and final separable geometry we will consider. The geometry for a spherical cavity of radius a is shown in Figure 4.1. Spherical cavities have the potential of use for making dielectric or permeability measurements [27], but are used less frequently than circular cylindrical cavities.

4.1 RESONANT MODES

In spherical coordinates (r, θ, ϕ), we cannot use the method of deriving modes that are transverse electric or magnetic to z as in Chapters 2 and 3. However, if we follow Tai [2] or Harrington [3], we can construct modes that are transverse electric or transverse magnetic to $\vec{r}$ (TE_r or TM_r). We begin by finding solutions to the scalar Helmholtz equation:

$$(\nabla^2 + k^2)\psi = 0 \tag{4.1}$$

By substituting the Laplacian in spherical coordinates into (4.1), we obtain:

$$\frac{1}{r^2}\frac{\partial}{\partial r}\left(r^2\frac{\partial \psi}{\partial r}\right) + \frac{1}{r^2 \sin\theta}\frac{\partial}{\partial \theta}\left(\sin\theta\frac{\partial \psi}{\partial \theta}\right) + \frac{1}{r^2 \sin^2\theta}\frac{\partial^2 \psi}{\partial \phi^2} + k^2\psi = 0 \tag{4.2}$$

We can use the method of separation of variables by writing the scalar potential ψ as:

$$\psi = R(r)H(\theta)\Phi(\phi) \tag{4.3}$$

By substituting (4.3) into (4.2), dividing by ψ, and multiplying by $r^2 \sin^2\theta$, we obtain:

$$\frac{\sin^2\theta}{R}\frac{\mathrm{d}}{\mathrm{d}r}\left(r^2\frac{\mathrm{d}R}{\mathrm{d}r}\right) + \frac{\sin\theta}{H}\frac{\mathrm{d}}{\mathrm{d}\theta}\left(\sin\theta\frac{\mathrm{d}H}{\mathrm{d}\theta}\right) + \frac{1}{\Phi}\frac{\mathrm{d}^2\Phi}{\mathrm{d}\phi^2} + k^2r^2\sin^2\theta = 0 \tag{4.4}$$

Electromagnetic Fields in Cavities: Deterministic and Statistical Theories, by David A. Hill

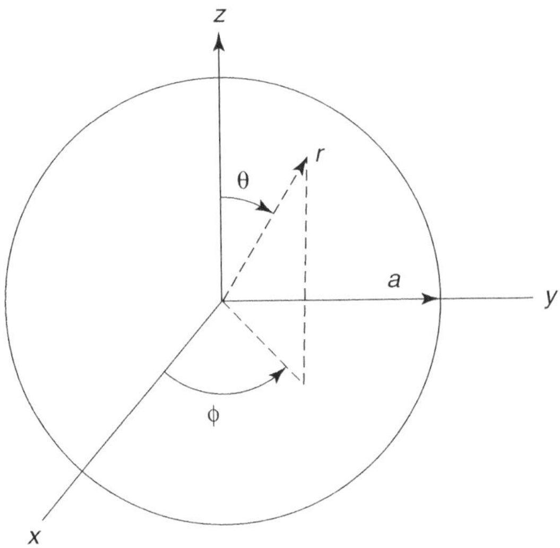

FIGURE 4.1 Spherical cavity.

The ϕ dependence in (4.4) is separated out by use of the integer m in the separation equation:

$$\frac{1}{\Phi}\frac{\mathrm{d}^2\Phi}{\mathrm{d}\phi^2} = -m^2 \tag{4.5}$$

If we substitute (4.5) into (4.4) and divide by $\sin^2\theta$, we obtain:

$$\frac{1}{R}\frac{\mathrm{d}}{\mathrm{d}r}\left(r^2\frac{\mathrm{d}R}{\mathrm{d}r}\right) + \frac{1}{H\sin\theta}\frac{\mathrm{d}}{\mathrm{d}\theta}\left(\sin\theta\frac{\mathrm{d}H}{\mathrm{d}\theta}\right) - \frac{m^2}{\sin^2\theta} + k^2r^2 = 0 \tag{4.6}$$

The θ dependence in (4.6) is separated out by use of the integer n in the following manner:

$$\frac{1}{H\sin\theta}\frac{\mathrm{d}}{\mathrm{d}\theta}\left(\sin\theta\frac{\mathrm{d}H}{\mathrm{d}\theta}\right) - \frac{m^2}{\sin^2\theta} = -n(n+1) \tag{4.7}$$

Substitution of (4.7) into (4.6) yields the final differential equation for R:

$$\frac{1}{R}\frac{\mathrm{d}}{\mathrm{d}r}\left(r^2\frac{\mathrm{d}R}{\mathrm{d}r}\right) - n(n+1) + k^2r^2 = 0 \tag{4.8}$$

We can now write (4.5), (4.7), and (4.8) in the following forms, which have solutions in terms of standard special functions:

$$\frac{\mathrm{d}^2\Phi}{\mathrm{d}\phi^2} + m^2\Phi = 0, \tag{4.9}$$

$$\frac{1}{\sin\theta}\frac{\mathrm{d}}{\mathrm{d}\theta}\left(\sin\theta\frac{\mathrm{d}H}{\mathrm{d}\theta}\right)+\left[n(n+1)-\frac{m^2}{\sin^2\theta}\right]H=0, \tag{4.10}$$

$$\frac{\mathrm{d}}{\mathrm{d}r}\left(r^2\frac{\mathrm{d}R}{\mathrm{d}r}\right)+\left[(kr)^2-n(n+1)\right]R=0 \tag{4.11}$$

The Φ equation in (4.9) is the familiar harmonic equation, which has even and odd solutions:

$$\Phi_{\substack{e\\o}}=\left\{\begin{array}{c}\cos m\phi\\ \sin m\phi\end{array}\right\} \tag{4.12}$$

The solutions of the H equation in (4.10) are the associated Legendre functions [25] of the first kind $\mathrm{P}_n^m(\cos\theta)$ and the second kind $\mathrm{Q}_n^m(\cos\theta)$. We will use only $\mathrm{P}_n^m(\cos\theta)$ because $\mathrm{Q}_n^m(\cos\theta)$ is not finite over the entire physical range of θ:

$$H(\theta)=\mathrm{P}_n^m(\cos\theta) \tag{4.13}$$

The associated Legendre functions are discussed in more detail in Appendix B. The solutions of the R equation in (4.11) are the spherical Bessel functions [25]. We require only the function that is finite at the origin ($r=0$):

$$R(kr)=\mathrm{j}_n(kr) \tag{4.14}$$

The spherical Bessel functions are discussed in more detail in Appendix C. Thus the elementary solutions for the scalar wave equation inside a spherical cavity are:

$$\psi_{\substack{e\\o}mn}=\mathrm{j}_n(kr)\mathrm{P}_n^m(\cos\theta)\left\{\begin{array}{c}\cos m\phi\\ \sin m\phi\end{array}\right\} \tag{4.15}$$

We can now write electric and magnetic vector potentials, $\vec{F}$ and $\vec{A}$, that are transverse to $\vec{r}$ as follows [3]:

$$\vec{F}=\vec{r}\psi_f,\quad \text{where}\quad \psi_f=f_{\substack{e\\o}mnp}\,\psi_{\substack{e\\o}mnp} \tag{4.16}$$

and:

$$\vec{A}=\vec{r}\psi_a,\quad \text{where}\quad \psi_a=a_{\substack{e\\o}mnp'}\,\psi_{\substack{e\\o}mnp'} \tag{4.17}$$

The constants, $f_{\substack{e\\o}mnp}$ and $a_{\substack{e\\o}mnp}$, are arbitrary, but $f_{\substack{e\\o}mnp}$ has units of V/m, and $a_{\substack{e\\o}mnp}$ has units of A/m. The p index relates to the cavity boundary condition, as indicated later.

The transverse (to $\vec{r}$) electric modes can be obtained from curl operations on $\vec{F}$:

$$\vec{E}^{TE} = -\nabla \times \vec{F} \quad \text{and} \quad \vec{H}^{TE} = \frac{-1}{i\omega\mu} \nabla \times \nabla \times \vec{F} \tag{4.18}$$

By requiring that the tangential components of the electric field be zero at $r = a$, we can write the radial component of $\vec{F}$ as:

$$\vec{F} = \hat{r} F_{r \substack{e\\o} mnp}, \text{ where}$$

$$F_{r \substack{e\\o} mnp} = \frac{f_{\substack{e\\o} mnp}}{k} kr \mathrm{j}_n\left(u_{np}\frac{r}{a}\right) \mathrm{P}_n^m(\cos\theta) \begin{matrix} \cos m\phi \\ \sin m\phi \end{matrix} \tag{4.19}$$

In (4.19), u_{np} is the pth zero of the spherical Bessel function:

$$\mathrm{j}_n(u_{np}) = 0 \tag{4.20}$$

Because r multiplies the spherical Bessel function in both the electric and magnetic scalar potentials, as seen in (4.16) and (4.17), it is convenient to introduce an alternative spherical Bessel function as defined by Harrington [3]:

$$\hat{\mathrm{J}}_n(kr) \equiv kr\mathrm{j}_n(kr) \tag{4.21}$$

Then the radial component of $\vec{F}$ in (4.19) can be written:

$$F_{r \substack{e\\o} mnp} = \frac{f_{\substack{e\\o} mnp}}{k} \hat{\mathrm{J}}_n\left(u_{np}\frac{r}{a}\right) \mathrm{P}_n^m(\cos\theta) \begin{matrix} \cos m\phi \\ \sin m\phi \end{matrix} \tag{4.22}$$

From (4.18), (4.19), and (4.22), we can write the scalar field components of the mnp TE modes as follows:

$$E^{TE}_{r \substack{e\\o} mnp} = 0, \tag{4.23}$$

$$E^{TE}_{\theta \substack{e\\o} mnp} = \frac{m f_{\substack{e\\o} mnp}}{kr\sin\theta} \hat{\mathrm{J}}_n\left(u_{np}\frac{r}{a}\right) \mathrm{P}_n^m(\cos\theta) \begin{matrix} \sin m\phi \\ -\cos m\phi \end{matrix}, \tag{4.24}$$

$$E^{TE}_{\phi \substack{e\\o} mnp} = \frac{f_{\substack{e\\o} mnp}}{kr} \hat{\mathrm{J}}_n\left(u_{np}\frac{r}{a}\right) \frac{\mathrm{d}}{\mathrm{d}\theta} \mathrm{P}_n^m(\cos\theta) \begin{matrix} \cos m\phi \\ \sin m\phi \end{matrix}, \tag{4.25}$$

$$H^{TE}_{r\,{}^{e}_{o}mnp} = \frac{-n(n+1)f_{{}^{e}_{o}mnp}}{i\omega\mu k r^2}\hat{\mathrm{J}}_n\left(u_{np}\frac{r}{a}\right)\mathrm{P}_n^m(\cos\theta)\begin{matrix}\cos m\phi\\ \sin m\phi\end{matrix}, \tag{4.26}$$

$$H^{TE}_{\theta\,{}^{e}_{o}mnp} = \frac{-f_{{}^{e}_{o}mnp}}{i\omega\mu r}\hat{\mathrm{J}}'_n\left(u_{np}\frac{r}{a}\right)\frac{\mathrm{d}}{\mathrm{d}\theta}\mathrm{P}_n^m(\cos\theta)\begin{matrix}\cos m\phi\\ \sin m\phi\end{matrix}, \tag{4.27}$$

$$H^{TE}_{\phi\,{}^{e}_{o}mnp} = \frac{-mf_{{}^{e}_{o}mnp}}{i\omega\mu r\sin\theta}\hat{\mathrm{J}}_n\left(u_{np}\frac{r}{a}\right)\mathrm{P}_n^m(\cos\theta)\begin{matrix}-\sin m\phi\\ \cos m\phi\end{matrix}. \tag{4.28}$$

In (4.27) and (4.28), $\hat{\mathrm{J}}'_n$ represents the derivative of $\hat{\mathrm{J}}_n$ with respect to the argument.

The resonant wavenumber k^{TE}_{mnp} of the TE_{mnp} mode is given by:

$$k^{TE}_{mnp} = u_{np}/a \tag{4.29}$$

Similarly, the resonant frequency f^{TE}_{mnp} is given by:

$$f^{TE}_{mnp} = \frac{u_{np}v}{2\pi a} \tag{4.30}$$

From (4.29) and (4.30), we see that the resonant frequencies are independent of the mode index m. This means that there are numerous degenerate modes (same resonant frequency) for spherical cavities. This is one reason why spherical cavities have not been used for reverberation chambers where it is desirable to have well spaced resonant modes [9].

We can treat the TM modes similarly. The transverse (to $\vec{r}$) magnetic modes can be obtained from curl operations on $\vec{A}$:

$$\vec{H}^{TM} = \nabla\times\vec{A} \quad \text{and} \quad \vec{E}^{TM} = \frac{-1}{i\omega\varepsilon}\nabla\times\nabla\times\vec{A} \tag{4.31}$$

By requiring that the tangential components of the electric field be zero at $r = a$, we can write the radial component of $\vec{A}$ as:

$$\vec{A} = \hat{r}A_{r\,{}^{e}_{o}mnp}, \text{ where}$$
$$A_{r\,{}^{e}_{o}mnp} = \frac{a_{{}^{e}_{o}mnp}}{k}\hat{\mathrm{J}}_n\left(u'_{np}\frac{r}{a}\right)\mathrm{P}_n^m(\cos\theta)\begin{matrix}\cos m\phi\\ \sin m\phi\end{matrix} \tag{4.32}$$

In (4.32), u'_{np} is the pth zero of the derivative of Harrington's spherical Bessel function [3]:

$$\hat{J}'_n(u'_{np}) = 0 \tag{4.33}$$

From (4.31) and (4.32), we can write the scalar field components of the mnp TM modes as follows:

$$H^{TM}_{r\,\substack{e\\o}\,mnp} = 0, \tag{4.34}$$

$$H^{TM}_{\theta\,\substack{e\\o}\,mnp} = \frac{m a_{\substack{e\\o}\,mnp}}{kr\sin\theta}\hat{J}_n\left(u'_{np}\frac{r}{a}\right)P_n^m(\cos\theta)\,\substack{-\sin m\phi\\ \cos m\phi}, \tag{4.35}$$

$$H^{TM}_{\phi\,\substack{e\\o}\,mnp} = \frac{-a_{\substack{e\\o}\,mnp}}{kr}\hat{J}_n\left(u'_{np}\frac{r}{a}\right)\frac{d}{d\theta}P_n^m(\cos\theta)\,\substack{\cos m\phi\\ \sin m\phi}, \tag{4.36}$$

$$E^{TM}_{r\,\substack{e\\o}\,mnp} = \frac{-n(n+1)a_{\substack{e\\o}\,mnp}}{i\omega\varepsilon kr^2}\hat{J}_n\left(u'_{np}\frac{r}{a}\right)P_n^m(\cos\theta)\,\substack{\cos m\phi\\ \sin m\phi}, \tag{4.37}$$

$$E^{TM}_{\theta\,\substack{e\\o}\,mnp} = \frac{-a_{\substack{e\\o}\,mnp}}{i\omega\varepsilon r}\hat{J}'_n\left(u'_{np}\frac{r}{a}\right)\frac{d}{d\theta}P_n^m(\cos\theta)\,\substack{\cos m\phi\\ \sin m\phi}, \tag{4.38}$$

$$E^{TM}_{\phi\,\substack{e\\o}\,mnp} = \frac{-m a_{\substack{e\\o}\,mnp}}{i\omega\varepsilon r\sin\theta}\hat{J}'_n\left(u'_{np}\frac{r}{a}\right)P_n^m(\cos\theta)\,\substack{-\sin m\phi\\ \cos m\phi} \tag{4.39}$$

The resonant wavenumber k^{TM}_{mnp} of the TM_{mnp} mode is given by:

$$k^{TM}_{mnp} = u'_{np}/a \tag{4.40}$$

Similarly, the resonant frequency f^{TM}_{mnp} is given by:

$$f^{TM}_{mnp} = \frac{u'_{np}v}{2\pi a} \tag{4.41}$$

From (4.41) we see that the resonant frequencies of the TM modes are also independent of m and hence have many degenerate modes.

TABLE 4.1 Ordered Zeros u_{np} of $\hat{J}_n(u)$ [3].

n/p	1	2	3	4	5	6	7	8
1	4.493	5.763	6.988	8.183	9.356	10.513	11.657	12.791
2	7.725	9.095	10.417	11.705	12.967	14.207	15.431	16.641
3	10.904	12.323	13.698	15.040	16.355	17.648	18.923	20.182
4	14.066	15.515	16.924	18.301	19.653	20.983	22.295	
5	17.221	18.689	20.122	21.525	22.905			
6	20.371	21.854						

The zeros u_{np} of (4.20) are given in Table 4.1 [3] for various values of n and p. These values can be used to obtain the resonant frequencies of the TE modes using (4.30). The zeros u'_{np} of (4.33) are given in Table 4.2 [3] for various values of n and p. These values can be used to obtain the resonant frequencies of the TM modes using (4.41). Tables of u_{np} and u'_{np} have also been published by Waldron [28].

From Tables 4.1 and 4.2, we see that the lowest-order mode is TM_{m11}, where m equals 0 or 1 and the resonant frequency is:

$$f_{m11}^{TM} = \frac{u'_{11} v}{2\pi a} \tag{4.42}$$

There are actually three degenerate modes ($TM_{e\,011}$, TM_{e111}, and TM_{o111}) at this frequency, and their field distributions are determined from the radial components of the magnetic vector potentials:

$$TM_{e\,011}: \quad A_{re\,011} = \frac{a_{e\,011}}{k} \hat{J}_1\left(u'_{11}\frac{r}{a}\right)\cos\theta, \tag{4.43}$$

$$TM_{e111}: \quad A_{re111} = \frac{a_{e111}}{k} \hat{J}_1\left(u'_{11}\frac{r}{a}\right)\sin\theta\cos\phi, \tag{4.44}$$

$$TM_{o111}: \quad A_{ro111} = \frac{a_{o111}}{k} \hat{J}_1\left(u'_{11}\frac{r}{a}\right)\sin\theta\sin\phi \tag{4.45}$$

The expressions for the field components of these modes can be obtained by taking the curl operations in (4.31) or by reducing the field expressions in (4.34)–(4.39) to the

TABLE 4.2 Ordered Zeros u'_{np} of $\hat{J}'_n(u')$ [3].

n/p	1	2	3	4	5	6	7	8
1	2.744	3.870	4.973	6.062	7.140	8.211	9.275	10.335
2	6.117	7.443	8.722	9.968	11.189	12.391	13.579	14.753
3	9.317	10.713	12.064	13.380	14.670	15.939	17.190	18.425
4	12.486	13.921	15.314	16.674	18.009	19.321	20.615	21.894
5	15.664	17.103	18.524	19.915	21.281	22.626		
6	18.796	20.272	21.714	23.128				
7	21.946							

particular mode indices for m, n, and p. In either case, the nonzero field components for the $\mathrm{TM}_{e\,011}$ mode are:

$$H_{\phi e\,011}^{TM} = \frac{a_{e\,011}}{kr}\hat{\mathrm{J}}_1\left(u'_{11}\frac{r}{a}\right)\sin\theta, \tag{4.46}$$

$$E_{re\,011}^{TM} = \frac{-2a_{e\,011}}{i\omega\varepsilon kr^2}\hat{\mathrm{J}}_1\left(u'_{11}\frac{r}{a}\right)\cos\theta, \tag{4.47}$$

$$E_{\theta e\,011}^{TM} = \frac{a_{e\,011}}{i\omega\varepsilon r}\hat{\mathrm{J}}'_1\left(u'_{11}\frac{r}{a}\right)\sin\theta \tag{4.48}$$

The nonzero field components for the TM_{e111} mode are:

$$H_{\theta e111}^{TM} = \frac{-a_{e111}}{kr}\hat{\mathrm{J}}_1\left(u'_{11}\frac{r}{a}\right)\sin\phi, \tag{4.49}$$

$$H_{\phi e111}^{TM} = \frac{-a_{e111}}{kr}\hat{\mathrm{J}}_1\left(u'_{11}\frac{r}{a}\right)\cos\theta\cos\phi, \tag{4.50}$$

$$E_{re111}^{TM} = -\frac{-2a_{e111}}{i\omega\varepsilon kr^2}\hat{\mathrm{J}}_1\left(u'_{11}\frac{r}{a}\right)\sin\theta\cos\phi, \tag{4.51}$$

$$E_{\theta e111}^{TM} = \frac{-a_{e111}}{i\omega\varepsilon r}\hat{\mathrm{J}}'_1\left(u'_{11}\frac{r}{a}\right)\cos\theta\cos\phi, \tag{4.52}$$

$$E_{\phi e111}^{TM} = \frac{a_{e111}}{i\omega\varepsilon r}\hat{\mathrm{J}}_1\left(u'_{11}\frac{r}{a}\right)\sin\phi \tag{4.53}$$

Similarly, the nonzero field components for the odd mode TM_{o111} are:

$$H_{\theta o111}^{TM} = \frac{a_{o111}}{kr}\hat{\mathrm{J}}_1\left(u'_{11}\frac{r}{a}\right)\cos\phi, \tag{4.54}$$

$$H_{\phi o111}^{TM} = \frac{-a_{o111}}{kr}\hat{\mathrm{J}}_1\left(u'_{11}\frac{r}{a}\right)\cos\theta\sin\phi, \tag{4.55}$$

$$E_{ro111}^{TM} = -\frac{-2a_{o111}}{i\omega\varepsilon kr^2}\hat{\mathrm{J}}_1\left(u'_{11}\frac{r}{a}\right)\sin\theta\sin\phi, \tag{4.56}$$

$$E_{\theta o111}^{TM} = \frac{-a_{o111}}{i\omega\varepsilon r}\hat{\mathrm{J}}'_1\left(u'_{11}\frac{r}{a}\right)\cos\theta\sin\phi, \tag{4.57}$$

$$E_{\phi o111}^{TM} = \frac{-a_{o111}}{i\omega\varepsilon r}\hat{\mathrm{J}}'_1\left(u'_{11}\frac{r}{a}\right)\cos\phi \tag{4.58}$$

It interesting that even though the $\mathrm{TM}_{e\,011}$, TM_{e111}, and TM_{o111} modes all have the same resonant frequency, the $\mathrm{TM}_{e\,011}$ mode has only three nonzero field components while the TM_{e111} and TM_{o111} modes have five nonzero field components. Actually this is due only to a rotation in space, and the three mode field patterns are actually the same. The field pattern is shown in Figure 4.2.

For use as single-mode resonators (filters or electromagnetic property measurements), the goal is to excite only a single mode at its resonant frequency or at its perturbed resonant frequency for material measurements [27]. However, for use of a

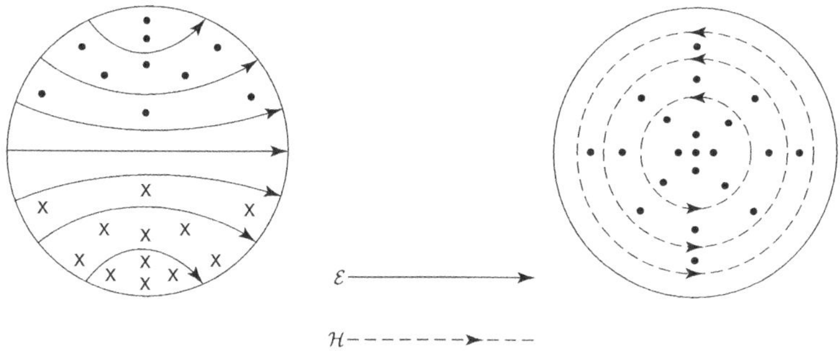

FIGURE 4.2 Instantaneous electric $\mathcal{E}$ and magnetic $\mathcal{H}$ field lines for the TM_{e011}, TM_{e111}, and TM_{o111} cavity modes [3].

spherical cavity as a reverberation chamber (mode-stirred chamber) [18,19], it is useful to know the number of modes available for stirring over a large bandwidth. The number of modes with eigenvalues $k_{\substack{e\\o}mnp}$ less than k can be approximated by (2.18) because that expression applies to cavities of arbitrary shape. The volume V of a cylindrical cavity is given by:

$$V = \frac{4}{3}\pi a^3 \tag{4.59}$$

If we substitute (4.59) in to (2.18), the Weyl approximation for the number of modes is:

$$N_W(k) = \frac{4a^3k^3}{9\pi} \tag{4.60}$$

If we wish to write the number of modes in terms of frequency f, we can replace k by $2\pi f/v$ in (4.60) to obtain:

$$N_W(f) = \frac{32\pi^2 a^3 f^3}{9v^3} \tag{4.61}$$

The mode density (modes/Hz) can be obtained by differentiating (4.61) with respect to f:

$$D_W(f) = \frac{dN_W(f)}{df} = \frac{32\pi^2 a^3 f^2}{3v^3} \tag{4.62}$$

However, as indicated previously, spherical cavities have not been popular shapes for reverberation chambers because of high mode degeneracy.

4.2 WALL LOSSES AND CAVITY Q

An expression for cavity Q due to wall losses of cavities of arbitrary shape was given in (1.41). For cylindrical cavities, the expressions for the magnetic field are known, and the integrals can be evaluated to determine Q for the various mode types and numbers.

Harrington [3, p. 312] has given the Q expressions for the TE and TM modes of arbitrary order.

To illustrate the details of the evaluation of Q, we will derive Q for the specific case of the $\mathrm{TM}_{e\,011}$ mode which, along with the TM_{e111} and TM_{o111} modes, has the lowest resonant frequency. We first write (1.41) in the following form:

$$Q_{e\,011}^{TM} = \frac{\omega_{e\,011}\mu}{R_s} \frac{\iiint\limits_V \vec{H}_{e\,011}^{TM} \bullet \vec{H}_{e\,011}^{TM*} \mathrm{d}V}{\oint\!\!\oint\limits_S \vec{H}_{e\,011}^{TM} \bullet \vec{H}_{e\,011}^{TM*} \mathrm{d}S}, \tag{4.63}$$

where the magnetic field (which includes only a ϕ component) is given by (4.46). From (4.46), the square of the magnetic field is:

$$|H_{\phi e\,011}^{TM}|^2 = \frac{a_{e\,011}^2}{k^2 r^2} \hat{\mathrm{J}}_1^2\left(u'_{11}\frac{r}{a}\right) \sin^2\theta \tag{4.64}$$

If we substitute (4.64) into the volume integral in the numerator of (4.63), the volume integral is:

$$\iiint\limits_V = \frac{a_{e\,011}^2}{k^2} \int_0^a \int_0^{2\pi} \int_0^{\pi} \hat{\mathrm{J}}_1^2\left(u'_{11}\frac{r}{a}\right) \sin^2\theta \sin\theta \, \mathrm{d}\theta \, \mathrm{d}\phi \, \mathrm{d}r \tag{4.65}$$

The θ and ϕ integrations in (4.65) are easily performed to yield

$$\iiint\limits_V = \frac{8\pi a_{e\,011}^2}{3k^2} \int_0^a \hat{\mathrm{J}}_1^2(kr) \, \mathrm{d}r, \tag{4.66}$$

where we have used the result from (4.40) that $k = u'_{11}/a$.

If we write the spherical Bessel function in (4.66) in terms of the corresponding cylindrical Bessel function [3, 25], then (4.66) becomes:

$$\iiint\limits_V = \frac{4\pi^2 a_{e\,011}^2}{3k} \int_0^a r\mathrm{J}_{3/2}^2(kr) \, \mathrm{d}r. \tag{4.67}$$

To evaluate the r integration in (4.67), the following integral [29, p. 146] is useful:

$$\int r\mathrm{J}_l^2(kr)\mathrm{d}r = \frac{r^2}{2}\left[\mathrm{J}_l^2(kr) - \mathrm{J}_{l-1}(kr)\mathrm{J}_{l+1}(kr)\right] \tag{4.68}$$

If we substitute (4.68) with $l = 3/2$ into (4.67), we obtain:

$$\iiint\limits_V = \frac{2\pi^2 a^2 a_{e\,011}^2}{3k}\left[\mathrm{J}_{3/2}^2(u'_{11}) - \mathrm{J}_{1/2}^2(u'_{11})\mathrm{J}_{5/2}^2(u'_{11})\right], \tag{4.69}$$

where we have used (4.40) in the arguments of the Bessel functions. We can simplify (4.69) further by using the following recurrence relations for Bessel functions [25, p. 361]:

$$J_{1/2}(u'_{11}) = J'_{3/2}(u'_{11}) + \frac{3/2}{u'_{11}} J_{3/2}(u'_{11}), \tag{4.70}$$

$$J_{5/2}(u'_{11}) = -J'_{3/2}(u'_{11}) + \frac{3/2}{u'_{11}} J_{3/2}(u'_{11}) \tag{4.71}$$

If we substitute (4.70) and (4.71) into (4.69), then only Bessel functions of order 3/2 remain. However, some Bessel function derivatives remain. From (4.33) we can derive the following relationship:

$$J'_{3/2}(u'_{11}) = \frac{-1}{2u'_{11}} J_{3/2}(u'_{11}) \tag{4.72}$$

Now if we substitute (4.70)–(4.72) into (4.69), we obtain:

$$\iiint_V = \frac{2\pi^2 a^2 a_{e\,011}^2}{3k}\left(1 - \frac{2}{u'^2_{11}}\right) J_{3/2}^2(u'_{11}) \tag{4.73}$$

The surface integral required in the denominator of (4.63) is simpler to evaluate because no r integration is required. Since the θ and ϕ integrations were required in (4.65), we can use the result (4.66) to obtain:

$$\oiint_S = \frac{4\pi^2 a_{e\,011}^2 u'_{11}}{3k^2} \hat{J}_{3/2}^2(u'_{11}) \tag{4.74}$$

If we substitute (4.73) and (4.74) into (4.63), we obtain the desired final result:

$$Q_{e\,011}^{TM} = \frac{\eta}{2R_s}\left(u'_{11} - \frac{2}{u'_{11}}\right) \tag{4.75}$$

From Table 4.2, we see that $u'_{11} = 2.744$. Thus, from (4.75) we have:

$$Q_{e\,011}^{TM} \approx 1.008\frac{\eta}{R_s} \tag{4.76}$$

For higher order modes, the Q expressions are derived by the same method, but more algebra is required. The general expressions have been given by Harrington [3, p. 312]:

$$Q_{mnp}^{TM} = \frac{\eta}{2R_s}\left(u'_{np} - \frac{n(n+1)}{u'_{np}}\right), \tag{4.77}$$

$$Q_{mnp}^{TE} = \frac{\eta u_{np}}{2R_s} \tag{4.78}$$

Comparing (4.75) and (4.77), we see that they agree for $n = p = 1$.

Since we have now analyzed rectangular, cylindrical, and spherical cavities, it is interesting to compare the cavity Q values for the three shapes. If we compare (4.76) with the lowest-order mode Q_r for a rectangular cavity with equal sides (cubic), the Q ratio is [3, p. 76]:

$$\frac{Q_{e\,011}^{TM}}{Q_r} \approx 1.36 \tag{4.79}$$

If we compare (4.76) with the lowest order mode Q_c for a cylindrical cavity with equal height and diameter, the Q ratio is [3, p. 216]:

$$\frac{Q_{e\,011}^{TM}}{Q_c} \approx 1.26 \tag{4.80}$$

4.3 DYADIC GREEN'S FUNCTIONS

Dyadic Green's functions for a spherical cavity have been derived by Tai [2]. The method of derivation is similar to, but somewhat different from, that for the rectangular and cylindrical cavities. Dyadic Green's functions are again useful in providing a compact notation for determining the electric and magnetic fields due to current sources.

The four sets of solenoidal eigenfunctions needed in the expansions are [2]:

$$\vec{M}_{\substack{e\\o}mn}(\kappa_p) = \nabla \times \left[\vec{r}\mathrm{j}_n(\kappa_p)P_n^m(\cos\theta)\begin{matrix}\cos m\phi\\ \sin m\phi\end{matrix}\right], \tag{4.81}$$

$$\vec{M}_{\substack{e\\o}mn}(\kappa_q) = \nabla \times \left[\vec{r}\mathrm{j}_n(\kappa_q)\mathrm{P}_n^m(\cos\theta)\begin{matrix}\cos m\phi\\ \sin m\phi\end{matrix}\right], \tag{4.82}$$

$$\vec{N}_{\substack{e\\o}mn}(\kappa_p) = \frac{1}{\kappa_p}\nabla \times \vec{M}_{\substack{e\\o}mn}(\kappa_p), \tag{4.83}$$

$$\vec{N}_{\substack{e\\o}mn}(\kappa_q) = \frac{1}{\kappa_q}\nabla \times \vec{M}_{\substack{e\\o}mn}(\kappa_q) \tag{4.84}$$

The quantities, κ_p and κ_q, are determined from mode equations that are equivalent to (4.20) and (4.33):

$$\mathrm{j}_n(\kappa_p a) = 0, \tag{4.85}$$

$$[\kappa_q a \mathrm{j}_n(\kappa_q a)]' = 0, \tag{4.86}$$

where the prime in (4.86) denotes differentiation with respect to the argument $\kappa_q a$. Hence, $\kappa_p a = u_{np}$ and $\kappa_q a = u'_{np}$.

The $\vec{M}$ and $\vec{N}$ vectors in (4.81)–(4.85) are proportional to the modal fields (within a constant factor) discussed in Section 4.1. Specifically, $\vec{M}_{\substack{e\\o}mn}(\kappa_p)$ corresponds to the electric fields of the TE modes, as given in (4.24) and (4.25), $\vec{M}_{\substack{e\\o}mn}(\kappa_q)$ corresponds to the magnetic fields of the TM modes as given in (4.35) and (4.36), $\vec{N}_{\substack{e\\o}mn}(\kappa_p)$ corresponds to the magnetic fields of the TE modes as given in (4.26)–(4.28), and $\vec{N}_{\substack{e\\o}mn}(\kappa_q)$ corresponds to the electric fields of the TM modes as given in (4.37)–(4.39).

The electric $\vec{G}_e$ and magnetic $\vec{G}_m$ dyadic Green's functions satisfy the differential equations given in (2.34) and (2.35). In addition to the differential equations, we need to specify boundary conditions to make the dyadic Green's functions unique. The electric dyadic Green's function needs to satisfy (2.38) at $r = a$, and the magnetic dyadic Green's function needs to satisfy (2.39) at $r = a$.

The solution for the magnetic dyadic Green's function from Tai [2] in shorthand summation form is:

$$\overleftrightarrow{G}_m(\vec{r},\vec{r}') = \sum_{l,m,n}\left[\frac{\kappa_p}{(\kappa_p^2-k^2)I_p}\vec{N}_p\vec{M}_p' + \frac{\kappa_q}{(\kappa_q^2-k^2)I_q}\vec{M}_q\vec{N}_q'\right], \tag{4.87}$$

where $\vec{M}_p'$ and $\vec{N}_q'$ are functions of the source coordinates (r', θ', ϕ') and l represents the discrete eigenvalues κ_p and κ_q. The quantities I_p and I_q are given by [2]:

$$I_p = \frac{a^3}{3}\left[\frac{\partial \mathrm{j}_n(\kappa_p a)}{\partial(\kappa_p a)}\right]^2, \tag{4.88}$$

$$I_q = \frac{a^3}{2}\left[1 - \frac{n(n+1)}{\kappa_q^2 a^2}\right]\mathrm{j}_n^2(\kappa_q a) \tag{4.89}$$

The electric dyadic Green's function can be obtained from the magnetic dyadic Green's function by the following curl operation [2]:

$$\overleftrightarrow{G}_e(\vec{r},\vec{r}') = \frac{1}{k^2}\left[\nabla \times \overleftrightarrow{G}_m(\vec{r},\vec{r}') - \overleftrightarrow{I}\delta(\vec{r}-\vec{r}')\right] \tag{4.90}$$

If we substitute (4.87) into (4.90), the result for $\overleftrightarrow{G}_e$ is [2]:

$$\overleftrightarrow{G}_e(\vec{r},\vec{r}') = -\frac{\overleftrightarrow{I}}{k^2}\delta(\vec{r}-\vec{r}') + \frac{1}{k^2}\sum_{l,m,n}\left[\frac{\kappa_p^2}{\kappa_p^2-k^2}\vec{M}_p\vec{M}_p' + \frac{\kappa_q^2}{\kappa_q^2-k^2}\vec{N}_q\vec{N}_q'\right] \tag{4.91}$$

When the excitation corresponds to that of a resonant mode such that:

$$k = \kappa_p \quad \text{or} \quad k = \kappa_q, \tag{4.92}$$

then $\overleftrightarrow{G}_m$ in (4.87) and $\overleftrightarrow{G}_e$ in (4.91) are singular because they have zeros in the denominators. However, if we include wall loss, as in Section 1.3, we can replace κ_p^2 or κ_q^2 by the following:

$$\kappa_p^2 \approx \kappa_p^2\left(1 - \frac{2i}{Q_{mnp}^{TE}}\right) \quad \text{or} \quad \kappa_q^2 \approx \kappa_q^2\left(1 - \frac{2i}{Q_{mnq}^{TM}}\right), \tag{4.93}$$

where Q_{mnp}^{TE} is given by (4.77) and Q_{mnq}^{TM} is given by (4.78). We have neglected the Q^{-2} terms in (4.93) because the Qs are large. For finite values of Q, the denominators in (4.87) and (4.91) cannot be zero for real k (or real frequency), and the singularities do not occur.

4.3.1 Fields in the Source-Free Region

Consider a volume current density $\vec{J}(\vec{r}')$ confined to a volume V' in a spherical cavity, as shown in Figure 4.3. The observation point $\vec{r}$ is located within the cavity, but outside the volume V'. The magnetic field can be written as an integral over the source volume [2]:

$$\vec{H}(\vec{r}) = \iiint_{V'} \overleftrightarrow{G}_m(\vec{r}, \vec{r}') \bullet \vec{J}(\vec{r}')\, dV', \tag{4.94}$$

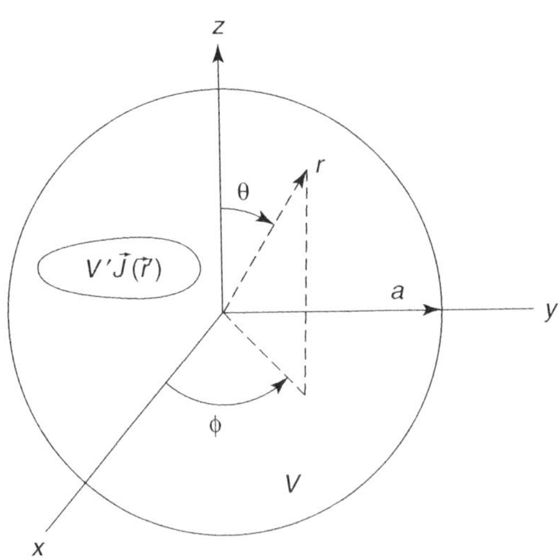

FIGURE 4.3 Current density $\vec{J}(\vec{r}')$ in a volume V' in a spherical cavity.

where $\overleftrightarrow{G}_m(\vec{r},\vec{r}\,')$ is given by (4.87). Similarly, the electric field can be written as an integral over the source volume [2]:

$$\vec{E}(\vec{r}) = i\omega\mu \iiint_{V'} \overleftrightarrow{G}_e(\vec{r},\vec{r}\,') \bullet \vec{J}(\vec{r}\,')\,\mathrm{d}V', \tag{4.95}$$

where $\overleftrightarrow{G}_e(\vec{r},\vec{r}\,')$ is given by (4.91). The volume integrals in (4.94) and (4.95) are well behaved because $\overleftrightarrow{G}_m(\vec{r},\vec{r}\,')$ and $\overleftrightarrow{G}_e(\vec{r},\vec{r}\,')$ are well behaved for $\vec{r} \neq \vec{r}\,'$.

4.3.2 Fields in the Source Region

In the source region, we must deal with the singularities in the Green's functions at $\vec{r} = \vec{r}\,'$. In evaluating the magnetic field, the singularity in $\overleftrightarrow{G}_m(\vec{r},\vec{r}\,')$ at $\vec{r} = \vec{r}\,'$ is integrable, and (4.94) can still be used to evaluate $\vec{H}$.

The evaluation of the electric field has been discussed in Section 2.3, and (4.95) needs to modified to (2.54) – (2.57). The only difference is that $\overleftrightarrow{G}_e$ for the rectangular cavity is replaced by $\overleftrightarrow{G}_e$ for the spherical cavity, as given by (4.91). The shape of the principle volume is arbitrary, but a logical shape is a sphere. In this case, $\overleftrightarrow{L}$ is given by [20]:

$$\overleftrightarrow{L} = \frac{\overleftrightarrow{I}}{3} \tag{4.96}$$

Note that the coefficient of the delta function in (4.91) also is proportional to $\overleftrightarrow{I}$. Further discussion of this term is contained in [15].

4.4 SCHUMANN RESONANCES IN THE EARTH-IONOSPHERE CAVITY

The earth-ionosphere cavity is very different from the cavities that have been covered to this point because it is so large, has very lossy boundaries, and is not simply connected. However, it is well worth studying because it can be analyzed by use of the formalism for the spherical cavity and is important in geophysical exploration [30] and extremely low frequency (ELF) communications [31]. The geometry of the cavity formed by the earth and ionosphere boundaries is shown in Figure 4.4. To begin with, the earth is modeled as a perfectly conducting sphere of radius a, and the lower boundary of the ionosphere is modeled as a perfect conductor of radius b. Because the cavity is so large, it supports extremely low resonant frequencies that are called Schumann resonances [32].

The lowest resonant frequencies are the most important and the most observable Schumann resonances. The lowest frequency modes are TM (to $\vec{r}$) and are independent of ϕ ($m = 0$). With this condition, the differential Equation (4.6) simplifies to:

$$\frac{1}{R}\frac{\mathrm{d}}{\mathrm{d}r}\left(r^2\frac{\mathrm{d}R}{\mathrm{d}r}\right) + \frac{1}{H\sin\theta}\frac{\mathrm{d}}{\mathrm{d}\theta}\left(\sin\theta\frac{\mathrm{d}H}{\mathrm{d}\theta}\right) + k^2r^2 = 0 \tag{4.97}$$

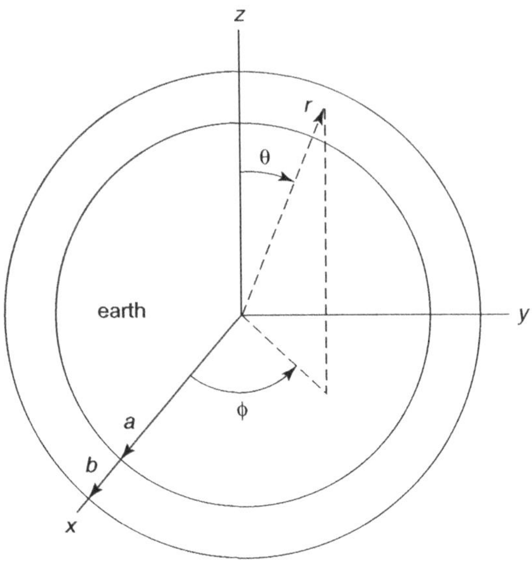

FIGURE 4.4 Geometry for the earth-ionosphere cavity which supports Schumann resonances.

Similarly, the separated equation for $H(\theta)$ in (4.7) simplifies to:

$$\frac{1}{H \sin\theta} \frac{\mathrm{d}}{\mathrm{d}\theta}\left(\sin\theta \frac{\mathrm{d}H}{\mathrm{d}\theta}\right) = -n(n+1) \tag{4.98}$$

The solution to (4.98) is given by (4.13) with $m = 0$:

$$H(\theta) = \mathrm{P}_n(\cos\theta) \tag{4.99}$$

If we substitute (4.98) into (4.97) and multiply by R, we obtain the following equation for $R(r)$:

$$\frac{\mathrm{d}}{\mathrm{d}r}\left(r^2 \frac{\mathrm{d}R}{\mathrm{d}r}\right) - n(n+1)R + k^2 r^2 R = 0 \tag{4.100}$$

In general, the solution of (4.100) can be written as a linear combination of two independent spherical Bessel functions, for example $\mathrm{j}_n(kr)$ and $\mathrm{y}_n(kr)$ [25]. However, an approximate solution to (4.100) is adequate for the special case of the earth-ionosphere cavity.

Simplifying (4.15), we first write the scalar potential as:

$$\psi_n = R(r)\mathrm{P}_n(\cos\theta) \tag{4.101}$$

To derive the TM modes, we follow (4.17) and write the magnetic vector potential as:

$$\vec{A} = \vec{r}\psi_n = \hat{r}rR(r)\mathrm{P}_n(\cos\theta) \tag{4.102}$$

As in (4.31), the magnetic field can be written as the curl of $\vec{A}$:

$$\vec{H}^{TM} = \nabla \times \vec{A} = -\hat{\phi}R(r)\mathrm{P}_n(\cos\theta) \tag{4.103}$$

Following (4.31), the electric field can be derived by taking a second curl operation on (4.103) and applying (4.10) to the θ component of the electric field:

$$\begin{aligned}\vec{E}^{TM} &= \frac{1}{i\omega\varepsilon}\nabla \times \vec{H}^{TM} \\ &= \frac{1}{i\omega\varepsilon}\left\{\hat{r}\frac{R(r)}{r}n(n+1)\mathrm{P}_n(\cos\theta) - \hat{\theta}\frac{1}{r}\frac{\mathrm{d}}{\mathrm{d}r}[rR(r)]\frac{\mathrm{dP}_n(\cos\theta)}{\mathrm{d}\theta}\right\}\end{aligned} \tag{4.104}$$

Before applying boundary conditions at the cavity walls, we can obtain an approximation to (4.100) for $R(r)$. We first make the following substitution for r:

$$r = a + h, \quad 0 < h < h_i, \tag{4.105}$$

where h is the height above the earth surface and $h_i = b - a$ is the height of the lower boundary of the ionosphere. The earth radius a is approximately 6400 km, and the height h_i of the ionosphere is approximately 100 km. So we can approximate r in (4.100) by a and derive the following approximate differential equation for R:

$$\frac{\mathrm{d}^2R}{\mathrm{d}h^2} + \left[k^2 - \frac{n(n+1)}{a^2}\right]R = 0 \tag{4.106}$$

Equation (4.106) is the well-known Helmholtz equation, which has sine and cosine solutions:

$$R(h) = \begin{cases} \cos\sqrt{k^2 - \dfrac{n(n+1)}{a^2}}h \\ \sin\sqrt{k^2 - \dfrac{n(n+1)}{a^2}}h \end{cases} \tag{4.107}$$

From (4.104) and (4.107), we can derive the following approximate expression for the θ component of the electric field:

$$E_\theta^{TM} = \frac{-1}{i\omega\varepsilon}\frac{\mathrm{d}R}{\mathrm{d}h}\frac{\mathrm{d}P_n(\cos\theta)}{\mathrm{d}\theta}$$

$$= \frac{1}{i\omega\varepsilon}\frac{\mathrm{d}P_n(\cos\theta)}{\mathrm{d}\theta}\left\{\begin{array}{l}\sqrt{k^2-\frac{n(n+1)}{a^2}}\sin\sqrt{k^2-\frac{n(n+1)}{a^2}}h \\ -\sqrt{k^2-\frac{n(n+1)}{a^2}}\cos\sqrt{k^2-\frac{n(n+1)}{a^2}}h\end{array}\right. \tag{4.108}$$

Since the tangential electric field must be zero at the cavity boundaries, the following conditions must be satisfied:

$$E_\theta^{TM}|_{h=0} = E_\theta^{TM}|_{h=h_i} = 0 \tag{4.109}$$

Equation (4.109) can be satisfied by setting the square root factor in (4.108) equal to zero:

$$\sqrt{k_n^2-\frac{n(n+1)}{a^2}} = 0 \quad \text{or} \quad \omega_n = \frac{c}{a}\sqrt{n(n+1)}, \tag{4.110}$$

where we assume that the cavity has free-space parameters $\left(c = 1/\sqrt{\mu_0\varepsilon_0}\right)$. Then the resonant frequencies are:

$$f_n = \frac{\omega_n}{2\pi} = \frac{c}{2\pi a}\sqrt{n(n+1)} \tag{4.111}$$

The same equation for f_n has been derived by Wait [33] and Jackson [34] by similar methods. With the earth radius $a = 6400$ km, Table 4.3 shows the first five Schumann resonances. The approximate field distributions for these modes are given by:

$$E_{\theta n}^{TM} \approx 0, \tag{4.112}$$

$$E_{rn}^{TM} \approx \frac{n(n+1)}{i\omega_n\varepsilon_0 a}\mathrm{P}_n(\cos\theta), \tag{4.113}$$

$$H_{\phi n}^{TM} \approx \mathrm{P}_n^1(\cos\theta) \tag{4.114}$$

TABLE 4.3 Approximate Schumann resonances f_n for the Earth-ionosphere cavity

f_1	10.6 Hz
f_2	18.3 Hz
f_3	25.8 Hz
f_4	33.4 Hz
f_5	40.9 Hz

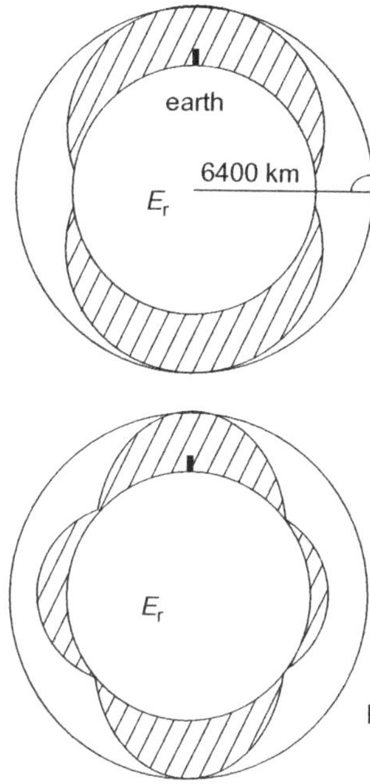

FIGURE 4.5 Radial electric field distributions, E_{r1}^{TM} and E_{r2}^{TM}, for the first two Schumann resonances as excited by a radial electric dipole at the pole ($\theta = 0$) [33].

Illustrations of E_{r1}^{TM} and E_{r2}^{TM} are shown in Figure 4.5 [33]. In both cases, the mode can be excited by a radial electric dipole, as shown. In nature, lightning constantly provides such an excitation somewhere on earth, and the electromagnetic noise caused by lightning is called atmospheric noise.

In reality, the earth and the ionosphere are far from perfect conductors. The conductivity of sea water is approximately 4 S/m, and the conductivity of the ionosphere is much lower yet (approximately 10^{-4} S/m [33]). These finite conductivities tend to reduce the actual resonant frequencies shown in Table 3.3 by about 20% [33]. In addition, the large loss results in Q values (determined by atmospheric noise measurements [35]) of only about 4 to 10 [34]. Such low Q values make it very difficult to measure Schumann resonances above about 40 Hz [35].

PROBLEMS

4-1 Consider a vacuum-filled spherical cavity as in Figure 3.1 with $a = 1$ cm. Determine the resonant frequencies of the TM_{m11} and TE_{m11} modes. Are they independent of m?

4-2 For copper walls (5.7×10^7 S/m), what are the Q values of the two modes in Problem 4-1?

4-3 Derive (4.77) from (1.41).

4-4 Derive (4.78) from (1.41).

4-5 Show that (4.87) satisfies (2.35).

4-6 Show that (4.87) satisfies (2.39) at $r = a$.

4-7 Show that (4.91) satisfies (2.34).

4-8 Show that (4.91) satisfies (2.38) at $r = a$.

4-9 Derive the approximation (4.106) from (4.100).

PART II

STATISTICAL THEORIES FOR ELECTRICALLY LARGE CAVITIES

CHAPTER 5

Motivation for Statistical Approaches

5.1 LACK OF DETAILED INFORMATION

For carefully designed cavities, such as microwave resonators for circuit applications [13] or cavities for material measurements [17, 23, 24], the cavity details (shape, size, dimensions, materials, etc.) are well known, and the cavity shape is generally a simple (separable) geometry. In such cases, deterministic theory (separation of variables and possibly perturbation techniques), as covered in Part I of this book, is appropriate.

However, for electrically large cavities that are not designed to perform a specific electromagnetic function (except possibly for shielding), the details of the cavity geometry and loading objects—such as cable bundles, scatterers, and absorbers—are not expected to be precisely known. Hence, for many applications in electromagnetic interference (EMI), compatibility (EMC) and in wireless communications, we are forced to deal with problems where we have only a partial knowledge of a large cavity and its interior loading. Gradually over the past two decades, techniques in statistical electromagnetics have been developed to deal with just such types of problems [36–39].

A good example of a structure with complex multiple cavities where EMI/EMC issues are important is an aircraft. A good description of aircraft cavities (crew cabin, main cabin, equipment bays, etc.) and their loading, electronic equipment, and apertures is given in [40, Sec 3.2.2]. The sources for aircraft EMI problems can be either external (such as a radar beam) or internal (inadvertent radiation from electronic devices). Clearly, all the information (cable bundle characteristics and routing, loading object characteristics and locations, etc.) will not be known in detail. The topological approach for EMI evaluation in [40] utilizes approximate deterministic solutions to individual, representative pieces of the entire structure. An alternative approach is to combine electromagnetic theory for a simplified aircraft cavity model with statistical estimates of quantities of interest (interior field strength, power coupled to a receiving antenna, etc.). A computer code using this combined method is included in [41].

Another example of a structure with complex multiple cavities is a large building where wireless communications [42] into or within the building is desired. Buildings

Electromagnetic Fields in Cavities: Deterministic and Statistical Theories, by David A. Hill

are particularly complicated because they change as doors are opened and closed, people move around, and furniture and other objects are moved. Ray tracing cannot possibly include all building features, but has been attempted [43]. More commonly, empirical models for indoor propagation attenuation [44, 45] have been proposed, but they have unknown parameters that are typically determined from experimental data [46]. Statistical models for angle of arrival have been found useful [47] for characterizing indoor multipath propagation.

5.2 SENSITIVITY OF FIELDS TO CAVITY GEOMETRY AND EXCITATION

It is well accepted that fields and resultant responses of almost any object located in an electrically large cavity are sensitive to geometrical parameters and excitation parameters [39,p.4]. This sensitivity has been seen in both frequency stirring [48, 49] and mechanical stirring [19] of reverberation chambers. There are also anecdotes of small geometrical changes, such as the position of a soda can in a large cavity, making large changes in field measurements [36, 39]. Sensitivity to geometry and excitation is one of the features of chaos that has been heavily studied for some time. The relevance of chaos to fields in complex cavities is discussed in Appendix D.

An easy way to quantify sensitivity to excitation is to examine the mode density of cavities which can infer the sensitivity of cavities to excitation frequency. The smoothed mode density $D_s(f)$ for an electrically large cavity was given in (1.33):

$$D_s(f) \cong \frac{8\pi f^2 V}{c^3} \tag{5.1}$$

Hence a typical frequency change Δf between adjacent modes is given by:

$$\Delta f \cong 1/D_s(f) \cong \frac{c^3}{8\pi V f^2} \tag{5.2}$$

The fractional frequency change between adjacent modes is obtained by dividing (5.2) by f:

$$\frac{\Delta f}{f} = \frac{c^3}{8\pi V f^3} = \frac{\lambda^3}{8\pi V}, \tag{5.3}$$

where V is the cavity volume and λ is the free-space wavelength.

Consider the following numerical example. The cavity is a 10m cube ($V = 10^3$ m^3), and the excitation frequency is 1 GHz ($\lambda = 0.3$ m). Then the fractional frequency change is approximately $\Delta f/f \approx 10^{-6}$. Thus the small relative frequency change of 10^{-6} will result in a totally different field structure. (In fact, the dominant mode will be orthogonal to the initially dominant mode.) Actually even a smaller relative frequency change could produce a substantial change in field by changing the mode coefficients.

It is also interesting to note in (5.3) that this sensitivity phenomenon depends only on volume and will occur for any cavity shape.

We can take a similar approach to determine the sensitivity of fields to cavity geometry. The smoothed mode number $N_s(f)$ for an electrically large cavity was given in (1.31):

$$N_s(f) \cong \frac{8\pi f^3 V}{3c^3} \cong \frac{8\pi V}{3\lambda^3} \tag{5.4}$$

If we make a small change ΔV in the cavity volume, the change in the smoothed mode number is:

$$\Delta N_s \cong \frac{8\pi}{3\lambda^3} \Delta V \tag{5.5}$$

To change the cavity volume by an amount sufficient to change the number of modes with resonant frequencies equal to or less than f by one, we can set ΔN_s in (5.5) equal to one. Then we can solve (5.5) for ΔV:

$$\Delta V \cong \frac{3\lambda^3}{8\pi} \tag{5.6}$$

We can obtain the relative change in volume by dividing both sides of (5.6) by V:

$$\frac{\Delta V}{V} \cong \frac{3\lambda^3}{8\pi V}. \tag{5.7}$$

If we consider the same parameters that we used in the frequency sensitivity example ($f = 1$ GHz and $V = 10^3$ m^3), then (5.7) yields $\Delta V/V \cong 3.22 \times 10^{-6}$. Thus a small relative change of cavity volume of 3.22×10^{-6} can shift the cavity mode to the mode of next higher order and completely change the field structure. This is a good example of the sensitivity of cavity fields to volume or geometry.

5.3 INTERPRETATION OF RESULTS

Even if it were possible to analyze a large, complex cavity accurately by use of modern computational techniques [50], the physical interpretation of the results (field strengths at all points within the cavity) would be difficult. Also, this is not generally the type of information desired. A typical question of practical interest is more of the flavor [39], "Given a cavity of approximately known parameters and some knowledge about the excitation, what is the probability that the performance of an electronic device located within that cavity will be degraded?" Such questions automatically take us out of the deterministic realm and require statistical treatments.

Analogous statistical approaches have been relied upon in other fields for many decades. For example, it is not productive to trace the complex path of every gas molecule in a large cavity. The averaged measurable quantities (such as temperature, pressure, and volume) are much more useful. Furthermore, it is fortunate that the ideal gas law does not depend on the details of the shape of the cavity. Similarly, the theory of room acoustics is really a statistical theory [51]. In fact, we will later show that some of the mathematics of room acoustics [52] are nearly identical to that of electromagnetic reverberation chambers [18].

Statistical methods have been used for some time in other electromagnetic applications. The theory in Ishimaru's classic book, *Wave Propagation and Scattering in Random Media* [53], is primarily statistical. Radiative transfer [54], a standard tool for analyzing propagation in random media is a statistical theory. The theory of optical coherence [55] is statistical. More recently in radar cross section (RCS) characterization, Mackay [56] has used statistical methods to deal with the chaotic behavior of electrically large ducts (open cavities) that has made deterministic RCS predictions difficult. The book by Holland and St. John, *Statistical Electromagnetics* [39], presents extensive comparisons of measured and analytical cumulative distributions for the responses of transmission lines located in cavities. Most of their experimental data are obtained by varying frequency, rather than cavity geometry as in a mechanically stirred reverberation chamber [18, 19], but their philosophy is the same in that the statistical results are more useful and easier to interpret than a measurement at a single frequency or a single stirrer position.

Although Part II of this book deals with statistical methods, the general philosophy is that solutions to Maxwell's equations are the desired starting point for the theory wherever possible. Then the statistics are introduced via unknown coefficients so that the general properties of electromagnetic fields in cavities as discussed in Part I are preserved.

PROBLEMS

5-1 Consider a large factory ($500\,\text{m} \times 250\,\text{m} \times 15\,\text{m}$) with metal walls with a communication frequency of 5 GHz. What is the mode separation as determined by the smoothed mode density in (5.3)?

5-2 For the same factory and communication frequency as in Problem 5-1, what is the relative change in volume that will change the smoothed mode number by one?

CHAPTER 6

Probability Fundamentals

6.1 INTRODUCTION

The remainder of Part II of this book makes frequent use of applied probability. Many good books [57–60] have been published on probability, statistics, and stochastic processes. The purpose of including this chapter on probability fundamentals is to attempt to make this book reasonably self-contained by covering the specific topics that will be used in Part II. However, for more complete knowledge of probability and related applications, the reader is advised to refer to a complete book, such as [57–60]. In addition, the four-volume set, *Principles of Statistical Radiophysics* [61], is of particular interest because of the applications to electromagnetic fields in random media. We will later see in Part II, that this area has several similarities to electromagnetic fields in large, complex cavities.

Probability theory deals with the mathematics of randomness. But how do we define randomness? An adequate definition for our purposes is "what happens in an experiment where we cannot predict the outcome with certainty." Some experiments have outcomes that appear to be truly random, as in quantum mechanics [62], but other experiments are less clear. For example, the flip of a coin is often cited as a simple case of a random process [60]. If we flip a coin, the outcome will be either heads or tails, but we cannot predict which. However, if we knew the exact initial conditions (position, velocity, rotation, etc.) of the coin flip and all other relevant parameters (coin weight, shape, and materials; table material and shape; etc.), then in theory we would be able to predict the outcome from the laws of physics. Hence the coin flip could be considered as an example of a way to use randomness to describe uncertainty due to lack of information. This is analogous to our discussion of large, complex cavities in Section 5.1 where we expect to lack detailed information. To continue on this line of reasoning, there are many complex deterministic processes where a random interpretation is actually clearer and more useful, as discussed previously in Section 5.3.

The next question is "What is probability." There are many definitions and interpretations of probability [57], but for our engineering purposes, the definitions are either "objective" or "subjective." The objective definition is statistical and

Electromagnetic Fields in Cavities: Deterministic and Statistical Theories, by David A. Hill

is sometimes called the *limit of relative frequencies*. The subjective definition usually requires some knowledge of the experiment or some reasoning and is sometimes called the *degree of belief*.

The statistical method for determining the probability P of an event E involves performing an experiment a large number of times N and recording the number of times M that the event E occurs. Then the statistical definition of P is:

$$P(E) = \lim_{N\to\infty} \frac{M}{N} \tag{6.1}$$

While the definition in (6.1) looks logical, it has some shortcomings. It assumes that the limit exists, and we will accept this assumption. It also does not tell us how many trials N are required because we cannot perform an infinite number of trials. This type of issue falls under the realm of statistics, and we will postpone it for now. An interesting experiment of flipping a coin a large number of times (N) and noting the number of heads (M) was performed by Karl Pearson (an eminent British statistician) about 100 hundred years ago [58]. He obtained $M = 12,012$ for $N = 24,000$. Hence, he obtained $P = 12,012/24,000 = 0.5005$, a value very close to our intuitive value of $\frac{1}{2}$.

Probability as a *degree of belief* is not as easily quantified, but sometimes it is the best that we can do, particularly if we do not have results from an experiment. If we return to the coin-flip example, we would expect that the probability that a coin flip gives heads is $\frac{1}{2}$ unless we have reason to believe that the coin is not fair. It is satisfying when the *degree of belief* probability agrees closely with the *limit of relative frequencies* probability, as with the coin flip. In the following chapters, we will use the *degree of belief* definition, but will follow that with experimental data that essentially generate *limit of relative frequencies* results and will generally find good agreement. For those in search of more rigor, a third method, the *axiomatic approach* is more satisfying [60, 63], but we will not need to pursue that approach.

6.2 PROBABILITY DENSITY FUNCTION

In this book, we deal primarily with random variables that can take continuous values. Typical examples are electric field strength, magnetic field strength, or received power. For a random variable g, the probability that g lies within a small range between g and $g + \mathrm{d}g$ can be written $f(g)\mathrm{d}g$. The function $f(g)$ is called the probability density function (PDF).

Since probabilities cannot be negative, all probability density functions must be positive or zero:

$$f(g) \geq 0, \quad \text{for all } g \tag{6.2}$$

Probability density functions need not be continuous or even finite. However, since the random variable g must lie between $-\infty$ and $+\infty$, the following integral relationship must hold:

$$\int_{-\infty}^{\infty} f(g)\mathrm{d}g = 1 \tag{6.3}$$

We will designate the mean value or ensemble average of g as $\langle g \rangle$. The mean value is also frequently designated μ, and it can be determined from the following integral involving the PDF:

$$\langle g \rangle = \mu = \int_{-\infty}^{\infty} gf(g)\mathrm{d}g \tag{6.4}$$

We define the variance of g as $\langle (g-\mu)^2 \rangle$. The variance is also frequently designated as σ^2, and it can also be determined from the PDF:

$$\langle (g-\mu)^2 \rangle = \sigma^2 = \int_{-\infty}^{\infty} (g-\mu)^2 f(g)\mathrm{d}g \tag{6.5}$$

The standard deviation σ is the square root of the variance.

Frequently we need to deal with two random variables, for example g and q. Here we introduce the joint PDF $f(g, q)$ such that $f(g, q)\mathrm{d}g\, \mathrm{d}q$ is the probability that g lies between g and $g + \mathrm{d}g$ and q lies between q and $q + \mathrm{d}q$. The two random variables are independent if their joint PDF equals the product of their individual PDFs:

$$f(g, q) = f_g(g) f_q(q) \tag{6.6}$$

Two random variables are uncorrelated if the expectation of their product is equal to the product of their expectations:

$$\langle gq \rangle = \langle g \rangle \langle q \rangle \tag{6.7}$$

We can show that if two random variables are independent, they are also uncorrelated [57]:

$$\begin{aligned} \langle gq \rangle &= \int_{-\infty}^{\infty}\int_{-\infty}^{\infty} gqf(g, q)\mathrm{d}g\, \mathrm{d}q \\ &= \int_{-\infty}^{\infty} gf_g(g)\mathrm{d}g \int_{-\infty}^{\infty} qf_q(q)\mathrm{d}q = \langle g \rangle \langle q \rangle \end{aligned} \tag{6.8}$$

The converse, that uncorrelated random variables are independent, is not generally true.

6.3 COMMON PROBABILITY DENSITY FUNCTIONS

In this section, we will define several specific probability functions that will appear later. The Gaussian PDF is:

$$f(g)=\frac{1}{\sigma\sqrt{2\pi}}\exp\left[-\frac{(g-\mu)^2}{2\sigma^2}\right], \tag{6.9}$$

where σ is the standard deviation and μ is the mean. This particular PDF is so common that it is also called the normal distribution.

The Rayleigh PDF is defined as [57, p. 104]:

$$f(g)=\frac{g}{\sigma^2}\exp\left(-\frac{g^2}{2\sigma^2}\right)\mathrm{U}(g), \tag{6.10}$$

where:

$$\mathrm{U}(g)=\begin{matrix}0, g<0\\ 1, g\geq 0\end{matrix} \tag{6.11}$$

The Rayleigh PDF is characterized by only one parameter, and the physical significance of σ^2 will be discussed in Chapter 7.

The Rice or Rice-Nakagami PDF [58, p. 252] is a generalization of the Rayleigh PDF:

$$f(g)=\frac{g}{\sigma^2}\exp\left[-\frac{g^2+s^2}{2\sigma^2}\right]\mathrm{I}_0\left(\frac{gs}{\sigma^2}\right)\mathrm{U}(g), \tag{6.12}$$

where I_0 is the zero-order, modified Bessel Function [25]. The Rice PDF is characterized by two parameters, σ^2 and s. The physical significance of s will be discussed in Chapter 9. For the case, $s/\sigma \ll 1$, the Rice PDF in (6.12) reduces to the Rayleigh PDF in (6.10).

The exponential PDF applies to a number of quantities in cavity problems [18]:

$$f(g)=\frac{1}{2\sigma^2}\exp\left(-\frac{g}{2\sigma^2}\right)\mathrm{U}(g) \tag{6.13}$$

Hence, the exponential is a one-parameter PDF, and its applications will be discussed in Chapter 7.

Chi and chi-square PDFs [57, p. 250] have several applications in cavity fields [18]. Suppose we have n independent, normal random variables g_i with zero mean and equal variances σ^2. We first form the random variable chi (or χ) as the square root of the sum of the squares of the normal random variables:

$$\chi=\sqrt{g_1^2+\ldots+g_n^2} \tag{6.14}$$

The random variable, chi squared ($q = \chi^2$), is also of interest. The chi and chi-squared PDFs are given by [57, p. 250]:

$$f_\chi(\chi) = \frac{2}{2^{n/2}\sigma^n\Gamma(n/2)}\chi^{n-1}\exp(-\chi^2/2\sigma^2)\mathrm{U}(\chi), \tag{6.15}$$

$$f_q(q) = \frac{1}{2^{n/2}\sigma^n\Gamma(n/2)}q^{(n-2)/2}\exp(-q/2\sigma^2)\mathrm{U}(q), \tag{6.16}$$

where Γ is the gamma function [25].

The special cases of chi and chi-square PDFs for $n = 2$ are of particular interest because they apply to the magnitude or magnitude squared of a complex scalar. If $n = 2$ is substituted into (6.15), the chi PDF simplifies to:

$$f_\chi(\chi)|_{n=2} = \frac{\chi}{\sigma^2}\exp(-\chi^2/2\sigma^2)\mathrm{U}(\chi) \tag{6.17}$$

The PDF in (6.17) is identical to (6.10) (with $\chi = g$). Hence the chi PDF with two degrees of freedom is frequently called a Rayleigh PDF. If $n = 2$ is substituted into (6.16), the chi-square PDF simplifies to:

$$f_q(q)|_{n=2} = \frac{1}{2\sigma^2}\exp(-q/\sigma^2)\mathrm{U}(q) \tag{6.18}$$

The PDF in (6.18) is identical to (6.13) (with $q = g$). Hence the chi-square PDF with two degrees of freedom is frequently called an exponential PDF. The special cases of chi and chi-square PDFs for $n = 6$ are of particular interest because they apply to the magnitude or magnitude squared of a complex vector. If $n = 6$ is substituted into (6.15) and (6.16), the chi and chi-square PDFs simplify to:

$$f_\chi(\chi)|_{n=6} = \frac{\chi^5}{8\sigma^6}\exp(-\chi^2/2\sigma^2)\mathrm{U}(\chi), \tag{6.19}$$

$$f_q(q)|_{n=6} = \frac{q^2}{16\sigma^6}\exp(-q/2\sigma^2)\mathrm{U}(q) \tag{6.20}$$

6.4 CUMULATIVE DISTRIBUTION FUNCTION

From the definition of the PDF in Section 6.2, we can write the probability P that the random variable G lies between a and b as an integral over f [58]:

$$P(a < G \leq b) = \int_a^b f(g)\mathrm{d}g \tag{6.21}$$

From (6.2) and (6.3), we can see that $P \leq 1$.

For the special case of $a = -\infty$, we can rewrite (6.21) in a way that allows us to define the cumulative distribution function (CDF), $F(g)$:

$$P(G \leq g) = \int_{-\infty}^{g} f(g')\mathrm{d}g' \equiv F(g). \tag{6.22}$$

From the properties of the PDF, the CDF must have the following properties [58]:

$$F(g) \text{ is a nondecreasing function of } g, \tag{6.23}$$

$$F(-\infty) = 0, \tag{6.24}$$

$$F(\infty) = 1 \tag{6.25}$$

To illustrate the derivation of F in (6.22) and the properties of F in (6.23) for a specific PDF, consider the exponential PDF in (6.13). If we substitute (6.13) into (6.22), we can evaluate the integral as follows:

$$\begin{aligned} F(g) &= \int_{-\infty}^{g} \frac{1}{2\sigma^2}\exp(-g'/2\sigma^2)\mathrm{U}(g')\mathrm{d}g' \\ &= -\exp(-g'/2\sigma^2)\mathrm{U}(g')|_0^g = [1-\exp(-g/2\sigma^2)]\mathrm{U}(g) \end{aligned} \tag{6.26}$$

It is clear that F in (6.26) satisfies (6.23)–(6.25).

6.5 METHODS FOR DETERMINING PROBABILITY DENSITY FUNCTIONS

Depending on the information given, there are many possibilities for determining or estimating the PDF for a random variable. In cases where only partial information is known, the PDF cannot be determined with complete certainty. However, the maximum entropy method [64, 65] has been found useful for deriving the PDF for underdetermined problems. The maximum entropy method selects the PDF $f(g)$ to maximize the entropy (uncertainty) given by the integral:

$$-\int_{-\infty}^{\infty} f(g)\ln[f(g)]\mathrm{d}g, \tag{6.27}$$

subject to the usual probability constraint in (6.3) and any other known constraints.

To illustrate the method, we consider the case where the mean μ and the variance σ^2 are given, but no other information is known about the pdf. Hence, the procedure is to select $f(g)$ to maximize the integral in (6.27) subject to the constraints given by (6.3), (6.4), and (6.5). This can be done by the method of Lagrange multipliers. We write the

Lagrangian L in the following form [65]:

$$L = -\int_{-\infty}^{\infty} f(g)\ln[f(g)] - (\lambda_0 - 1)\left[\int_{-\infty}^{\infty} f(g)\mathrm{d}g - 1\right] - \lambda_1\left[\int_{-\infty}^{\infty} f(g)g\mathrm{d}g - \mu\right] - \lambda_2\left[\int_{-\infty}^{\infty} f(g)(g-\mu)^2 - \sigma^2\right], \tag{6.28}$$

where λ_0, λ_1, and λ_2 are unknown constants. An extremum (maximum) of L can be obtained from the following derivate relation:

$$\frac{\partial L}{\partial f(g)} = 0 \tag{6.29}$$

If we substitute (6.28) into (6.29), we obtain:

$$-\ln[f(g)] - \lambda_0 - \lambda_1 g - \lambda_2(g-\mu)^2 = 0 \tag{6.30}$$

Equation (6.30) can be converted to the following exponential form:

$$f(g) = \exp[-\lambda_0 - \lambda_1 g - \lambda_2(g-\mu)^2] \tag{6.31}$$

Equation (6.31) gives us the general form of $f(g)$, but we still need to determine the constants, λ_0, λ_1, and λ_2. We first choose to write (6.31) in the following equivalent form:

$$f(g) = a\exp[-b(g-c)^2], \tag{6.32}$$

where a, b, and c are now the unknown constants. If we substitute (6.32) into the three constraint equations, (6.3) to (6.5), and carry out the g integrations, we obtain the following three equations in three unknowns:

$$a\sqrt{\frac{\pi}{b}} = 1, \tag{6.33}$$

$$ac\sqrt{\frac{\pi}{b}} = \mu, \tag{6.34}$$

$$a\left[\frac{1}{2}\sqrt{\frac{\pi}{b^3}} + (c-\mu)^2\sqrt{\frac{\pi}{b}}\right] = \sigma^2 \tag{6.35}$$

Simultaneous solution of (6.33)–(6.35) yields the following values for the constants:

$$a = \frac{1}{\sqrt{2\pi}\sigma}, \quad b = \frac{1}{2\sigma^2}, \quad \text{and} \quad c = \mu \tag{6.36}$$

If we substitute (6.36) into (6.32), we obtain:

$$f(g) = \frac{1}{\sigma\sqrt{2\pi}} \exp\left[-\frac{(g-\mu)^2}{2\sigma^2}\right] \tag{6.37}$$

Equation (6.37) is recognized as the Gaussian (or normal) PDF previously discussed and given in (6.9).

An alternative way to state the result in (6.37) is that if the mean and variance are specified for a PDF over the range from $-\infty$ to ∞, then the maximum entropy method predicts a Gaussian PDF. Even though there are other PDFs that would satisfy the constraints in (6.3) to (6.5) over the range from $-\infty$ to ∞, the normal PDF maximizes the entropy (uncertainty) in (6.27) and is the least biased. Any other PDF would have to be based on additional information that is not provided by the constraints, (6.3) to (6.5). The maximum entropy method has been used to determine PDFs for a number of other combinations of constraints and ranges of g, and some are listed in [65].

Since the Gaussian PDF is so common and is encountered here in the following chapter, the central limit theorem [57, pp. 266–268] is also worth mentioning. It states that if a random variable is the sum of a large number of independent random variables of the continuous type, then the PDF approaches a Gaussian as the number of random variables increases. Both the central limit theorem and the maximum entropy method can be used for determining a Gaussian PDF for certain quantities in cavities, as will be seen in the following chapter.

PROBLEMS

6-1 For the Rayleigh PDF in (6.10), show that the integral of the PDF equals 1: $\int\limits_0^\infty \frac{g}{\sigma^2}\exp\left(-\frac{g^2}{2\sigma^2}\right)dg = 1.$

6-2 For the Rayleigh PDF in (6.10), show that the mean value is $\mu = \langle g \rangle = \sigma\sqrt{\pi/2}.$

6-3 Using the Rayleigh PDF result for the mean value in Problem 6-2, show that the variance is $\langle (g-\mu)^2 \rangle = \sigma^2\left(2-\frac{\pi}{2}\right).$

6-4 For the exponential PDF in (6.13), show that the integral of the PDF equals 1: $\int\limits_0^\infty \frac{1}{2\sigma^2}\exp\left(-\frac{g}{2\sigma^2}\right)dg = 1.$

6-5 For the exponential PDF in (6.13), show that the mean value is $\mu = \langle g \rangle = 2\sigma^2.$

6-6 Using the exponential result for the mean value in Problem 6-5, show that the variance is $\langle (g-\mu)^2 \rangle = 4\sigma^4.$

6-7 For the chi PDF with $n = 6$ in (6.19), show that the integral of the PDF equals 1: $\int_0^\infty \frac{\chi^5}{8\sigma^2} \exp\left(-\frac{\chi^2}{2\sigma^2}\right) d\chi = 1$.

6-8 For the chi PDF with $n = 6$ in (6.19), show that the mean value is $\mu = \langle\chi\rangle = 15\sigma\sqrt{2\pi}/16$.

6-9 Using the chi PDF with $n = 6$ result for the mean value in Problem 6-8, show that the variance is $\langle(\chi-\mu)^2\rangle = \sigma^2[6-(225\pi/128)]$.

6-10 For the chi-square pdf with $n = 6$ in (6.20), show that the integral of the PDF equals 1: $\int_0^\infty \frac{q^2}{16\sigma^2} \exp\left(-\frac{q}{2\sigma^2}\right) dq = 1$.

6-11 For the chi-square PDF with $n = 6$ in (6.20), show that the mean value is $\mu = \langle q\rangle = 6\sigma^2$.

6-12 Using the chi-square with $n = 6$ result in Problem 6-11, show that the variance is $\langle(q-\mu)^2\rangle = 12\sigma^4$.

6-13 For the Rice pdf of (6.12), show that the integral of the PDF equals 1: $\int_0^\infty \frac{g}{\sigma^2} \exp\left(-\frac{g^2+s^2}{2\sigma^2}\right) \mathrm{I}_0\left(\frac{gs}{2\sigma^2}\right) dg = 1$.

6-14 Consider a PDF $f(x)$ which is zero for negative x. If we specify only the mean value m, show that the maximum entropy method yields an exponential PDF: $f(x) = \frac{1}{m} \exp\left(-\frac{x}{m}\right) \mathrm{U}(x)$.

CHAPTER 7

Reverberation Chambers

The primary electrically large cavity that we choose to cover is the reverberation chamber. Reverberation chambers have been well studied theoretically [18] and experimentally [19, 66]. The use of reverberation chambers (also called mode-stirred chambers) for electromagnetic compatibility (EMC) measurements was first proposed in 1968 [67]. It took some time for reverberation chamber measurements to gain acceptance, but by the 1980s their use was well established in EMC measurements [68,19]. Reverberation chambers are electrically large, high-Q cavities that obtain statistically uniform fields by either mechanical stirring [19,68] or frequency stirring [48,49]. This chapter will be devoted to the theory of reverberation chambers [18] that use mechanical stirring. Frequency stirring will be covered in Chapter 9.

7.1 PLANE-WAVE INTEGRAL REPRESENTATION OF FIELDS

A typical rectangular-cavity reverberation chamber with a rotating stirrer is shown in Figure 7.1. As discussed in Chapter 5, deterministic mode theory is not convenient for predicting the field properties or the response of antennas and test objects in electrically large, complex cavities. Since many stirrer positions are employed in reverberation chamber measurements, some type of statistical method [37,39] is required to determine the statistics of the fields and test object response. At the same time, it is important to ensure that the associated electromagnetic theory is consistent with Maxwell's equations.

We choose a plane-wave integral representation for the electric and magnetic fields that satisfies Maxwell's equations and also includes the statistical properties expected for a well-stirred field [69]. The statistical nature of the fields is introduced through the plane-wave coefficients that are taken to be random variables with fairly simple statistical properties. Because the theory uses only propagating plane waves, it is fairly easy to use to calculate the responses of test objects or reference antennas.

As shown in Figure 7.1, a transmitting antenna radiates cw fields, and the mechanical stirrer (or multiple stirrers [66]) is rotated to generate a statistically

Electromagnetic Fields in Cavities: Deterministic and Statistical Theories, by David A. Hill

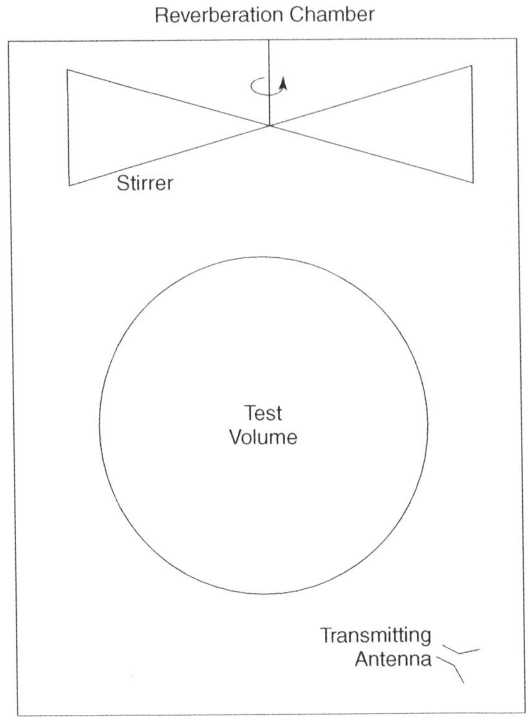

FIGURE 7.1 Transmitting antenna in a reverberation chamber with a mechanical stirrer.

uniform field. The test volume can occupy a fairly large portion of the chamber volume. The electric field $\vec{E}$ at location $\vec{r}$ in a source-free, finite volume can be represented as an integral of plane waves over all real angles [70]:

$$\vec{E}(\vec{r}) = \iint_{4\pi} \vec{F}(\Omega)\exp(i\vec{k} \bullet \vec{r})\mathrm{d}\Omega, \tag{7.1}$$

where the solid angle Ω is shorthand for the elevation and azimuth angles, α and β, and $\mathrm{d}\Omega = \sin\alpha\, \mathrm{d}\alpha\, \mathrm{d}\beta$. The vector wavenumber $\vec{k}$ is:

$$\vec{k} = -k(\hat{x}\sin\alpha\ \cos\beta + \hat{y}\sin\alpha\ \sin\beta + \hat{z}\ \cos\alpha) \tag{7.2}$$

The geometry for a plane-wave component is shown in Figure 7.2. So (7.1) could be written more explicitly as:

$$\vec{E}(\hat{r}) = \int_0^{2\pi}\int_0^{\pi} \vec{F}(\alpha,\beta)\exp(i\vec{k} \bullet \vec{r})\sin\alpha\, \mathrm{d}\alpha\, \mathrm{d}\beta \tag{7.3}$$

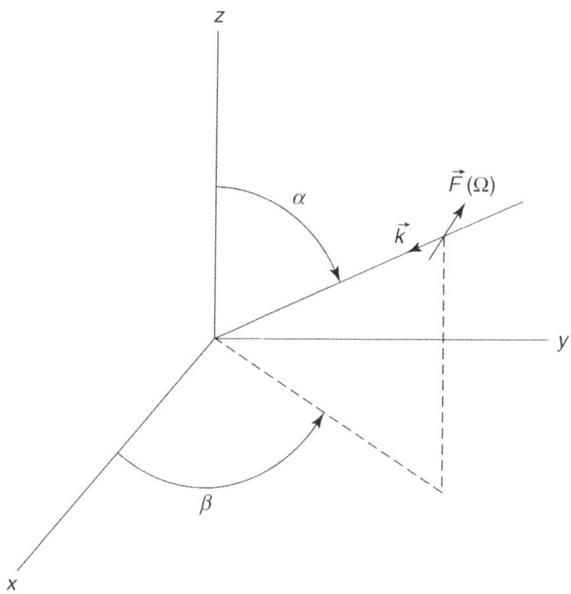

FIGURE 7.2 Plane-wave component $\vec{F}(\Omega)$ of the electric field with wavenumber $\vec{k}$.

The angular spectrum $\vec{F}(\Omega)$can be written:

$$\vec{F}(\Omega) = \hat{\alpha}F_\alpha(\Omega) + \hat{\beta}F_\beta(\Omega), \tag{7.4}$$

where $\hat{\alpha}$ and $\hat{\beta}$ are unit vectors that are orthogonal to each other and to $\hat{k}$. Both F_α and F_β are complex and can be written in terms of their real and imaginary parts:

$$F_\alpha(\Omega) = F_{\alpha r}(\Omega) + iF_{\alpha i}(\Omega) \quad \text{and} \quad F_\beta(\Omega) = F_{\beta r}(\Omega) + iF_{\beta i}(\Omega) \tag{7.5}$$

The electric field in (7.1) satisfies Maxwell's equations because each plane-wave component satisfies Maxwell's equations. For a spherical volume, the representation in (7.1) can be shown to be complete because it is equivalent to the rigorous spherical-wave expansion [71]. For a non-spherical volume, the plane-wave expansion can be analytically continued outward from a spherical volume, but the general conditions under which the analytical continuation holds have yet to be established. In this chapter, we assume that the volume is selected so that (7.1) is valid.

Up to this point, the angular spectrum $\vec{F}(\Omega)$ in (7.1) is general and could be either deterministic or random. However, for a statistical field as generated in a reverberation chamber, we take $\vec{F}(\Omega)$ to be a random variable (that depends on stirrer position). For the derivation of many of the important field quantities, the probability density function of the angular spectrum is not required, and it is sufficient to specify certain means and variances. In a typical reverberation chamber measurement, the statistical ensemble is generated by rotating the stirrer (or stirrers). For general cavities, the same statistical ensemble could also be thought of as being generated from a large number of cavities of different shapes. In the rest of this book, we use $\langle\,\rangle$ to represent an ensemble average. The starting point for the statistical analysis is to select statistical properties for the angular spectrum that are representative of a well-stirred field that would be obtained in an electrically large, multimode chamber with a large effective stirrer [19]. Appropriate statistical assumptions for such a field are as follows:

$$\langle F_\alpha(\Omega)\rangle = \langle F_\beta(\Omega)\rangle = 0, \tag{7.6}$$

$$\begin{aligned}\langle F_{\alpha r}(\Omega_1)F_{\alpha i}(\Omega_2)\rangle &= \langle F_{\beta r}(\Omega_1)F_{\beta i}(\Omega_2)\rangle = \\ \langle F_{\alpha r}(\Omega_1)F_{\beta r}(\Omega_2)\rangle &= \langle F_{\alpha r}(\Omega_1)F_{\beta i}(\Omega_2)\rangle = \\ \langle F_{\alpha i}(\Omega_1)F_{\beta r}(\Omega_2)\rangle &= \langle F_{\alpha i}(\Omega_1)F_{\beta i}(\Omega_2)\rangle = 0,\end{aligned} \tag{7.7}$$

$$\begin{aligned}\langle F_{\alpha r}(\Omega_1)F_{\alpha r}(\Omega_2)\rangle &= \langle F_{\alpha i}(\Omega_1)F_{\alpha i}(\Omega_2)\rangle = \\ \langle F_{\beta r}(\Omega_1)F_{\beta r}(\Omega_2)\rangle &= \langle F_{\beta i}(\Omega_1)F_{\beta i}(\Omega_2)\rangle = C_E\delta(\Omega_1-\Omega_2),\end{aligned} \tag{7.8}$$

where δ is the Dirac delta function and C_E is a constant with units of $(\mathrm{V/m})^2$.

The mathematical reasons for the assumptions, (7.6)–(7.8), will become clear when the field properties are derived, but the physical justifications are as follows. Since the angular spectrum is a result of many rays or bounces with random phases, the mean value should be zero, as indicated in (7.6). Since multipath scattering changes the phase and rotates the polarization many times, angular spectrum components with orthogonal polarizations or quadrature phase ought to be uncorrelated, as indicated in (7.7). Since angular spectrum components arriving from different directions have taken very different multiple scattering paths, they ought to be uncorrelated, as indicated by the delta function on the right side of (7.8). The coefficient C_E of the delta function is proportional to the square of the electric field strength, as will be shown later. The following useful relationships can be derived from (7.7) and (7.8):

$$\langle F_\alpha(\Omega_1)F_\beta^*(\Omega_2)\rangle = 0, \tag{7.9}$$

$$\langle F_\alpha(\Omega_1)F_\alpha^*(\Omega_2)\rangle = \langle F_\beta(\Omega_1)F_\beta^*(\Omega_2)\rangle = 2C_E\delta(\Omega_1-\Omega_2), \tag{7.10}$$

where * denotes complex conjugate.

7.2 IDEAL STATISTICAL PROPERTIES OF ELECTRIC AND MAGNETIC FIELDS

A number of field properties can be derived from (7.1) and (7.6)–(7.10). Consider first the mean value of the electric field $\langle \vec{E}(\vec{r})\rangle$, which can be derived from (7.1) and (7.6):

$$\langle \vec{E}(\vec{r})\rangle = \iint_{4\pi} \langle \vec{F}(\Omega)\rangle \exp(i\vec{k}\bullet\vec{r})\,\mathrm{d}\Omega = 0 \tag{7.11}$$

Thus the mean value of the electric field is zero because the mean value of the angular spectrum is zero. This result is expected for a well-stirred field which is the sum of a large number of multipath rays with random phases.

The square of the absolute value of the electric field is important because it is proportional to the electric energy density [38]. From (7.1), the square of the absolute value of the electric field can be written as a double integral:

$$|\vec{E}(\vec{r})|^2 = \iint_{4\pi}\iint_{4\pi} \vec{F}(\Omega_1) \bullet \vec{F}^*(\Omega_2)\exp[i(\vec{k}_1-\vec{k}_2) \bullet \vec{r}\}\mathrm{d}\Omega_1\mathrm{d}\Omega_2 \tag{7.12}$$

The mean value of (7.12) can be derived by applying (7.9) and (7.10) to the integrand:

$$\langle|\vec{E}(\vec{r})|^2\rangle = 4C_E \iint_{4\pi}\iint_{4\pi} \delta(\Omega_1-\Omega_2)\exp[i(\vec{k}_1-\vec{k}_2) \bullet \vec{r}]\mathrm{d}\Omega_1\mathrm{d}\Omega_2 \tag{7.13}$$

One integration in (7.13) can be evaluated by use of the sampling property of the delta function, and the second integration is easily evaluated to obtain the final result:

$$\langle|\vec{E}(\vec{r})|^2\rangle = 4C_E \iint_{4\pi} \mathrm{d}\Omega_2 = 16\pi C_E \equiv E_0^2 \tag{7.14}$$

Thus the mean-square value of the electric field is E_0^2 and is independent of position. This is the spatial uniformity property of an ideal reverberation chamber; it applies to the ensemble average of the squared electric field and has been verified experimentally with an array of three-axis, electric-field probes [19,66]. For convenience from here on, C_E is defined in terms of the mean-square value of the electric field as indicated in (7.14). For now, we postpone the dependence of E_0^2 on chamber properties and excitation.

By a similar derivation, the mean-square values of the rectangular components of the electric field can be derived:

$$\langle|E_x|^2\rangle = \langle|E_y|^2\rangle = \langle|E_z|^2\rangle = \frac{E_0^2}{3} \tag{7.15}$$

This is the isotropy property of an ideal reverberation chamber, and it has been verified with three-axis, electric-field probes [19,66]. Both isotropy and spatial uniformity are demonstrated experimentally in Figure 7.3 for frequencies from 80 MHz to 18 GHz [66]. The measurements were taken with 10 three-axis probes (equivalent to 30 single-axis probes) spaced at least one meter apart. So there are 30 measurements at each frequency. The results are best (least spread) above about 200 MHz where the chamber has sufficient electrical size.

The magnetic field $\vec{H}$ can be derived by applying Maxwell's curl equation (1.1) to (7.1):

$$\vec{H}(\vec{r}) = \frac{1}{i\omega\mu}\nabla \times \vec{E}(\vec{r}) = \frac{1}{\eta}\iint_{4\pi} \hat{k} \times \vec{F}(\Omega)\exp(i\vec{k} \bullet \vec{r})\mathrm{d}\Omega, \tag{7.16}$$

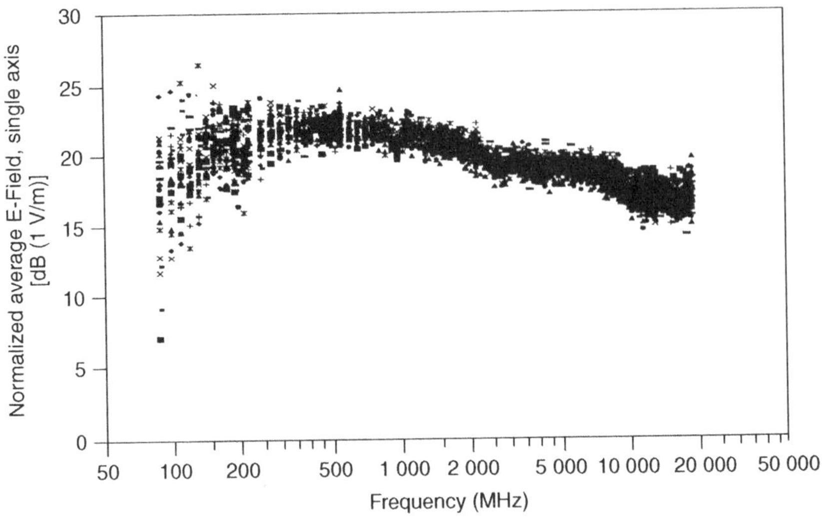

FIGURE 7.3 The average measured electric field (rectangular component) for each of 30 short dipoles. Field values are for a constant net input power of 1 W [66].

where η is the characteristic impedance of free space. Applying (7.6) to (7.16) shows that the mean value of the magnetic field is zero:

$$\langle \vec{H}(\vec{r}) \rangle = \frac{1}{\eta} \iint_{4\pi} \hat{k} \times \langle \vec{F}(\Omega) \rangle \exp(i\vec{k} \bullet \vec{r}) \mathrm{d}\Omega = 0 \tag{7.17}$$

The square of the magnitude of the magnetic field can be written:

$$|\vec{H}(\vec{r})|^2 = \frac{1}{\eta^2} \iint_{4\pi} \iint_{4\pi} \left[\hat{k}_1 \times \vec{F}(\Omega_1)\right] \bullet \left[\hat{k}_2 \times \vec{F}^*(\Omega_2)\right] \exp\left[i(\vec{k}_1 - \vec{k}_2) \bullet \vec{r}\right] \mathrm{d}\Omega_1 \mathrm{d}\Omega_2 \tag{7.18}$$

The derivation of the mean-square value follows closely that of the electric field, and the result is:

$$\langle |\vec{H}(\vec{r})|^2 \rangle = \frac{E_0^2}{\eta^2} \tag{7.19}$$

Thus the mean-square magnetic field also exhibits spatial uniformity, and the value is related to the mean-square electric field by the square of the free-space impedance:

$$\langle |\vec{H}(\vec{r}_1)|^2 \rangle = \frac{\langle |\vec{E}(\vec{r}_2)|^2 \rangle}{\eta^2}, \tag{7.20}$$

where $\vec{r}_1$ and $\vec{r}_2$ are arbitrary locations. This free-space relationship has been demonstrated experimentally by use of electric and magnetic field probes [19].

By using the previous formalism, we can also derive the isotropy relationship for the magnetic field:

$$\langle |H_x|^2 \rangle = \langle |H_y|^2 \rangle = \langle |H_z|^2 \rangle = \frac{E_0^2}{3\eta^2} \tag{7.21}$$

The energy density W can be written [3]:

$$W(\vec{r}) = \frac{1}{2}\left[\varepsilon|\vec{E}(\vec{r})|^2 + \mu|\vec{H}(r)|^2\right] \tag{7.22}$$

The mean value can be obtained from (7.14), (7.20), and (7.22):

$$\langle W(r) \rangle = \frac{1}{2}\left[\varepsilon\langle|\vec{E}(\vec{r})|^2\rangle + \mu\langle|\vec{H}^*(\vec{r})|^2\rangle\right] = \varepsilon E_0^2 \tag{7.23}$$

Thus the average value of the energy density is also independent of position.

The power density or Poynting vector $\vec{S}$ can be written [3]:

$$\vec{S}(\vec{r}) = \vec{E}(\vec{r}) \times \vec{H}^*(\vec{r}) \tag{7.24}$$

From (7.1), (7.16), and (7.24), the mean power density can be written:

$$\langle \vec{S}(\vec{r}) \rangle = \frac{1}{\eta} \iint_{4\pi} \iint_{4\pi} \langle \vec{F}(\Omega_1) \times [\vec{k}_2 \times \hat{F}^*(\Omega_2)] \rangle \exp[i(\vec{k}_1 - \vec{k}_2) \bullet \vec{r}] \mathrm{d}\Omega_1 \mathrm{d}\Omega_2 \tag{7.25}$$

The expectation in the integrand can be evaluated from vector identities and (7.9) and (7.10):

$$\langle \vec{F}(\Omega_1) \times [\hat{k}_2 \times \vec{F}(\Omega_2)] \rangle = \vec{k}_2 \frac{E_0^2}{4\pi} \delta(\Omega_1 - \Omega_2) \tag{7.26}$$

The right side of (7.25) can now be evaluated from (7.26) and the sampling property of the delta function:

$$\langle \vec{S}(\vec{r}) \rangle = \frac{E_0^2}{4\pi\eta} \iint_{4\pi} \hat{k}_2 \mathrm{d}\Omega_2 = 0 \tag{7.27}$$

A physical interpretation of (7.27) is that each plane wave carries equal power in a different direction so that the vector integration of 4π steradians is zero. This result is important because it shows that the power density is not the proper quantity for characterizing field strength in reverberation chambers. The mean value of energy density as given by (7.23) is an appropriate positive scalar quantity that could be used.

Another possibility is to define a positive scalar quantity S that has units of power density and is proportional to the mean energy density:

$$S = v\langle W \rangle = \frac{E_0^2}{\eta}, \tag{7.28}$$

where $v = 1/\sqrt{\mu\varepsilon}$. For lack of a better term, S will be called scalar power density from here on. This quantity could be used to compare with uniform-field, plane-wave testing where power density, rather than field strength, is sometimes specified.

7.3 PROBABILITY DENSITY FUNCTIONS FOR THE FIELDS

The statistical assumptions for the angular spectrum in (7.6)–(7.8) have been used to derive a number of useful ensemble averages in Section 7.2. These results have not required a knowledge of the particular form of the probability density functions. However, such knowledge would be very useful for analysis of measured data which is always based on some limited number of samples (stirrer positions). For example, the probability density function is needed to determine the expectation of maximum field strength strength for a given number of samples [66]. This maximum is important in immunity testing of electronic equipment.

The starting point for deriving electric-field probability density functions is to write the rectangular components in terms of their real and imaginary parts:

$$E_x = E_{xr} + iE_{xi}, \quad E_y = E_{yr} + iE_{yi}, \quad E_z = E_{zr} + iE_{zi} \tag{7.29}$$

(The dependence on $\vec{r}$ will be omitted where convenient because all of the results in this section are independent of $\vec{r}$.) The mean values of all the real and imaginary parts in (7.29) are zero, as shown in (7.11):

$$\langle E_{xr} \rangle = \langle E_{xi} \rangle = \langle E_{yr} \rangle = \langle E_{yi} \rangle = \langle E_{zr} \rangle = \langle E_{zi} \rangle = 0 \tag{7.30}$$

The variances of the real and imaginary parts can be shown to equal half the result for the complex components in (7.15):

$$\langle E_{xr}^2 \rangle = \langle E_{xi}^2 \rangle = \langle E_{yr}^2 \rangle = \langle E_{yi}^2 \rangle = \langle E_{zr}^2 \rangle = \langle E_{zi}^2 \rangle = \frac{E_0^2}{6} \equiv \sigma^2 \tag{7.31}$$

The mean and variance of the real and imaginary parts in (7.30) and (7.31) are all the information that can be derived from the initial statistical assumptions in (7.6) to (7.8).

However, as shown in Section 6.5, if the mean and variance are specified for a PDF over the range from $-\infty$ to ∞, then the maximum entropy method predicts a Gaussian PDF. So from (6.37) the PDF $f(E_{xr})$ is:

$$f(E_{xr}) = \frac{1}{\sqrt{2\pi}\sigma} \exp\left[-\frac{E_{xr}^2}{2\sigma^2}\right], \tag{7.32}$$

where σ is defined in (7.31). The same pdf also applies to the other real and imaginary parts of the electric components.

Equations (7.1), (7.10) and (7.11) can be used to show that the real and imaginary parts of the electric-field components are uncorrelated. Only the derivation for $\langle E_{xr}E_{xi}\rangle$ will be shown, but the derivations for the other correlations are similar. From (7.1) to (7.5), the real and imaginary parts of E_x can be written:

$$E_{xr} = \iint_{4\pi} \{[\cos\alpha\cos\beta\, F_{\alpha r}(\Omega) - \sin\beta\, F_{\beta r}(\Omega)]\cos(\vec{k}\bullet\vec{r}) - [\cos\alpha\cos\beta\, F_{\alpha i}(\Omega) - \sin\beta\, F_{\beta i}(\Omega)]\sin(\vec{k}\bullet\vec{r})\}\mathrm{d}\Omega, \tag{7.33}$$

$$E_{xi} = \iint_{4\pi} \{[\cos\alpha\cos\beta\, F_{\alpha i}(\Omega) - \sin\beta\, F_{\beta i}(\Omega)]\cos(\vec{k}\bullet\vec{r}) + [\cos\alpha\cos\beta\, F_{\alpha r}(\Omega) - \sin\beta\, F_{\beta r}(\Omega)]\sin(\vec{k}\bullet\vec{r})\}\mathrm{d}\Omega \tag{7.34}$$

The average value of the product of (7.33) and (7.34) can be evaluated by use of (7.7) and (7.8) inside the double integral and making use of the delta function to evaluate one integration. Then the remaining integrand (and hence the integral) is zero:

$$\langle E_{xr}(\vec{r})E_{xi}(\vec{r})\rangle = \frac{E_0^2}{16\pi}\iint_{4\pi}[\cos^2\alpha_2\cos^2\beta_2][\cos(\vec{k}_2\bullet\vec{r})\sin(\vec{k}_2\bullet\vec{r}) - \cos(\vec{k}_2\bullet\vec{r})\sin(\vec{k}_2\bullet\vec{r})]\mathrm{d}\Omega_2 = 0 \tag{7.35}$$

Similar evaluations show that the real and imaginary parts of all three rectangular components of the electric field are uncorrelated. Since they are Gaussian, they are also independent [57].

Since the real and imaginary parts of the rectangular components of the electric field have been shown to be normally distributed with zero mean and equal variances and are independent, the probability density functions of various electric magnitudes or squared magnitudes are chi or chi-square distributions with appropriate number of degrees of freedom. The magnitude of any of the electric field components, for example $|E_x|$, is chi distributed with two degrees of freedom and consequently has a Rayleigh distribution [57]:

$$f(|E_x|) = \frac{|E_x|}{\sigma^2}\exp\left[-\frac{|E_x|^2}{2\sigma^2}\right] \tag{7.36}$$

Figure 7.4 shows a comparison of (7.36) with measured data taken at 1 GHz in the NASA Chamber A [66]. The chamber has two stirrers, and the total number of samples (stirrer positions) is 225. The data were taken with a small electric-field probe that was calibrated at NIST [66]. The agreement is about as good as can be expected for 225 samples.

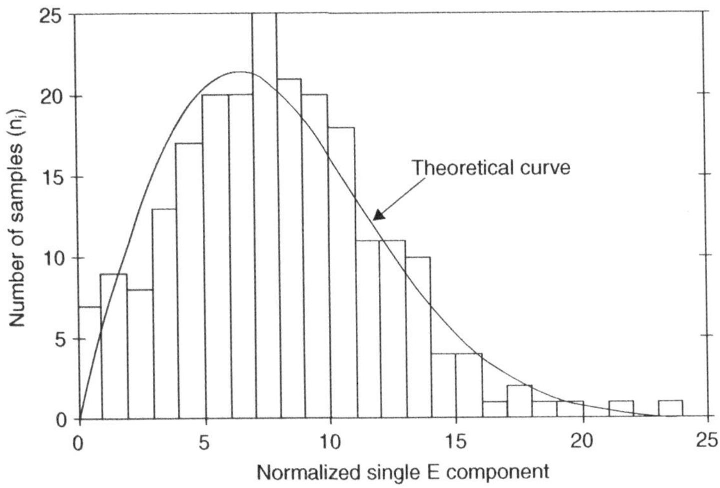

FIGURE 7.4 Comparison of the measured probability density function of the magnitude of a single rectangular component of the electric field with theory (Rayleigh distribution) [18].

The squared magnitude of any of the electric field components, for example $|E_x|^2$, is chi-square distributed with two degrees of freedom, and consequently it has an exponential distribution [57]:

$$f(|E_x|^2) = \frac{1}{2\sigma^2}\exp\left[-\frac{|E_x|^2}{2\sigma^2}\right] \tag{7.37}$$

The probability density functions in (7.36) and (7.37) agree with Kostas and Boverie [72]. They suggest the exponential distribution in (7.37) is also applicable to the power received by a small, linearly polarized antenna, but it was shown that the exponential distribution applies to the power received by any type of antenna [18]. The exponential distribution has been confirmed experimentally for a horn antenna [18].

The total electric field magnitude is chi distributed with six degrees of freedom and has the following probability density function [57]:

$$f(|\vec{E}|) = \frac{|\vec{E}|^5}{8\sigma^6}\exp\left[-\frac{|\vec{E}|^2}{2\sigma^2}\right] \tag{7.38}$$

Figure 7.5 shows a comparison of (7.38) with measured data taken under the same conditions as in Figure 7.4. In this case a three-axis, electric-field probe was used to take the data [72]. Again the agreement is about as good as can be expected for 225 samples. The squared magnitude of the total electric field is chi-square distributed with six degrees of freedom and has the following probability density function [57]:

$$f(|\vec{E}|^2) = \frac{|\vec{E}|^4}{16\sigma^6}\exp\left[-\frac{|\vec{E}|^2}{2\sigma^2}\right] \tag{7.39}$$

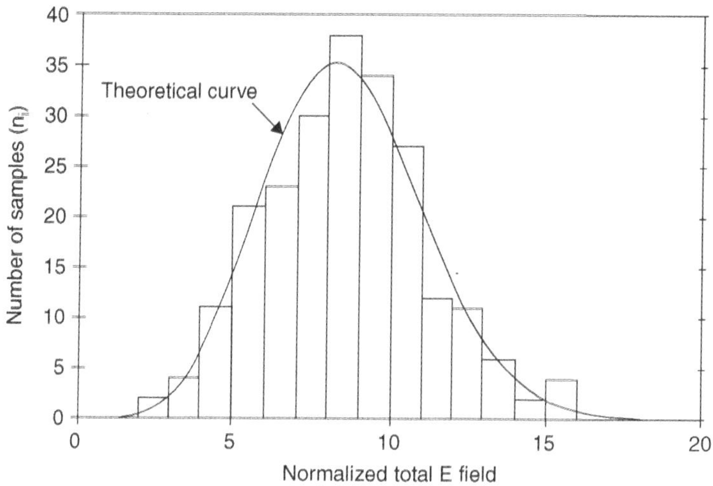

FIGURE 7.5 Comparison of the measured probability density function of the total electric field with theory (chi distribution with six degrees of freedom) [18].

The dual probability density functions for the magnetic field can be obtained by starting with the variance of the real or imaginary parts of one of the magnetic field components, for example H_{xr}:

$$\langle H_{xr}^2 \rangle = \frac{E_0^2}{6\eta^2} \equiv \sigma_H^2 \tag{7.40}$$

Now the dual of the results in (7.36)–(7.39) can be obtained by replacing E by H and σ by σ_H.

7.4 SPATIAL CORRELATION FUNCTIONS OF FIELDS AND ENERGY DENSITY

In the previous section, field properties at a point were considered. Real antennas and test objects have significant spatial extent, and the correlation functions of the fields [73] are important in understanding responses of extended objects in reverberation chambers [74].

7.4.1 Complex Electric or Magnetic Field

We begin by deriving the spatial correlation function $\rho(\vec{r}_1, \vec{r}_2)$ for the total complex electric field in a reverberation chamber. Without loss of generality, we can locate $\vec{r}_1$ at the origin and $\vec{r}_2$ on the z axis:

$$\vec{r}_1 = 0 \quad \text{and} \quad \vec{r}_2 = \hat{z}r \tag{7.41}$$

Now we can write the correlation function ρ as a function of the separation r of the two field points [75]:

$$\rho(r) \equiv \frac{\langle \vec{E}(0) \bullet E^*(\hat{z}r)\rangle}{\sqrt{\langle |\vec{E}(0)|^2\rangle \langle |\vec{E}(\hat{z}r)|^2\rangle}} \tag{7.42}$$

The numerator in (7.42) is the correlation function (or mutual coherence function), which has been used to describe wave propagation in random media [53].

The expectations in the denominator of (7.42) have been evaluated in (7.14):

$$\langle |\vec{E}(0)|^2\rangle = \langle |\vec{E}(\hat{z}r)|^2\rangle = E_0^2 \tag{7.43}$$

The numerator in (7.42) can be rewritten using (7.1):

$$\langle \vec{E}(0) \bullet \vec{E}^*(\hat{z}r)\rangle = \iint_{4\pi}\iint_{4\pi} \langle \vec{F}(\Omega_1) \bullet \vec{F}^*(\Omega_2)\rangle \exp(-i\vec{k}_2 \bullet \hat{z}r)\mathrm{d}\Omega_1\mathrm{d}\Omega_2 \tag{7.44}$$

One of the integrations in (7.44) can be performed using (7.9), (7.10), and (7.14):

$$\langle \vec{E}(0) \bullet \vec{E}^*(\hat{z}r)\rangle = \frac{E_0^2}{4\pi}\iint_{4\pi} \exp(-i\vec{k}_2 \bullet \hat{z}r)\mathrm{d}\Omega_2 \tag{7.45}$$

By writing $\vec{k}_2$ and $\mathrm{d}\Omega_2$ explicitly as in (7.2) and (7.3), we can write (7.45) in the following form:

$$\langle \vec{E}(0) \bullet \vec{E}^*(\hat{z}r)\rangle = \frac{E_0^2}{4\pi}\int_0^{2\pi}\int_0^{\pi} \exp(-ikr\cos\alpha_2)\sin\alpha_2 \,\mathrm{d}\alpha_2 \,\mathrm{d}\beta_2 \tag{7.46}$$

The β_2 integration in (7.46) contributes a 2π factor, and the α_2 integrand is a perfect differential so that (7.46) reduces to:

$$\langle \vec{E}(0) \bullet \vec{E}(\hat{z}r)\rangle = E_0^2\frac{\sin(kr)}{kr} \tag{7.47}$$

By substituting (7.43) and (7.47) into (7.42), we can write the correlation function $\rho(r)$ as:

$$\rho(r) = \frac{\sin(kr)}{kr} \tag{7.48}$$

It is perhaps surprising that the spatial correlation function in (7.48) decays in an oscillatory manner as kr increases, but the identical result has been obtained independently [37,76]. The same correlation function can be derived for the magnetic field, and it also applies to acoustic reverberation chambers [77]. A correlation length l_c can be defined as the separation corresponding to the first zero in (7.48):

$$kl_c = \pi \quad \text{or} \quad l_c = \pi/k = \lambda/2, \tag{7.49}$$

where λ is the wavelength in the medium (usually free space).

An angular correlation function $\rho(\hat{s}_1, \hat{s}_2)$ can be defined as:

$$\rho(\hat{s}_1, \hat{s}_2) = \frac{\langle E_{s1}(\vec{r}) \bullet E_{s2}^*(\vec{r})\rangle}{\sqrt{\langle |E_{s1}(\vec{r})|^2\rangle\langle |E_{s2}(\vec{r})|^2\rangle}}, \tag{7.50}$$

where the two electric field components are defined as:

$$E_{s1}(\vec{r}) = \hat{s}_1 \bullet \vec{E}(\vec{r}) \quad \text{and} \quad E_{s2}(\vec{r}) = \hat{s}_2 \bullet \vec{E}(\vec{r}), \tag{7.51}$$

and $\hat{s}_1$ and $\hat{s}_2$ are unit vectors separated by an angle γ, as shown in Figure 7.6. From (7.15), the denominator of (7.50) is $E_0^2/3$. The numerator of (7.50) is evaluated from (7.1), (7.9), and (7.10), and the result for the angular correlation is:

$$\rho(\hat{s}_1, \hat{s}_2) = \hat{s}_1 \bullet \hat{s}_2 = \cos\gamma \tag{7.52}$$

This result is independent of $\vec{r}$. The same angular correlation applies to the magnetic field components. For the case of $\cos\gamma = 0$, (7.52) is in agreement with (7.31) and the theory of Kostas and Boverie [72].

We now turn to spatial correlation functions for the linear components of the electric field. The spatial correlation function $\rho_l(r)$ for the longitudinal electrical field can be defined as:

$$\rho_l(r) = \frac{\langle E_z(0)E_z^*(\hat{z}r)\rangle}{\sqrt{\langle |E_z(0)|^2\rangle\langle |E_z(\hat{z}r)|^2\rangle}} \tag{7.53}$$

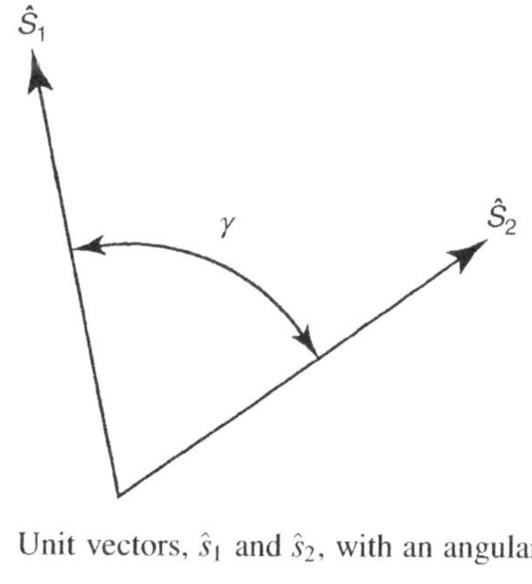

FIGURE 7.6 Unit vectors, $\hat{s}_1$ and $\hat{s}_2$, with an angular separation γ.

From (7.15), the denominator of (7.53) is $E_0^2/3$. The evaluation of the numerator in (7.53) has been studied in [74]:

$$\langle E_z(0)E_z^*(\hat{z}r)\rangle = \iint\limits_{4\pi}\iint\limits_{4\pi} \sin\alpha_1 \sin\alpha_2 \langle F_\alpha(\Omega_1)F_\alpha^*(\Omega_2)\rangle \exp(ikr\cos\alpha_2)\mathrm{d}\Omega_1\,\mathrm{d}\Omega_2 \quad (7.54)$$

One of the integrations in (7.54) can be evaluated by use of (7.14):

$$\langle E_z(0)E_z^*(\hat{z}r)\rangle = \frac{E_0^2}{8\pi}\iint\limits_{4\pi} \sin^2\alpha_2 \exp(ikr\cos\alpha_2)\mathrm{d}\Omega_2 \quad (7.55)$$

The Ω_2 integration can be written explicitly in the following form:

$$\langle E_z(0)E_z^*(\hat{z}r)\rangle = \frac{E_0^2}{8\pi}\int_0^{2\pi}\int_0^{\pi} \sin^2\alpha_2 \exp(ikr\cos\alpha_2)\sin\alpha_2\,\mathrm{d}\alpha_2\,\mathrm{d}\beta_2 \quad (7.56)$$

The β_2 integration in (7.56) contributes a 2π factor, and the α_2 integration can be performed by substituting $u = \cos\alpha_2$ and using integration by parts [74]:

$$\langle E_z(0)E_z^*(\hat{z}r)\rangle = \frac{E_0^2}{(kr)^2}\left[\frac{\sin(kr)}{kr} - \cos(kr)\right] \quad (7.57)$$

Hence we can now write the final result for ρ_l [74]:

$$\rho_l(r) = \frac{3}{(kr)^2}\left[\frac{\sin(kr)}{kr} - \cos(kr)\right] \quad (7.58)$$

Similarly, a spatial correlation function $\rho_t(r)$ for the transverse electric field, such as E_x or E_y, can be defined as [73]:

$$\begin{aligned}\rho_t(r) &\equiv \frac{\langle E_x(0)E_x^*(\hat{z}r)\rangle}{\sqrt{\langle|E_x(0)|^2\rangle\langle|E_x(\hat{z}r)|^2\rangle}} \\ &= \frac{\langle E_y(0)E_y^*(\hat{z}r)\rangle}{\sqrt{\langle|E_y(0)|^2\rangle\langle|E_y(\hat{z}r)|^2\rangle}}\end{aligned} \quad (7.59)$$

The results are identical, but we will choose to deal with E_x rather than E_y. As with (7.53), the denominator of (7.59) is $E_0^2/3$. The evaluation of the numerator in (7.59) has been studied in [73]:

$$\begin{aligned}\langle E_x(0)E_x^*(\hat{z}r)\rangle = &\iint\limits_{4\pi}\iint\limits_{4\pi}\langle[\cos\alpha_1\cos\beta_1 F_{\alpha1} - \sin\beta_1 F_{\beta1}(\Omega_1)] \\ &[\cos\alpha_2\cos\beta_2 F_{\alpha2}^*(\Omega_2) - \sin\beta_2 F_{\beta2}^*(\Omega_2)]\rangle \\ &\exp(ikr\cos\alpha_2)\mathrm{d}\Omega_1\mathrm{d}\Omega_2\end{aligned} \quad (7.60)$$

The expectation in the integrand of (7.60) can be evaluated by use of (7.9) and (7.10). Then the Ω_1 integration can be done by using the sampling property of the delta function so that (7.60) reduces to:

$$\langle E_x(0)E_x^*(\hat{z}r)\rangle = \frac{E_0^2}{8\pi}\iint_{4\pi}(\cos^2\alpha_2\cos^2\beta_2 + \sin^2\beta_2)\exp(ikr\cos\alpha_2)\mathrm{d}\Omega_2 \qquad (7.61)$$

The β_2 integration (0 to 2π) and the α_2 integration (0 to π) can be done analytically to obtain [73]:

$$\langle E_x(0)E_x^*(\hat{z}r)\rangle = \frac{E_0^2}{2}\left[\frac{\sin(kr)}{kr} - \frac{1}{(kr)^2}\left(\frac{\sin(kr)}{kr} - \cos(kr)\right)\right] \qquad (7.62)$$

Hence we can now write the final expression for ρ_t [73]:

$$\rho_t(r) = \frac{3}{2}\left[\frac{\sin(kr)}{kr} - \frac{1}{(kr)^2}\left(\frac{\sin(kr)}{kr} - \cos(kr)\right)\right] \qquad (7.63)$$

The spatial correlation functions, ρ, ρ_l, and ρ_t, all have the following three properties: (1) they equal one for $r = 0$, (2) they are even in r, and (3) they decay to zero in an oscillatory manner for increasing kr. The first property can be seen by performing Taylor series expansions in kr [74,75]:

$$\rho(kr) = 1 - \frac{1}{6}(kr)^2 + O(kr)^4, \qquad (7.64)$$

$$\rho_l(kr) = 1 - \frac{1}{10}(kr)^2 + O(kr)^4, \qquad (7.65)$$

$$\rho_t(kr) = 1 - \frac{1}{5}(kr)^2 + O(kr)^4 \qquad (7.66)$$

From the definitions of ρ, ρ_l, and ρ_t, the following consistency relation can be derived:

$$\rho(kr) = \frac{1}{3}[2\rho_t(kr) + \rho_l(kr)] \qquad (7.67)$$

The derived expressions in (7.48), (7.58), and (7.63) satisfy (7.67). Also, the Taylor series expansions in (7.66) satisfy (7.67). The results given here for ρ, ρ_l, and ρ_t are consistent with the results in [78] derived by a volume average of a mode sum.

Although the correlation functions were defined for field points at the origin and on the z axis, the results are invariant to translation and rotation. The general results are a function of the separation r, the longitudinal correlation function ρ_l is a function of the longitudinal field component E_l, and the transverse correlation function ρ_t is a function of the transverse electric field E_t. The geometry is shown in Figure 7.7.

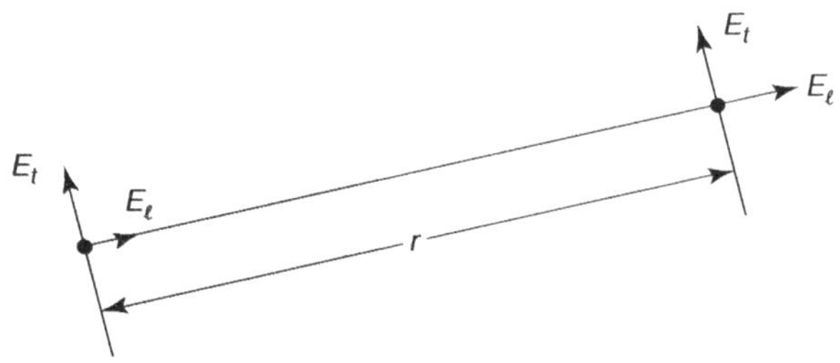

FIGURE 7.7 Geometry for correlation functions for general field locations [73].

7.4.2 Mixed Electric and Magnetic Field Components

Most of the electric and magnetic components are uncorrelated. Without loss of generality, we can consider the correlations of electric field components at the origin and magnetic field components on the z axis. For example, the following ensemble averages (and hence correlations) are all zero [73]:

$$\begin{aligned}&\langle E_x(0)H_x^*(\hat{z}r)\rangle = \langle E_x(0)H_z^*(\hat{z}r)\rangle = \langle E_y(0)H_y^*(\hat{z}r)\rangle = \langle E_y(0)H_z^*(\hat{z}r)\rangle = \\ &\langle E_z(0)H_x^*(\hat{z}r)\rangle = \langle E_z(0)H_y^*(\hat{z}r)\rangle = \langle E_z(0)H_z^*(\hat{z}r)\rangle = 0\end{aligned} \tag{7.68}$$

The results in (7.68) indicate that most of the electric and magnetic field components are uncorrelated at all separations r.

However, the orthogonal transverse components of $\vec{E}$ and $\vec{H}$ are correlated for $r \neq 0$. For this case, we define the correlation function:

$$\rho_{xy}(r) = \frac{\langle E_x(0)H_y^*(\hat{z}r)\rangle}{\sqrt{\langle|E_x(0)|^2\rangle\langle|H_y(\hat{z}r)|^2\rangle}} \tag{7.69}$$

The denominator of (7.69) can be evaluated from the known mean-square values of the electric and magnetic field components in (7.15) and (7.21):

$$\sqrt{\langle|E_x(0)|^2\rangle\langle|H_y(\hat{z}r)|^2\rangle} = \frac{E_0^2}{3\eta} \tag{7.70}$$

By substituting (7.11) and (7.16) into (7.69), the numerator of (7.69) can be written:

$$\begin{aligned}\langle E_x(0)H_y^*(\hat{z}r)\rangle = \ &\frac{1}{\eta}\iint\limits_{4\pi}\iint\limits_{4\pi}\langle\left[\cos\alpha_1\cos\beta_1 F_{\alpha_1}(\Omega_1) - \sin\beta_1 F_{\beta_1}(\Omega_1)\right] \\ &\left[\cos\alpha_2\sin\beta_2 F_{\beta_2}^*(\Omega_2) - \cos\beta_2 F_{\alpha_2}^*(\Omega_2)\right]\rangle\exp(ikr\cos\alpha_2)\mathrm{d}\Omega_1\,\mathrm{d}\Omega_2\end{aligned} \tag{7.71}$$

The expectation $\langle\ \rangle$ in the integrand can be evaluated using (7.8). Then the Ω_1 integration can be done by use of the sampling property of the delta function so that (7.71) reduces to:

$$\langle E_x(0)H_y^*(\hat{z}r)\rangle = \frac{-E_0^2}{8\pi\eta}\iint_{4\pi} \cos\alpha_2 \exp(ikr\cos\alpha_2)\mathrm{d}\Omega_2 \tag{7.72}$$

The Ω_2 (β_2 and α_2) integration can be done analytically to obtain:

$$\langle E_x(0)H_y^*(\hat{z}r)\rangle = \frac{-iE_0^2}{2\eta(kr)^2}(\sin(kr) - kr\cos(kr)) \tag{7.73}$$

Substitution of (7.70) and (7.73) into (7.69) yields the final result for ρ_{xy}:

$$\rho_{xy}(r) = \frac{-3i}{2(kr)^2}[\sin(kr) - kr\cos(kr)] \tag{7.74}$$

For small kr, the leading term in (7.74) is:

$$\rho_{xy}(r) \cong \frac{-ikr}{2} \tag{7.75}$$

Equation (7.75) shows that $\rho_{xy}(0) = 0$. Hence the following two correlations are zero:

$$\langle E_x(0)H_y^*(0)\rangle = \langle E_y(0)H_x^*(0)\rangle = 0 \tag{7.76}$$

Equations (7.68) and (7.76) show that all electric and magnetic field components are uncorrelated when evaluated at the same point.

7.4.3 Squared Field Components

In this section, we consider correlations of squared field quantities. These quantities are of interest because they appear in expressions for power and energy. The simplest way to handle squared field quantities is to write them in terms of the squares of the real and imaginary parts. For example, the square of the magnitude of the electric field at an arbitrary point $\vec{r}$ can be written:

$$|E(\vec{r})|^2 = E_{xr}^2(\vec{r}) + E_{xi}^2(\vec{r}) + E_{yr}^2(\vec{r}) + E_{yi}^2(\vec{r}) + E_{zr}^2(\vec{r}) + E_{zi}^2(\vec{r}) \tag{7.77}$$

As shown previously in (7.32), each real and imaginary part of the electric field is Gaussian. They are also independent with zero means and equal variances as shown in Section 7.3.

The correlation function ρ_{ll} for the square of the longitudinal field component is defined as:

$$\rho_{ll}(r) = \frac{\langle [|E_z(0)|^2 - \langle |E_z(0)|^2 \rangle][|E_z(\hat{z}r)|^2 - \langle |E_z(\hat{z}r)|^2 \rangle] \rangle}{\sqrt{\langle [|E_x(0)|^2 - \langle |E_z(\hat{z}r)|^2 \rangle]^2 \rangle \langle [|E_z(\hat{z}r)|^2 - \langle |E_z(\hat{z}r)|^2 \rangle]^2 \rangle}} \tag{7.78}$$

In (7.78), the mean values of the squares of the fields are subtracted according to the usual definition of correlation function [57]. This was not necessary in (7.53), (7.59), and (7.69) because the mean values of the fields are zero. If the squared magnitudes in (7.78) are written as the sums of the real and imaginary parts, then the evaluation of (7.78) involves expectations of terms of the type $\langle g^2 h^2 \rangle$, where g and h represent real or imaginary parts of E_z. Since the real and imaginary parts of field components are Gaussian variables with zero mean, the expectations can all be evaluated by use of the following relationship [57]:

$$\langle g^2 h^2 \rangle = \langle g^2 \rangle \langle h^2 \rangle + 2 \langle gh \rangle^2 \tag{7.79}$$

Then the result for ρ_{ll} is:

$$\rho_{ll}(r) = \rho_l^2(r), \tag{7.80}$$

where ρ_l is given in (7.63). Thus, ρ_{ll} has the same nulls as ρ_l, but is never negative.

The correlation function ρ_{tt} for the square of the transverse field component is similarly defined as:

$$\rho_{tt}(r) = \frac{\langle [|E_x(0)|^2 - \langle |E_x(0)|^2 \rangle][|E_x(\hat{z}r)|^2 - \langle |E_x(\hat{z}r)|^2 \rangle] \rangle}{\sqrt{\langle [|E_x(0)|^2 - \langle |E_x(\hat{z}r)|^2 \rangle]^2 \rangle \langle [|E_x(\hat{z}r)|^2 - \langle |E_x(\hat{z}r)|^2 \rangle]^2 \rangle}} \tag{7.81}$$

The expectations can again be evaluated by use of (7.79), and the result is:

$$\rho_{tt}(r) = \rho_t^2(r), \tag{7.82}$$

where ρ_t is given by (7.63).

The correlation function ρ_{EE} of the square of the magnitude of the electric field can be defined as:

$$\rho_{EE}(r) = \frac{\langle [|\vec{E}(0)|^2 - \langle |\vec{E}(0)|^2 \rangle][|\vec{E}(\hat{z}r|^2 - \langle |\vec{E}(\hat{z}r)|^2 \rangle] \rangle}{\sqrt{\langle [|\vec{E}(0)|^2 - \langle |\vec{E}(0)|^2 \rangle]^2 \rangle \langle [|\vec{E}(\hat{z}r)|^2 - \langle |\vec{E}(\hat{z}r)|^2 \rangle]^2 \rangle}} \tag{7.83}$$

The expectations can be evaluated by using (7.79), and the result is:

$$\rho_{EE}(r) = \frac{2\rho_{tt}(r) + \rho_{ll}(r)}{3} \tag{7.84}$$

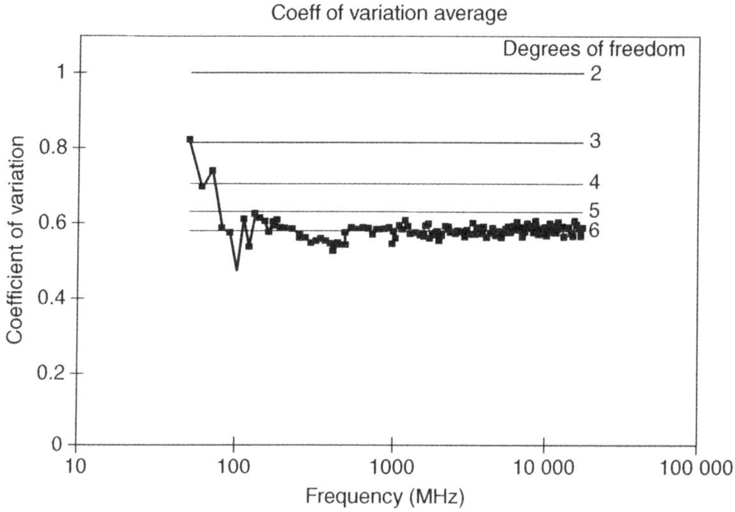

FIGURE 7.8 Measured ratio of the standard deviation to the root-mean-square electric field averaged over a large number of field probe locations for a large frequency range [73].

The result for ρ_{EE} in [78] includes a combination of ρ_{tt} and ρ_{ll} plus a constant term. The constant arises because the mean value of the square of the electric field was not subtracted out in the definition, as it is in (7.81). There are some other differences in the results of [78] because those results were based on real, single-mode fields of an unstirred cavity. Our results are for complex, multi-mode fields that result from stirring and ensemble averaging. Hence, our electric field has six degrees of freedom [73], as shown in Figure 7.8, rather than three degrees as found in [78]. All of the correlations in this section are valid for magnetic fields as well as electric fields.

There is a shortage of measured correlations in three-dimensional cavities, but some correlation results have been reported with monopole receiving antennas [79, 80]. The experiment was done by measuring received power with short monopoles in a transverse geometry, and the range of kr values was obtained by varying the frequency for a fixed separation r. Since the received power is proportional to the square of the magnitude of the transverse electric field, the relevant correlation function is ρ_{tt}. Mitra and Trost [79,80] compared their experimental data with the square ρ^2 of the correlation function given in (7.48) because the transverse correlation functions ρ_t and ρ_{tt} were not known at that time. A comparison of of measurements with both ρ_{tt} and ρ^2 is given in Figure 7.9. Even though there is a good deal of scatter in the experimental data, two important features (the slope for $kr < 2$ and the maximum near $kr = 4$) agree better with ρ_{tt} than with ρ^2. The experimental data were taken for $r = 1.5$ cm with frequency varying from 1.0 to 13.5 GHz, but more data are available in [80].

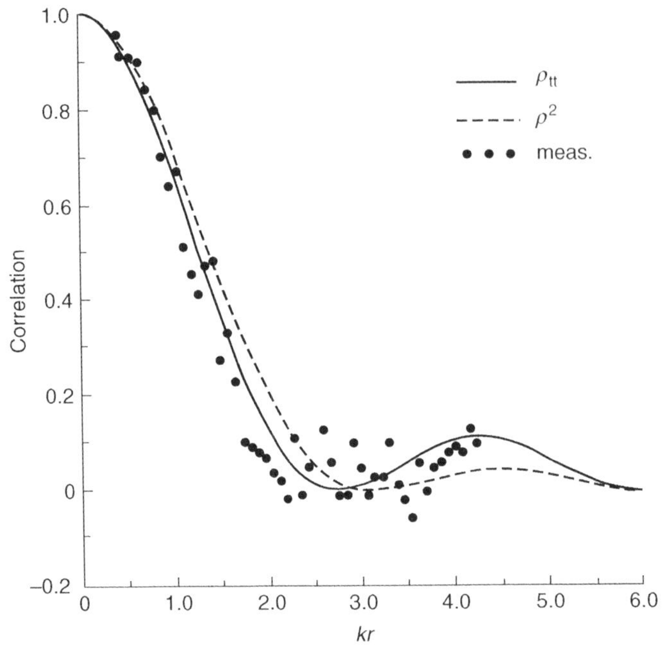

FIGURE 7.9 Measured correlation for power received by transverse monopole antennas compared to ρ_{tt} and ρ^2 [73].

7.4.4 Energy Density

The energy density W can be written as the sum of electric and magnetic energy densities [3]:

$$W(\vec{r}) = W_E(\vec{r}) + W_H(\vec{r}), \tag{7.85}$$

where:

$$W_E(\vec{r}) = \frac{\varepsilon}{2}|\vec{E}(\vec{r})|^2 \quad \text{and} \quad W_H(\vec{r}) = \frac{\mu}{2}|\vec{H}(\vec{r})|^2 \tag{7.86}$$

The spatial properties of the electric energy density are of interest in applications such as heating of electric conductors. Similarly, the spatial properties of magnetic energy density are of interest in applications such as heating of materials with magnetic loss (such as ferrites). Without loss of generality, we again perform our derivations for locations on the z axis.

The correlation function ρ_{W_E} of the electric energy density is defined as:

$$\rho_{W_E}(r) \equiv \frac{\langle [W_E(0) - \langle W_E(0)\rangle][W_E(\hat{z}r) - \langle W_E(\hat{z}r)\rangle]\rangle}{\sqrt{\langle [W_E(0) - \langle W_E(0)\rangle]^2\rangle \langle [W_E(\hat{z}r) - \langle W_E(\hat{z}r)\rangle]^2\rangle}} \tag{7.87}$$

When the definition of W_E is substituted into (7.87), the result is equal to that for the square of the electric field in (7.84):

$$\rho_{W_E}(r) = \rho_{EE}(r) \tag{7.88}$$

The result for the correlation function ρ_{W_H} of the magnetic energy density is the same:

$$\rho_{W_H}(r) \equiv \frac{\langle [W_H(0) - \langle W_H(0)\rangle][W_H(\hat{z}r) - \langle W_H(\hat{z}r)\rangle]\rangle}{\sqrt{\langle [W_H(0) - \langle W_H(0)\rangle]^2\rangle \langle [W_H(\hat{z}r) - \langle W_H(\hat{z}r)\rangle]^2\rangle}} = \rho_{EE}(r) \tag{7.89}$$

The correlation function ρ_W of the total energy density is defined as:

$$\rho_W(r) \equiv \frac{\langle [W(0) - \langle W(0)\rangle][W(\hat{z}r) - \langle W(\hat{z}r)\rangle]\rangle}{\sqrt{\langle [W(0) - \langle W(0)\rangle]^2\rangle \langle [W(\hat{z}r) - \langle W(\hat{z}r)\rangle]^2\rangle}} \tag{7.90}$$

When (7.85) and (7.86) are substituted into (7.90), the result for ρ_W is:

$$\rho_W(r) = \rho_{EE}(r) + \frac{2}{3}|\rho_{xy}(r)|^2, \tag{7.91}$$

where ρ_{xy} is given by (7.74). The first term on the right side is the same as the correlation function for W_E and W_H, and the second term is a result of the correlation of the orthogonal transverse components of $\vec{E}$ and $\vec{H}$. Since $\rho_{xy}(0) = 0$ and $\rho_{EE}(0) = 1$, we have the necessary result that $\rho_W(0) = 1$.

The mean values of the electric, magnetic, and total energy densities are also of interest and are given by:

$$\langle W_E(\vec{r})\rangle = \langle W_H(\vec{r})\rangle = \frac{\varepsilon}{2}E_0^2 \quad \text{and} \quad \langle W(\vec{r})\rangle = \varepsilon_0 E_0^2 \tag{7.92}$$

The mean energy values in (7.92) are independent of position, and E_0^2 is the mean-square electric field, and indicated in (7.14).

7.4.5 Power Density

As indicated in (7.27), the mean of the power density or Poynting vector $\vec{S}$ is zero Even though the mean of the Poynting vector is zero, the variance is not. The real part of the Poynting vector, $\text{Re}(\vec{S})$, gives the real power flow and can be written:

$$\text{Re}(\vec{S}) = \hat{x}S_{xr} + \hat{y}S_{yr} + \hat{z}S_{zr} \tag{7.93}$$

The x component S_{xr} can be written in terms of the real and imaginary parts of electric and magnetic field components:

$$S_{xr} = E_{yr}H_{zr} + E_{yi}H_{zi} - E_{zr}H_{yr} - E_{zi}H_{yi} \tag{7.94}$$

The variance of S_{xr} is equal to the variances of S_{yr} and S_{zr}, and can be determined by use of (7.79) because the field components in (7.94) are Gaussian. The result is:

$$\langle S_{xr}^2 \rangle = \langle S_{yr}^2 \rangle = \langle S_{zr}^2 \rangle = \left(\frac{E_0^2}{3\eta} \right)^2, \tag{7.95}$$

where E_0^2 is the mean-square electric field, which is independent of position. The factor of three in the denominator of (7.95) is a result of the variance being distributed between three components. The spatial correlation of the Poynting vector is difficult to derive and is generally of little interest anyway because it has a zero mean. Therefore, we will not pursue it.

7.5 ANTENNA OR TEST-OBJECT RESPONSE

Now that we have characterized the fields in reverberation chambers, we can consider the response of a receiving antenna or a test object placed in a reverberation chamber. The simplest case of a lossless, impedance-matched antenna will be considered first. The received signal can be written as an integral over incidence angle by analogy with Kern's plane-wave, scattering-matrix theory [81]. The received signal could be a current, a voltage, or a waveguide mode coefficient, but the general formulation remains the same. Consider the received signal to be a current I induced in a matched load. For an antenna located at the origin, the current can be written as a dot product of the angular spectrum with a receiving function $\vec{S}_r(\Omega)$ integrated over angle:

$$I = \iint_{4\pi} \vec{S}_r(\Omega) \bullet \vec{F}(\Omega) \mathrm{d}\Omega, \tag{7.96}$$

where the receiving function can be written in terms of two components,

$$\vec{S}_r(\Omega) = \hat{\alpha} S_{r\alpha}(\Omega) + \hat{\beta} S_{r\beta}(\Omega) \tag{7.97}$$

In general, $S_{r\alpha}$ and $S_{r\beta}$ are complex, so the antenna can have arbitrary polarization, such as linear or circular. For example, a z-directed linear antenna with linear polarization would have $S_{r\beta}(\Omega) = 0$. A circularly polarized antenna would have $S_{r\beta}(\Omega_b) = \pm S_{r\alpha}(\Omega_b)$ for right- or left-hand circular polarization, where Ω_b is the direction of the main beam.

The mean value of the current I can be shown to be zero from (7.6) and (7.96):

$$\langle I \rangle = \iint_{4\pi} \vec{S}_r(\Omega) \bullet \langle \vec{F}(\Omega) \rangle \, \mathrm{d}\Omega = 0 \tag{7.98}$$

The absolute value of the square of the current is important because it is proportional to received power P_r:

$$P_r = |I|^2 R_r = R_r \iint_{4\pi} \iint_{4\pi} [\vec{S}_r(\Omega_1) \bullet \vec{F}(\Omega_1)][\vec{S}_r^*(\Omega_2) \bullet \vec{F}^*(\Omega_2)] \mathrm{d}\Omega_1 \mathrm{d}\Omega_2, \tag{7.99}$$

where the radiation resistance R_r of the antenna is also equal to the real part of the matched load impedance. The mean value of the received power can be determined from (7.9), (7.10), and (7.99):

$$\langle P_r \rangle = \langle |I|^2 \rangle R_r = \frac{E_0^2}{2} \frac{R_r}{4\pi} \iint_{4\pi} [|S_{r\alpha}(\Omega_2)|^2 + |S_{r\beta}(\Omega_2)|^2] \mathrm{d}\Omega_2 \tag{7.100}$$

The physical interpretation of (7.100) is that the ensemble average of received power is equal to an average over incidence angle (Ω_2) and polarization (α and β components).

The integrand of (7.100) can be related to the effective area of an isotropic antenna $\lambda^2/4\pi$ and the antenna directivity $D(\Omega_2)$ by [82]:

$$\eta R_r \left[|S_{r\alpha}(\Omega_2)|^2 + |S_{r\beta}(\Omega_2)|^2 \right] = \frac{\lambda^2}{4\pi} D(\Omega_2) \tag{7.101}$$

Substitution of (7.101) into (7.100) yields:

$$\langle P_r \rangle = \frac{1}{2} \frac{E_0^2}{\eta} \frac{\lambda^2}{4\pi} \frac{1}{4\pi} \iint_{4\pi} D(\Omega_2) \mathrm{d}\Omega_2 \tag{7.102}$$

The integral in (7.102) is known because the average (over Ω_2) of D is 1. Thus the final result for the average received power is:

$$\langle P_r \rangle = \frac{1}{2} \frac{E_0^2}{\eta} \frac{\lambda^2}{4\pi} \tag{7.103}$$

The physical interpretation of (7.103) is that the average received power is the product of the scalar power density E_0^2/η and the effective area $\lambda^2/4\pi$ of an isotropic antenna times a polarization mismatch factor of one half [83]. This result is independent of the antenna directivity and is consistent with the reverberation chamber analysis [68] of Corona et al. Some of the earlier data indicated that (7.103) was in better agreement with measurements if the one-half polarization mismatch factor was omitted [19]. However, more recent comparisons of antenna received power with field-probe data [66] and with a well-characterized test object [84] support the inclusion of the factor of one-half. Consequently, the polarization mismatch factor needs to be included to be in agreement with theory and with most measured data. Traditionally, linearly polarized antennas have been used as reference antennas in reverberation chambers, but this result suggests that circularly polarized antennas are also appropriate. Experimental data with circularly polarized antennas would be useful for confirming this theoretical result. The special cases of an electrically short dipole (electric-field probe) and an electrically small loop (magnetic-field probe) are discussed in Appendices D and E, respectively.

The preceding analysis can be extended to the case of a real antenna with loss and impedance mismatch by use of Tai's theory [83]. The effective area A_e can be generalized to:

$$A_e(\Omega) = \frac{\lambda^2}{4\pi} D(\Omega) p m \eta_a, \tag{7.104}$$

where p is the polarization mismatch, m is the impedance mismatch, and η_a is the antenna efficiency. All three quantities, p, m, and η_a, are real and can vary between 0 and 1. The average of A_e over incidence angle and polarization can be written [83]:

$$\langle A_e \rangle = \frac{\lambda^2}{8\pi} m \eta_a \tag{7.105}$$

The average received power is:

$$\langle P_r \rangle = \frac{E_0^2}{\eta} \langle A_e \rangle, \tag{7.106}$$

where E_0^2/η can again be interpreted as the average scalar power density.

Test objects can be thought of as lossy, impedance-mismatched antennas, so (7.106) also applies to test objects as long as terminals with linear loads can be identified. This theory has been used to predict the coupling to an apertured coaxial line [85], an apertured rectangular box [38], and a microstrip transmission line [84,86]

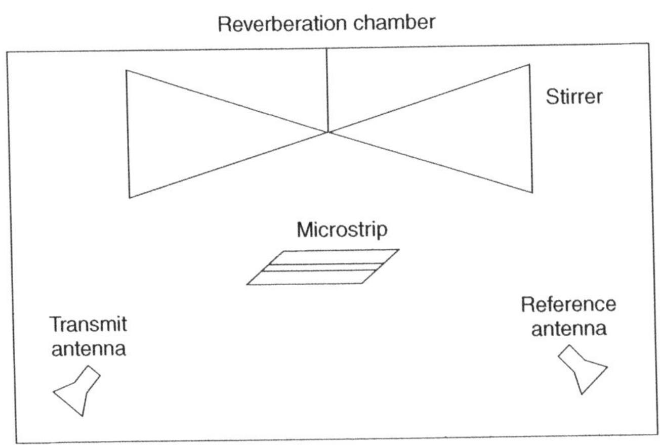

FIGURE 7.10 Reverberation chamber configuration for emissions or immunity measurements of a microstrip transmission line [86].

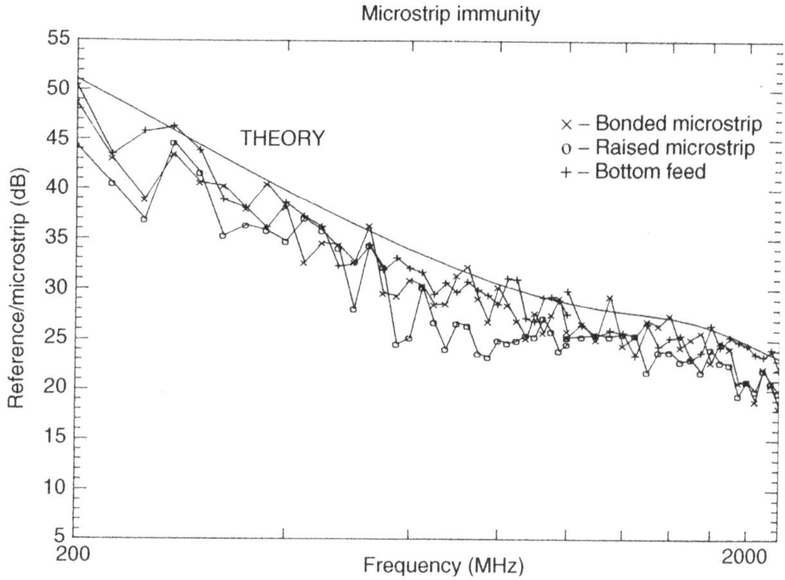

FIGURE 7.11 Theory (smooth curve) and measurements for microstrip transmission line immunity [86].

when compared to a reference antenna in a reverberation chamber. Good agreement with measurements has been obtained in each case.

The microstrip line example is a good illustration of the use of the above theory. The response of a terminated microstrip line was computed by use of the above theory and measured in the NIST reverberation chamber [86] with the setup shown in Figure 7.10. A comparison of theory and measurements is shown in Figure 7.11 for frequencies from 200 to 2000 MHz. The plotted quantity is the ratio of the average power received by the reference antenna to the average power received by the microstrip line in decibels. (This ratio is sometimes called shielding effectiveness in decibels.) The theoretical ratio is $20\log_{10}[(\lambda^2/8\pi)/\langle A_e\rangle]$, where $\lambda^2/8\pi$ is the theoretical average effective area of the reference antenna, and $\langle A_e\rangle$ is the average effective area of the microstrip transmission line. The measurements were performed on three different physical models, and the "bottom feed" microstrip line best fits the theoretical model. Even that measured curve has a small negative bias which is probably due to impedance mismatch in the reference antenna, which was not taken into account. The actual reference antenna was a log periodic dipole array below 1000 MHz and a broadband ridged horn above 1000 MHz.

7.6 LOSS MECHANISMS AND CHAMBER Q

In (7.14), E_0^2 was introduced as the mean-square value of the electric field, which was shown to be independent of position. This constant can be related to the power P_t

transmitted and the chamber Q by conservation of power [38,41]. The starting equation is the definition of quality factor (Q):

$$Q = \frac{\omega U}{P_d}, \tag{7.107}$$

where U is the energy stored in the cavity and P_d is the power dissipated. Since the average energy density was shown to be independent of position in (7.92), the stored energy can be written as the product of the average energy density and the chamber volume V:

$$U = \langle W \rangle V \tag{7.108}$$

For steady state conditions, conservation of power requires that the dissipated power P_d equals the transmitted power P_t. Then (7.92), (7.107), and (7.108) can be used to derive:

$$E_0^2 = \frac{QP_t}{\omega \varepsilon V} \tag{7.109}$$

This analysis can be carried further to relate the transmitted power to the power received by a receiving antenna located in the chamber. If (7.109) is substituted into (7.103), the power received by a matched, lossless antenna is found to be:

$$\langle P_r \rangle = \frac{\lambda^3 Q}{16\pi^2 V} P_t \tag{7.110}$$

Equations (7.109) and (7.110) show the importance of the Q enhancement in determining the field strength or the received power in the chamber. The most popular method of measuring Q is based on the solution of (7.110) for Q:

$$Q = \frac{16\pi^2 V}{\lambda^3} \frac{\langle P_r \rangle}{P_t} \tag{7.111}$$

Equation (7.111) is applicable to an impedance-matched, lossless receiving antenna, but dissipative or mismatch loss can be accounted for by modifying the effective area as shown in (7.105).

The calculation of chamber Q requires that all losses are accounted for in evaluating P_d in (7.107). A theory has been developed for including the following four types of loss [38]:

$$P_d = P_{d1} + P_{d2} + P_{d3} + P_{d4}, \tag{7.112}$$

where P_{d1} is the power dissipated in the cavity walls, P_{d2} is the power absorbed in loading objects within the cavity, P_{d3} is the power lost through aperture leakage, and

P_{d4} is the power dissipated in the loads of receiving antennas. By substituting (7.112) into (7.107), we can write the following expression for the inverse of Q:

$$Q^{-1} = Q_1^{-1} + Q_2^{-1} + Q_3^{-1} + Q_4^{-1}, \tag{7.113}$$

where:

$$Q_1 = \frac{\omega U}{P_{d1}}, \ Q_2 = \frac{\omega U}{P_{d2}}, \ Q_3 = \frac{\omega U}{P_{d3}}, \quad \text{and} \quad Q_4 = \frac{\omega U}{P_{d4}} \tag{7.114}$$

The four loss mechanisms can be analyzed as follows. Wall loss is usually dominant, so it will be covered in most detail.

For highly conducting walls, the plane-wave integral representation can be analytically continued all the way to the wall surfaces, and the reflected fields are related to the incident fields via plane-wave reflection coefficients as shown in Figure 7.12. Then P_{d1} in (7.112) can be evaluated in terms of the wall area A and the wall reflection coefficient [11].

The power P_{d1} dissipated in the walls can be written:

$$P_{d1} = \frac{1}{2} SA \langle (1 - |\Gamma|^2) \cos\theta \rangle_\Omega, \tag{7.115}$$

where Γ is the plane wave reflection coefficient, θ is the incidence angle shown in Figure 7.12, and $\langle\,\rangle_\Omega$ indicates average over incidence angle and polarization. The factor $\frac{1}{2}$ arises because only half of the plane waves are propagating toward the wall. From (7.114), Q_1 can then be written:

$$Q_1 = \frac{\omega U}{P_{d1}} = \frac{2kV}{A\langle (1 - |\Gamma|^2) \cos\theta \rangle_\Omega} \tag{7.116}$$

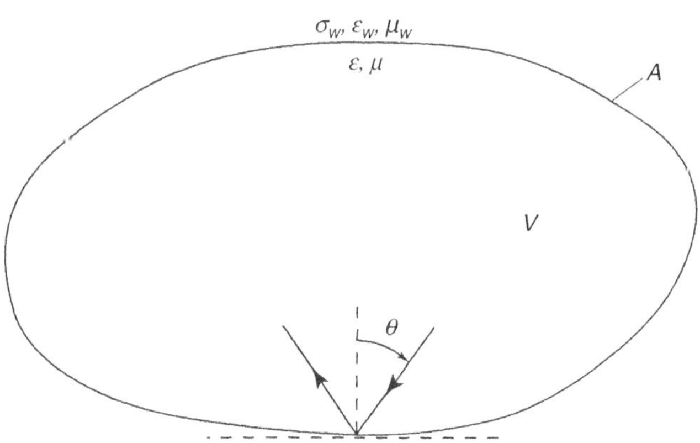

FIGURE 7.12 Plane-wave reflection from an imperfectly conducting wall of a reverberation chamber.

Equation (7.116) is a general result for highly reflecting walls where $1-|\Gamma|^2 \ll 1$. The next step is the evaluation of the average value in the denominator of (7.116).

The reflection coefficients for TE (perpendicular) polarization Γ_{TE} and vertical (parallel) polarization Γ_{TM} are given by [9]:

$$\Gamma_{TE} = \frac{\mu_w k \cos\theta - \mu\sqrt{k_w^2 - k^2\sin^2\theta}}{\mu_w k \cos\theta + \mu\sqrt{k_w^2 - k^2\sin^2\theta}} \tag{7.117}$$

and:

$$\Gamma_{TM} = \frac{\mu k_w^2 \cos\theta - \mu_w k\sqrt{k_w^2 - k^2\sin^2\theta}}{\mu k_w^2 \cos\theta - \mu_w k\sqrt{k_w^2 - k^2\sin^2\theta}}, \tag{7.118}$$

where $k_w = \omega\sqrt{\mu_w(\varepsilon_w + i\sigma_w/\omega)}$, σ_w is the wall conductivity, ε_w is the wall permittivity, and μ_w is the wall permeability. To account equally for both polarizations in (7.115), the average quantity can be written:

$$\begin{aligned}\langle(1-|\Gamma|^2)\cos\theta\rangle_\Omega &= \left\langle\left[1-\frac{1}{2}(|\Gamma_{TE}|^2+|\Gamma_{TM}|^2)\right]\cos\theta\right\rangle_\Omega \\ &= \int_0^{\pi/2}\left[1-\frac{1}{2}(|\Gamma_{TE}|^2+|\Gamma_{TM}|^2)\right]\cos\theta\sin\theta\,d\theta\end{aligned} \tag{7.119}$$

For $|k_w/k| \gg 1$, the squares of the reflection coefficients can be approximated as:

$$|\Gamma_{TE}|^2 \approx 1 - \frac{4\mu_w k\,\mathrm{Re}(k_w)\cos\theta}{\mu|k_w|^2} \tag{7.120}$$

and

$$|\Gamma_{TM}|^2 \approx 1 - \frac{4\mu_w k\,\mathrm{Re}(k_w)}{\mu|k_w|^2\cos\theta}, \tag{7.121}$$

where Re indicates real part. The approximation in (7.121) does not hold for θ close to $\pi/2$ because of the $\cos\theta$ factor in the denominator, but it can still be used in approximating (7.119) because of the $\cos\theta$ factor in (7.119). Substitution of (7.119)–(7.121) into (7.116) yields:

$$Q_1 \approx \frac{3|k_w|^2 V}{4A\mu_r\mathrm{Re}(k_w)}, \tag{7.122}$$

where $\mu_r = \mu_w/\mu$.

Equation (7.122) does not require that the walls be highly conducting. However, if the walls are highly conducting and conduction currents dominate displacement currents, $\sigma_w/(\omega\varepsilon_w) \gg 1$, then Q_1 simplifies to:

$$Q_1 \approx \frac{3V}{2\mu_r \delta A}, \tag{7.123}$$

where $\delta = 2/\sqrt{\omega\mu_w\sigma_w}$. This is the usual expression for metal wall reverberation chamber Q for the case where wall losses are dominant. A related derivation has employed the skin depth approximation from the start, followed by an average over an ensemble of plane waves [87]. For the case of nonmagnetic walls ($\mu_r = 1$), (7.123) agrees with the result for a single mode, given in (1.48). For the case of a rectangular cavity with $\mu_r = 1$ where the modes are known, this has been derived by averaging the modal Q values for modes whose resonant frequencies are in the vicinity of the excitation frequency [9]. A correction term was derived for rectangular cavities [9], but it is important only at low frequencies.

If the cavity contains absorbers (lossy objects distinct from the walls), the absorption loss P_{d2} can be written in terms of the absorption cross section σ_a [88] which is generally a function of incidence angle and polarization:

$$P_{d2} = S\langle\sigma_a\rangle_\Omega \tag{7.124}$$

The appropriate average is over 4π steradians and both (TE and TM) polarizations [38]:

$$\langle\sigma_a\rangle_\Omega = \frac{1}{8\pi}\iint_{4\pi} (\sigma_{a\mathrm{TE}} + \sigma_{a\mathrm{TM}})\mathrm{d}\Omega \tag{7.125}$$

The absorption cross section in (7.125) can be that of a single object or a summation for multiple absorbers. For example, for M absorbers $\langle\sigma_a\rangle_\Omega$ is replaced by:

$$\langle\sigma_a\rangle_\Omega = \sum_{m=1}^{M} \langle\sigma_{am}\rangle_\Omega, \tag{7.126}$$

where $\langle\sigma_{am}\rangle_\Omega$ is the averaged absorption cross-section of the mth absorber. From (7.114) and (7.124), the result for Q_2 is [38]:

$$Q_2 = \frac{2\pi V}{\lambda\langle\sigma_a\rangle_\Omega} \tag{7.127}$$

The formulation for leakage loss P_{d3} is similar to that of absorption loss because apertures can be characterized by a transmission cross section σ_l [89]. However, only plane waves that propagate toward the wall aperture(s) contribute to leakage power. So the expression for Q_3 is modified from (7.127) by a factor of 2 [38]:

$$Q_3 = \frac{4\pi V}{\lambda\langle\sigma_l\rangle_\Omega} \tag{7.128}$$

Also, the angular average is over 2π steradians ($0 \leq \theta \leq \pi/2$):

$$\langle \sigma_l \rangle_\Omega = \frac{1}{4\pi} \iint_{2\pi} (\sigma_{\mathrm{TE}} + \sigma_{\mathrm{TM}}) \mathrm{d}\Omega \tag{7.129}$$

For the case of N apertures, $\langle \sigma_l \rangle_\Omega$ in (7.129) is replaced by a summation:

$$\langle \sigma_l \rangle_\Omega = \sum_{n=1}^{N} \langle \sigma_{ln} \rangle_\Omega, \tag{7.130}$$

where $\langle \sigma_{ln} \rangle_\Omega$ is the averaged transmission cross section of the nth aperture. For electrically large apertures, $\langle \sigma_l \rangle_\Omega$ is independent of frequency and Q_3 is proportional to frequency. For small or resonant apertures, the frequency dependence of Q_3 is more complicated. The Q of a cavity with a circular aperture [38] will be studied in detail in the following chapter.

The power dissipated in the load of a receiving antenna was covered in Section 7.5. For a lossless receiving antenna, P_{d4} can be written:

$$P_{d4} = \frac{m\lambda^2}{8\pi} S, \tag{7.131}$$

where m is the impedance mismatch. From (7.15) and (7.131), Q_4 can be written:

$$Q_4 = \frac{16\pi^2 V}{m\lambda^3} \tag{7.132}$$

If there are multiple receiving antennas, (7.131) and (7.132) can be modified accordingly. For example, if there are N identical receiving antennas, P_{d4} is multiplied by N and Q_4 is divided by N. For a matched load ($m = 1$), Q_4 is proportional to frequency cubed. This means that Q_4 is small for low frequencies and is the dominant contributor to the total Q in (7.113). The effect of antenna loading on the Q of reverberation chambers has been observed experimentally [90]. At high frequencies, Q_4 becomes large and contributes little to the total Q.

A comparison of measured and calculated Q [38] is shown in Figure 7.13 for a rectangular aluminum cavity of dimensions $0.514\,\mathrm{m} \times 0.629\,\mathrm{m} \times 1.75\,\mathrm{m}$. The Q measurements were performed by the power ratio method of (7.111) and the decay-time method [91], as discussed in the following chapter. Standard-gain, K_u-band horn antennas were used to cover the frequency range from 12 to 18 GHz. The measured Q values fall below the theoretical Q, but agreement is much better than that obtained in earlier comparisons [19]. The decay-time measurement [91] generally agrees better with theory than the power-ratio method because it is less affected by antenna efficiency and impedance mismatch.

A second comparison of theory and measurement in Figure 7.14 shows the effect of loading the cavity with three spheres of radius 0.066m filled with salt water [92]. In this case, the absorption loss as described by (7.127) decreases the Q dramatically.

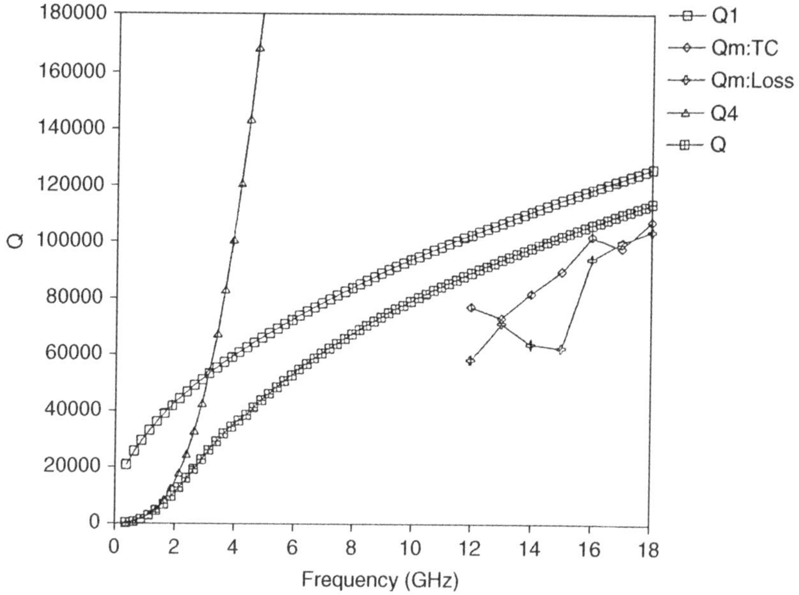

FIGURE 7.13 Comparison of Q measured by power ratio (Qm: Loss) and decay time (Qm:TC) with Q calculated from (7.113) for an aluminium cavity [41]. The theoretical values for wall loss (Q1) and receiving antennas (Q4) are also shown.

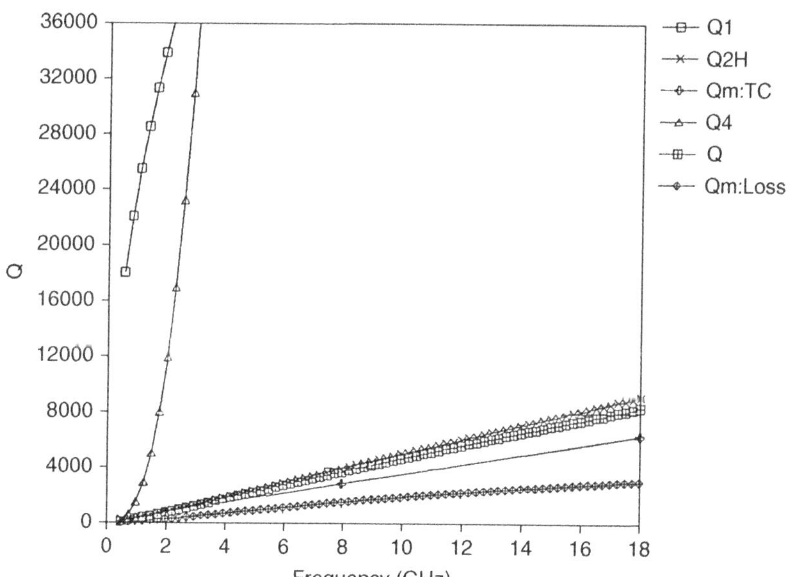

FIGURE 7.14 Comparison of Q measured by power ratio (Qm:Loss) and decay time (Qm:TC) with Q calculated from (7.113) for an absorber-loaded aluminium cavity [41]. The theoretical values for wall loss (Q1), absorption by salt-water spheres (Q2H), and receiving antennas (Q4) are also shown.

Broadband ridged horns were used, and the agreement with theory is not as good. However, the decay-time measurement is again a significant improvement over the power-ratio measurement.

7.7 RECIPROCITY AND RADIATED EMISSIONS

Reverberation chambers have been primarily used for radiated immunity measurements, and as a result a great deal of research has been done in characterizing chamber fields. However, reverberation chambers are reciprocal devices, and can and have been used for radiated emissions measurements [84]. The quantity measured is the total radiated power, and the measurement can be explained by either power conservation [38] or reciprocity [92,93].

7.7.1 Radiated Power

If the equipment under test (EUT) radiates (transmits) power P_{tEUT}, (7.110) can be used to determine the average power $\langle P_{rEUT} \rangle$ received by a matched, lossless reference antenna. Equation (7.110) is based on conservation of power and can be solved for P_{tEUT}:

$$P_{tEUT} = \frac{16\pi^2 V}{\lambda^3 Q} \langle P_{rEUT} \rangle \tag{7.133}$$

In theory this equation could be used directly for measurement of P_{tEUT}. However, (7.133) requires that the chamber volume V and (loaded) Q be known. It also requires that the receiving antenna be impedance-matched and lossless, or that the received power be corrected for antenna effects.

A better way to determine P_{tEUT} is to perform a separate reference measurement under the same chamber conditions. If a known power P_{tref} is transmitted and an average power $\langle P_{rref} \rangle$ is received, the coefficient on the right side of (7.133) can be determined:

$$\frac{16\pi^2 V}{\lambda^3 Q} = \frac{P_{tref}}{\langle P_{rref} \rangle} \tag{7.134}$$

Then P_{tEUT} can be determined by the ratio:

$$P_{tEUT} = \frac{P_{tref}}{\langle P_{rref} \rangle} \langle P_{rEUT} \rangle \tag{7.135}$$

If the same receiving antenna is used for both the EUT and the reference measurement, this method has the additional advantage of approximately canceling efficiency and impedance mismatch effects of the receiving antenna.

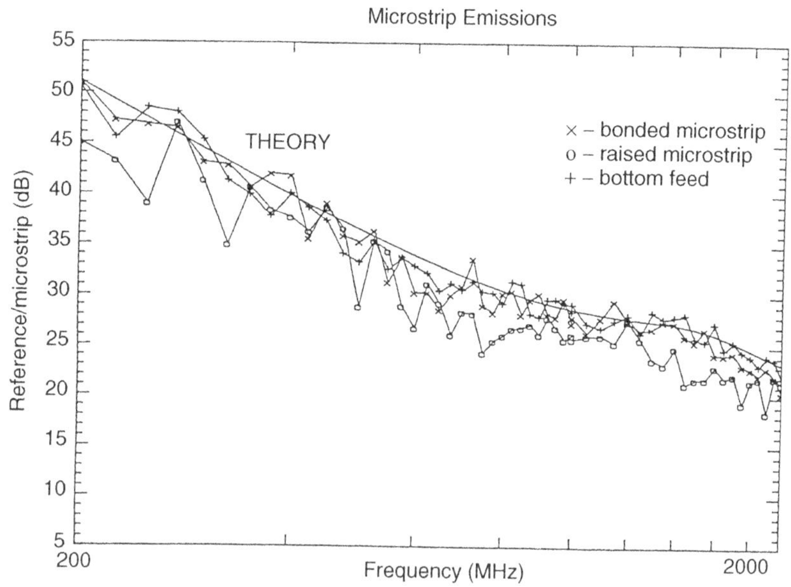

FIGURE 7.15 Comparison of theory with three measurements of the radiated emissions of a microstrip transmission line [86] in the NIST reverberation chamber.

This was done in the measurement of radiated power (emission) from a microstrip line [84], and the agreement between theory and measurement as shown in Figure 7.15 was good. The actual quantity plotted was the following power ratio:

$$\frac{\langle P_{rref} \rangle}{\langle P_{rEUT} \rangle} = \frac{P_{tref}}{P_{tEUT}} \tag{7.136}$$

Because the same input power was fed to the reference antenna and the microstrip line, the ratio in Figure 7.15 can be interpreted as either a shielding effectiveness or the reciprocal of the radiation efficiency of the microstrip line.

7.7.2 Reciprocity Relationship to Radiated Immunity

Electromagnetic reciprocity has many mathematical forms, and it can be applied to fields, circuits, or a mixture of the two [94]. Since reciprocity involves interchanging the source and receiver, it provides a method for relating radiated emissions and immunity. Consider an EUT located at the center of a spherical volume as shown in Figure 7.16. In an immunity measurement, the EUT is illuminated by incident electric and magnetic fields, $\vec{E}_i$ and $\vec{H}_i$, due to sources located outside the spherical surface S_r. In an emissions measurement, the EUT radiates (transmits) electric and magnetic fields, $\vec{E}_t$ and $\vec{H}_t$.

A typical EUT is very complex, and de Hoop and Quak [92] have developed a multi-port reciprocity formulation to relate emissions and immunity. Here we consider the

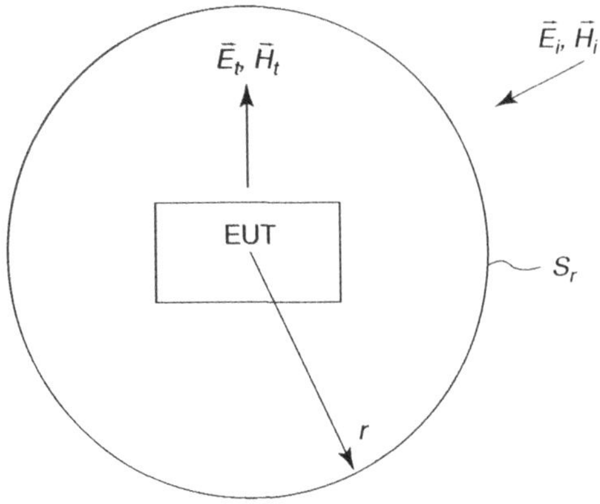

FIGURE 7.16 Equipment under test (EUT) radiating fields $\vec{E}_t, \vec{H}_t$ (emissions measurement) or illuminated by fields $\vec{E}_i, \vec{H}_i$ (immunity measurement).

simpler special case of a single port within the EUT, as shown in Figure 7.17. In an immunity measurement, the incident fields induce an open-circuit voltage V_i, and Z_t is the impedance of the Thevenin equivalent circuit. An arbitrary load impedance Z_l is connected across the terminals. In an emissions measurement, V_i is zero and a current I_t flows in the loop. The radiated fields are proportional to I_t and can be normalized as follows:

$$\vec{E}_t(\vec{r}) = I_t\vec{e}_n(\vec{r}) \quad \text{and} \quad H_t(\vec{r}) = I_t\vec{h}_n(\vec{r}), \tag{7.137}$$

where $\vec{e}_n$ and $\vec{h}_n$ are the electric and magnetic fields that are radiated when $I_t = 1$ A. If reciprocity is applied at the circuit terminals and the spherical surface, the following expression is obtained for V_i [92]:

$$V_i = -\iint_{S_r} \hat{r} \bullet [\vec{e}_n(\vec{r}) \times \vec{H}_i(r) - \vec{E}_i(\vec{r}) \times \vec{h}_n(\vec{r})]\mathrm{d}S_r \tag{7.138}$$

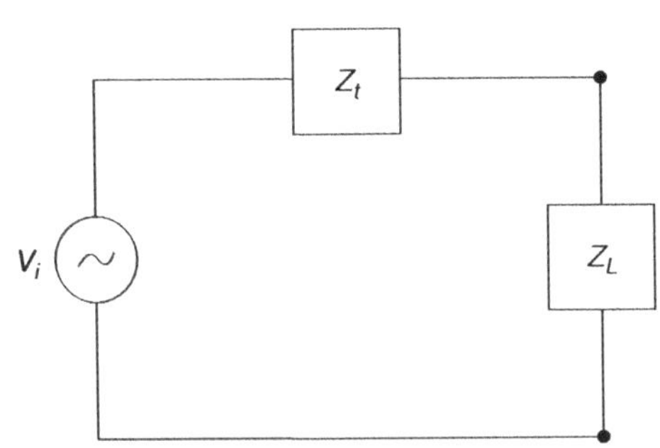

FIGURE 7.17 Thevenin equivalent circuit for a single port in equipment under test [18].

Up to this point, (7.138) is fairly general because there are no restrictions on the sphere radius r or the incident fields. If the surface integral in (7.138) is performed in the far field of the EUT ($kr \gg 1$), the normalized EUT fields can be written in the following forms:

$$\begin{aligned} \vec{e}_n(\vec{r}) &= \vec{e}_t(\theta, \phi)\frac{\exp(ikr)}{r}, \\ \vec{h}_n(\vec{r}) &= \hat{r} \times \vec{e}_t(\theta, \phi)\frac{\exp(ikr)}{\eta r}, \end{aligned} \tag{7.139}$$

where $\vec{e}_t(\theta, \phi) \bullet \vec{r} = 0$ and θ and ϕ are standard spherical coordinates. To apply (7.139) to reverberation chamber measurements, the incident electric and magnetic fields are replaced by plane-wave integral representations from (7.1) and (7.16). Then (7.138) can be rewritten as:

$$\begin{aligned} V_i = -\iint\limits_{S_r} \frac{\exp(ikr)}{r}\hat{r} \bullet \Bigg\{ \vec{e}_t(\theta, \phi) \times \left[\frac{1}{\eta} \iint\limits_{4\pi} \hat{k} \times \vec{F}(\Omega)\exp(i\vec{k} \bullet \vec{r})\mathrm{d}\Omega \right. \\ - \left[\iint\limits_{4\pi} \vec{F}(\Omega)\exp(i\vec{k} \bullet \vec{r})\mathrm{d}\Omega \right] \times \left[\frac{1}{\eta}\hat{r} \times \vec{e}_t(\theta, \phi) \right] \Bigg\} \mathrm{d}\Omega \end{aligned} \tag{7.140}$$

To evaluate the surface integration, it is written explicitly in terms of spherical coordinates:

$$\iint\limits_{S_r} \{\quad\}\mathrm{d}S_r = \int\limits_0^{2\pi}\int\limits_0^{\pi} \{\quad\} r^2 \sin\theta \,\mathrm{d}\theta\,\mathrm{d}\phi \tag{7.141}$$

The exponential factor $\exp(i\vec{k} \bullet \vec{r})$ in (7.140) is a rapidly oscillating function of θ and ϕ except at the stationary point $\hat{r} = -\hat{k}$. A stationary-phase [95] evaluation of (7.140) yields:

$$\begin{aligned} V_i = \frac{2\pi i}{k\eta} \iint\limits_{4\pi} \hat{k} \bullet \{\vec{e}_t(\alpha, \beta) \times [\hat{k} \times \vec{F}(\Omega)] \\ + \vec{F}(\Omega) \times [\hat{k} \times \vec{e}_t(\alpha, \beta)]\}\mathrm{d}\Omega \end{aligned} \tag{7.142}$$

Because the reciprocity integral in (7.138) is independent of the surface over which it is evaluated, the result in (7.142) is an exact, rather than an asymptotic result. (This is consistent with the observation that (7.142) is independent of r.) Vector identities can be used to reduce (7.142) to:

$$V_i = \frac{4\pi i}{k\eta} \iint\limits_{4\pi} \vec{e}_t(\alpha, \beta) \bullet \vec{F}(\Omega)\mathrm{d}\Omega \tag{7.143}$$

This is as far as the expression for V_i can be simplified. It shows that the open-circuit voltage induced when the EUT is illuminated in an immunity test is proportional to a weighted integral of the transmitted far field $\vec{e}_t$ when the EUT is transmitting. Equation (7.143) is similar to the earlier receiving response in (7.96), except that the receiving function in (7.96) was not derived in terms of the transmission properties of the antenna. Another interpretation of (7.143) is that the transmitting and receiving patterns of an antenna or an EUT are the same.

The statistical properties of the plane-wave spectrum $\vec{F}(\Omega)$ were discussed in Section 7.1, and they can be used to derive the statistical properties of V_i. For example, (7.6) and (7.143) can be used to show that the average value of V_i is zero:

$$\langle V_i \rangle = \frac{4\pi i}{k\eta} \iint_{4\pi} \vec{e}_t(\alpha, \beta) \bullet \langle \vec{F}(\Omega) \rangle \, d\Omega = 0 \tag{7.144}$$

The mean square value of V_i is the most useful quantity because it is proportional to the received power in an emissions measurement. The squared magnitude $|V_i|^2$ can be written:

$$|V_i|^2 = \left(\frac{4\pi}{k\eta}\right)^2 \iint_{4\pi} \iint_{4\pi} [\vec{e}_t(\alpha_1, \beta_1) \bullet \vec{F}(\Omega_1)][\vec{e}_t^{\,*}(\alpha_2, \beta_2) \bullet \vec{F}^*(\Omega_2)] d\Omega_1 \, d\Omega_2 \tag{7.145}$$

The average value $\langle |V_i|^2 \rangle$ can be determined by applying the properties of $\vec{F}$ in (7.9) and (7.10) to (7.145):

$$\langle |V_i|^2 \rangle = \frac{2\pi E_0^2}{k^2\eta^2} \iint_{4\pi} |\vec{e}_t(\alpha_1, \beta_1)|^2 d\Omega \tag{7.146}$$

Equation (7.139) shows that the total radiated power in an emissions measurement is proportional to the mean-square, induced voltage in an immunity measurement. For an arbitrary current I in the transmitting (emissions) case, the radiated power P_{rad} is given by:

$$P_{rad} = |I|^2 R_{rad}, \tag{7.147}$$

where R_{rad} is the radiation resistance part of the transmitting impedance Z_t in Figure 7.17. For $I = 1$ A, we have $P_{rad1} = R_{rad}$. If we substitute for P_{rad1} and k $(= 2\pi/\lambda)$, (7.146) can be rewritten:

$$\frac{\langle |V_i|^2 \rangle / (4R_{rad})}{E_0^2/\eta} = \frac{\lambda^2}{8\pi} \tag{7.148}$$

The numerator of the left side of (7.148) is the received power for the case of a matched load ($Z_L = Z_t^*$) with no dissipative loss in the circuit ($\mathrm{Re}(Z_t) = R_{rad}$) in Figure 7.17, and the denominator is the scalar power density. This ratio is the average effective area, and it is equal to $\lambda^2/8\pi$, as shown previously in (7.103).

If the circuit in Figure 7.17 has loss ($\mathrm{Re}(Z_t) = R_{rad} + R_{loss}$), but is still impedance matched ($Z_L = Z_t^*$), (7.148) can be manipulated into the following form:

$$\frac{\{\langle |V_i|^2\rangle/[4(R_{rad}+R_{loss})]\}/\{E_0^2/\eta\}}{\lambda^2/8\pi} = \frac{R_{rad}}{R_{rad}+R_{loss}} \tag{7.149}$$

In (7.149), the numerator is the average received power divided by the scalar power density, which equals the average effective area. The denominator $\lambda^2/8\pi$ is the maximum effective area for any antenna in a well-stirred field. Kraus [96] has termed this ratio the "effectiveness ratio, α_i" for the simpler case where the incident field is a plane wave that can be polarization matched by the receiving antenna to yield a maximum effective area of $\lambda^2/4\pi$. The right side of (7.149) is the radiation efficiency η_a for the emissions case. Thus we can rewrite (7.149):

$$\alpha_i(\text{immunity}) = \eta_a(\text{emissions}) \tag{7.150}$$

The theoretical and experimental results in Figures 7.11 and 7.15 provide a verification of (7.150) for the specific case of a microstrip transmission line [84]. Typically, in the electromagnetic compatibility (EMC) community, the left side of (151) is called shielding effectiveness and is given in decibels. If there is impedance mismatch, both sides of (7.150) can be multiplied by the same mismatch factor to provide a comparison with ideal receivers or transmitters.

7.8 BOUNDARY FIELDS

Because of the electromagnetic boundary conditions at highly conducting metal walls (tangential electric field and normal magnetic field equal zero), statistical field uniformity and isotropy cannot be established in the vicinity of reverberation chamber walls [97]. Consequently, the useful test volume for EMC measurements must exclude the region near the chamber walls, with the possible exception of test objects that are intended to operate on a ground plane [84].

Dunn's theory [87] describes electric and magnetic field transitions from a planar interface (chamber wall) to free space (where the fields are statistically uniform). In this section we confirm Dunn's results and analyze the fields near right-angle bends and right-angle corners. All three geometries (planar interface, right-angle bend, and right-angle corner) are important in determining the useful test volume in rectangular chambers (the usual shape), and all three cases can be analyzed by use of the plane-wave integral representation described in Section 7.1 for predicting field properties and test object responses away from chamber walls. A typical rectangular-cavity reverberation chamber is shown in Figure 7.18. It includes a mechanical stirrer, but the fields near the stirrer are not discussed here.

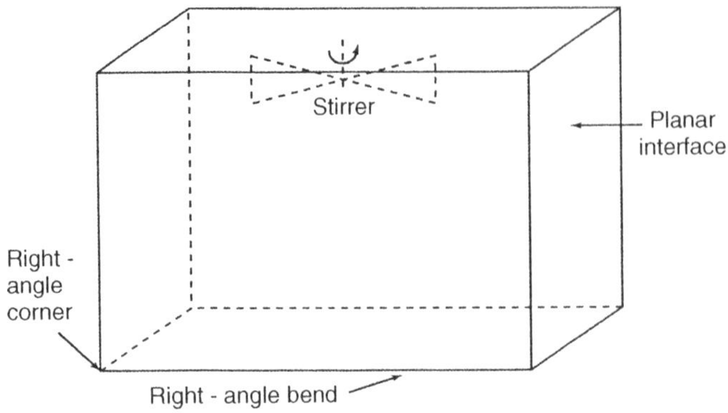

FIGURE 7.18 Rectangular reverberation chamber with mechanical stirring [97].

7.8.1 Planar Interface

The geometry of a planar interface in Figure 7.19 applies to the case where the field point is close to one wall, but distant from all other walls. In fact, there is no assumption needed regarding the geometry of the other chamber walls. We assume here and throughout this section that the walls are perfectly conducting because we are interested only in the field distributions and not the wall losses, as in a calculation of chamber Q. In the analysis of fields far from walls in Section 7.1, the fields in the source-free region included plane waves propagating at all real angles. In this section, we include only propagation directions toward the wall(s) for the incident field and the reflected field from the boundary conditions at one, two, or three walls.

The incident electric field $\vec{E}^i$ at location $\vec{r}$ follows the plane-wave integral form for the total electric field in free space as in (7.1), except for the integration limits:

$$\vec{E}^i(\vec{r}) = \iint_{2\pi} \vec{F}(\Omega)\exp(i\vec{k}^i \bullet \vec{r})\mathrm{d}\Omega, \tag{7.151}$$

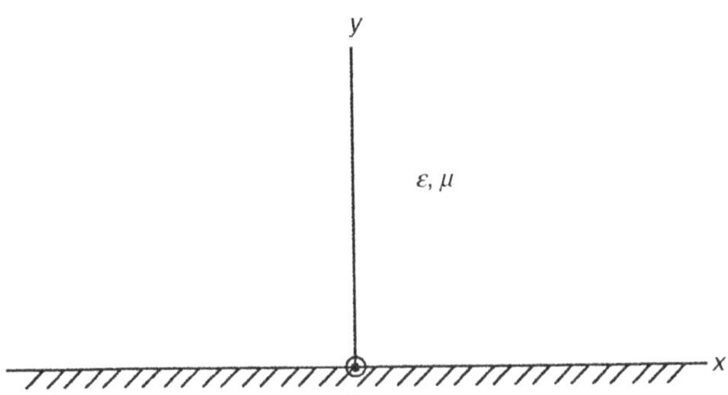

FIGURE 7.19 Single planar wall in a reverberation chamber [97].

where the incident vector wavenumber $\vec{k}^i$ is:

$$\vec{k}^i = -k(\hat{x}\sin\alpha\cos\beta + \hat{y}\sin\alpha\sin\beta + \hat{z}\cos\alpha) \tag{7.152}$$

The coordinates in (7.152) are essentially the same as shown in Figure 7.2. The integral over solid angle 2π steradians in (7.151) actually represents the following double integral:

$$\iint_{2\pi} [\quad] d\Omega = \int_{\beta=0}^{\pi} \int_{\alpha=0}^{\pi} [\quad] \sin\alpha \, d\alpha \, d\beta \tag{7.153}$$

The range of β is only 0 to π, rather than 0 to 2π, because the incident field includes only plane waves propagating toward the interface, $y = 0$.

To use image theory to determine the reflected field, we can first write the incident field in rectangular components as a function of rectangular coordinates:

$$\vec{E}^i(x, y, z) = \hat{x}E_x^i(x, y, z) + \hat{y}E_y^i(x, y, z) + \hat{z}E_z^i(x, y, z) \tag{7.154}$$

The reflected field $\vec{E}^r$ can be determined by image theory:

$$\vec{E}^r(x, y, z) = -\hat{x}E_x^i(x, -y, z) + \hat{y}E_y^i(x, -y, z) - \hat{z}E_z^i(x, -y, z) \tag{7.155}$$

The expressions in this planar interface section are valid for $y \geq 0$. The total field $\vec{E}^t$ is the sum of the incident and reflected fields:

$$\begin{aligned} \vec{E}^t(x, y, z) = {} & \hat{x}[E_x^i(x, y, z) - E_x^i(x, -y, z)] + \hat{y}[E_n^i(x, y, z) \\ & + E_y^i(x, -y, z)] + \hat{z}[E_z^i(x, y, z) - E_z^i(x, -y, z)] \end{aligned} \tag{7.156}$$

At the interface, $y = 0$, the total electric field is:

$$\vec{E}^t(x, 0, z) = 2\hat{y}E_y^i(x, 0, z) \tag{7.157}$$

Thus the tangential electric field is zero and the normal incident electric field is doubled, as expected at a perfectly conducting plane.

The magnetic field analysis is very similar [97], and we can again use image theory to derive the total magnetic field $\vec{H}^t$ in terms of the rectangular components of the incident field:

$$\begin{aligned} \vec{H}^t(x, y, z) = {} & \hat{x}[H_x^i(x, y, z) + H_x^i(x, -y, z)] + \hat{y}[H_y^i(x, y, z) \\ & - H_y^i(x, -y, z)] + \hat{z}[H_z^i(x, y, z) + H_z^i(x, -y, z)] \end{aligned} \tag{7.158}$$

At the interface, $y = 0$, the total magnetic field is:

$$\vec{H}^t(x,0,z) = 2[\hat{x}H_x^i(x,0,z) + \hat{z}H_z^i(x,0,z)] \tag{7.159}$$

Thus the normal magnetic field is zero, and the tangential magnetic field is doubled, as expected at a perfectly conducting plane.

The statistical properties of the angular spectrum $\vec{F}$ have been used in Section 7.2 to derive various ensemble averages at locations away from chamber walls. Here we can use the same methods to obtain ensemble averages for field quantities near chamber walls. For example, the average values of the fields are zero:

$$\langle \vec{E}^t(x,y,z)\rangle = \langle \vec{H}^t(x,y,z)\rangle = 0 \tag{7.160}$$

The result in (7.160) is due to the average value of the angular spectrum $\langle \vec{F}\rangle$ being zero as in (7.6).

The averages of the squares of the field components were shown to be independent of position in Section 7.2 for positions far from the chamber walls. Here the averages evolve from required boundary conditions at the wall ($y = 0$) to uniformity for large ky. Consider first the normal component E_y^t of the electric field. From the two-term y component of (7.156), the magnitude of the square can be written:

$$\begin{aligned}|E_y^t(x,y,z)|^2 &= |E_y^i(x,y,z)|^2 + |E_y^i(x,-y,z)|^2 \\ &+ E_y^i(x,y,z)E_y^{i*}(x,-y,z) + E_y^i(x,-y,z)E_y^{i*}(x,y,z)\end{aligned} \tag{7.161}$$

In determining the average value of (7.161), the first two terms can be determined from the uniformity result in (7.15), and the last two terms can be obtained from the longitudinal correlation function described in (7.53) to (7.58):

$$\left\langle |E_y^t(x,y,z)|^2\right\rangle = \frac{E_0^2}{3}[1 + \rho_l(2y)], \tag{7.162}$$

where E_0^2 is the mean-square of the total electric field at large distances from the wall where the field is spatially uniform, as shown in (7.14). The result in (7.162) agrees exactly with Dunn's result [87]. The result is independent of x and z as expected by translational symmetry. For large ky, ρ_l decays as $(ky)^{-2}$. So the limit of (7.162) for large ky is:

$$\lim\left\langle |E_y^t(x,y,z)|^2\right\rangle = \frac{E_0^2}{3}, \quad \text{for} \quad ky\to\infty \tag{7.163}$$

This is the known result far from the chamber walls, as shown in (7.15).

At the wall boundary ($y = 0$), (7.162) reduces to:

$$\left\langle |E_y^t(x,0,z)|^2\right\rangle = \frac{2E_0^2}{3} \tag{7.164}$$

Thus the mean-square value of the normal component of the electric field at the wall is twice that of the value far from the chamber wall.

Consider next the tangential components E_x^t and E_z^t of the electric field. The results are the same for both tangential components; so we consider only E_x^t. The square of the magnitude can be written:

$$\begin{aligned}&|E_x^t(x,y,z)|^2 = |E_x^i(x,y,z)|^2 + |E_x^i(x,-y,z)|^2\\&-E_x^i(x,y,z)E_x^{i*}(x,-y,z) - E_x^i(x,-y,z)E_x^{i*}(x,y,z)\end{aligned} \tag{7.165}$$

In determining the average value of (7.165), the first two terms can be determined from the uniformity result in (7.15), and the last two terms can be obtained from the transverse correlation function described in (7.59)–(7.63):

$$\left\langle |E_x^t(x,y,z)|^2 \right\rangle = \frac{E_0^2}{3}[1-\rho_t(2y)] \tag{7.166}$$

The result in (7.166) also agrees with Dunn's result [87]. Again, the result is independent of x and z. For large ky, ρ_t decays as $(ky)^{-1}$. So the limit of (7.166) for large ky is:

$$\lim \left\langle |E_x^t(x,y,z)|^2 \right\rangle = \frac{E_0^2}{3}, \quad \text{for} \quad ky \to \infty \tag{7.167}$$

At the wall boundary ($y = 0$), (7.167) reduces to:

$$\left\langle |E_x^t(x,0,z)|^2 \right\rangle = 0 \tag{7.168}$$

This is the expected result because the tangential electric field must be zero at the wall.

The analysis of the square of the magnetic field components is similar to that of the electric field components. Consider first the normal component H_y^t of the magnetic field. The square of the magnitude can be written:

$$\begin{aligned}&|H_y^t(x,y,z)|^2 = |H_y^i(x,y,z)|^2 + |H_y^i(x,-y,z)|^2\\&-H_y^i(x,y,z)H_y^{i*}(x,-y,z)|^2 - H_y^i(x,-y,z)H_y^{i*}(x,y,z)\end{aligned} \tag{7.169}$$

The procedure for determining the average value of (7.169) follows that for the normal electric field. The first two terms can be determined from the uniformity results in (7.21), and the last two terms can be determined from the longitudinal correlation function in (7.58):

$$\left\langle |H_y^t(x,y,z)|^2 \right\rangle = \frac{E_0^2}{3\eta^2}[1-\rho_l(2y)] \tag{7.170}$$

The result in (7.170) agrees with Dunn's result [87]. The limit of (7.170) for large ky is:

$$\lim \left\langle |H_y^t(x,y,z)|^2 \right\rangle = \frac{E_0^2}{3\eta^2}, \quad \text{for} \quad ky \to \infty \tag{7.171}$$

This is the result for a uniform, well stirred magnetic field far from the chamber walls, as shown in (7.21). At the wall boundary ($y = 0$), (7.170) reduces to:

$$\left\langle |H_y^t(z, 0, z)|^2 \right\rangle = 0 \tag{7.172}$$

Thus the mean-square value of the normal component of the magnetic field is zero.

Consider next the tangential components, H_x^t and H_z^t, of the magnetic field. The results are the same for both components; so we consider only H_x^t. The square can be written:

$$\begin{aligned} |H_x^t(x, y, z)|^2 &= |H_x^i(x, y, z)|^2 + |H_x^i(x, -y, z)|^2 \\ &+ H_x^i(x, y, z)H_x^{i*}(x, -y, z)|^2 + H_x^i(x, -y, z)H_x^{i*}(x, y, z) \end{aligned} \tag{7.173}$$

In determining the average value of (7.173), the first two terms can again be determined from the uniformity results in (7.21), and the last two terms can be determined from the transverse correlation function in (7.63):

$$\left\langle |H_x^t(x, y, z)|^2 \right\rangle = \frac{E_0^2}{3\eta^2}[1 + \rho_t(2y)] \tag{7.174}$$

The result in (7.715) agrees with Dunn's result [87]. The limit of (7.715) for large ky is

$$\lim \left\langle |H_x^t(x, y, z)|^2 \right\rangle = \frac{E_0^2}{3\eta^2}, \quad \text{for } ky \to \infty \tag{7.175}$$

As with (7.171), this is the known result for a uniform, well stirred magnetic field far from the chamber walls as shown in (7.21). At the wall boundary ($y = 0$), (7.174) reduces to:

$$\left\langle |H_x^t(x, 0, z)|^2 \right\rangle = \frac{2E_0^2}{3\eta^2} \tag{7.176}$$

Thus the mean-square value of the tangential magnetic field at the chamber wall is twice that of the value far from the chamber walls.

7.8.2 Right-Angle Bend

The geometry of a right-angle in Figure 7.20 applies to the case where the field point is close to two mutually perpendicular walls, but distant from all other walls. The expression for the incident electric field is similar to that in (7.151) except that the solid angle integration is now over only π steradians:

$$\vec{E}^i(\vec{r}) = \iint_{\pi} \vec{F}(\Omega)\exp(i\vec{k}^i \bullet \vec{r})\mathrm{d}\Omega \tag{7.177}$$

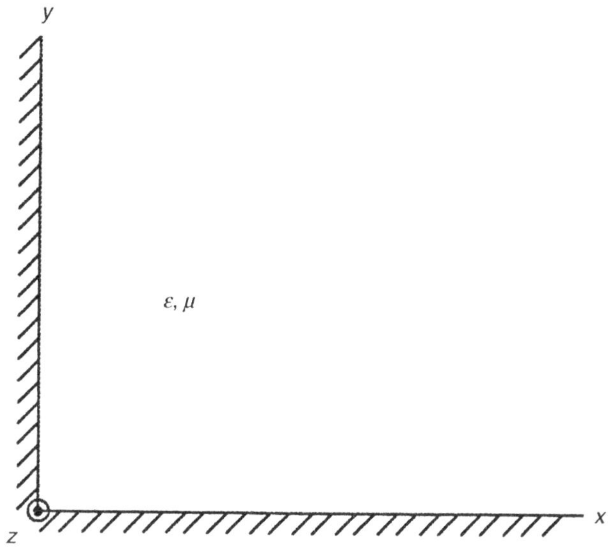

FIGURE 7.20 Junction of two planar walls (right-angle bend) in a reverberation chamber [97].

The integral over solid angle π steradians in (7.177) actually represents the following double integral:

$$\iint_{\pi} [\quad] d\Omega = \int_{\beta=0}^{\pi/2} \int_{\alpha=0}^{\pi} [\quad] \sin\alpha \, d\alpha \, d\beta \tag{7.178}$$

The range of β is 0 to $\pi/2$ because the incident field includes only plane waves propagating toward the two walls of the right-angle bend.

The incident field is again written in rectangular coordinates as in (7.154). The reflected field is more complicated than that given in (7.155), because three images rather than one are needed to satisfy the boundary conditions on both walls ($y = 0$ and $x = 0$). Hence, the reflected field is written:

$$\begin{aligned} \vec{E}^r(x, y, z) = \; & \hat{x}[-E^i_x(x, -y, z) - E^i_x(-x, -y, z) + E^i_x(-x, y, z)] \\ & + \hat{y}[F^i_y(x, -y, z) - E^i_y(-x, -y, z) - E^i_y(-x, y, z)] \\ & + \hat{z}[-E^i_z(x, -y, z) + E^i_z(-x, -y, z) - E^i_z(-z, y, z)] \end{aligned} \tag{7.179}$$

The expressions in this section are valid for $x, y \geq 0$. The total electric field is the sum of the incident and reflected fields:

$$\begin{aligned} \vec{E}^t(x, y, z) = \; & \hat{x}[E^i_x(x, y, z) - E^i_x(x, -y, z) - E^i_x(-x, -y, z) + E^i_x(-x, y, z)] \\ & + \hat{y}[E^i_y(x, y, z) + E^i_y(x, -y, z) - E^i_y(-x, -y, z) - E^i_y(-x, y, z)] \\ & + \hat{z}[E^i_z(x, y, z) - E^i_z(x, -y, z) + E^i_z(-x, -y, z) - E^i_z(-x, y, z)] \end{aligned} \tag{7.180}$$

At the interface, $x = 0$, the total electric field is:

$$\vec{E}^t(0,y,z) = 2\hat{x}[E_x^i(0,y,z) - E_x^i(0,-y,z)] \tag{7.181}$$

Thus, the tangential electrical field is zero as expected on a perfect conductor, and the normal electric field is the difference of two doubled terms. An analogous result occurs on the interface, $y = 0$:

$$\vec{E}^t(x,0,z) = 2\hat{y}[E_y^i(x,0,z) - E_y^i(-x,0,z)] \tag{7.182}$$

The magnetic field analysis is very similar, and we can again use double-image theory to derive the total magnetic field in terms of the rectangular components of the incident field:

$$\begin{aligned} \vec{H}^t(x,y,z) = {} & \hat{x}[H_x^i(x,y,z) + H_x^i(x,-y,z) - H_x^i(-x,-y,z) - H_x^i(-x,y,z) \\ & + \hat{y}[H_y^i(x,y,z) - H_y^i(x,-y,z) - H_y^i(-x,-y,z) + H_y^i(-x,y,z)] \\ & + \hat{z}[H_z^i(x,y,z) + H_z^i(x,-y,z) + H_z^i(-x,-y,z) + H_z^i(-x,y,z)] \end{aligned} \tag{7.183}$$

At the interface, $x = 0$, the total magnetic field is:

$$\begin{aligned} \vec{H}^t(0,y,z) = {} & 2\hat{y}[H_y^i(0,y,z) - H_y^i(0,-y,z)] \\ & + 2\hat{z}[H_z^i(0,y,z) - H_z^i(0,-y,z)] \end{aligned} \tag{7.184}$$

Thus the normal magnetic field is zero, as expected on a perfect conductor, and the tangential magnetic field is the difference of two doubled terms. An analogous result occurs on the interface, $y = 0$:

$$\begin{aligned} \vec{H}^t(x,0,z) = {} & 2\hat{x}[H_x^i(x,0,z) - H_x^i(-x,0,z)] \\ & + 2\hat{z}[H_x^i(x,0,z) - H_z^i(-x,0,z)] \end{aligned} \tag{7.185}$$

As with the previous analysis of the planar interface, the average values of the total electric and magnetic fields are zero because the average value of the angular spectrum $\langle \vec{F} \rangle$ is zero. We can follow the previous method of determining the average values of the squared magnitudes of the field components, except that there are more terms involved because of the additional image terms. Consider first the z (tangential) component E_z^t of the total electric field. Its squared magnitude can be written as:

$$\begin{aligned} |E_z^t(x,y,z)|^2 = {} & |E_z^i(x,y,z)|^2 + |E_z^i(x,-y,z)|^2 + |E_z^i(-x,-y,z)|^2 + |E_z^i(-x,y,z)|^2 \\ & + E_z^i(x,y,z)[-E_z^{i*}(x,-y,z) + E_z^{i*}(-x,-y,z) - E_z^{i*}(-x,y,z)] \\ & - E_z^i(x,-y,z)[E_z^{i*}(x,y,z) + E_z^{i*}(-x,-y,z) - E_z^{i*}(-x,y,z)] \\ & + E_z^i(-x,-y,z)[E_z^{i*}(x,y,z) - E_z^{i*}(x,-y,z) - E_z^{i*}(-x,y.z)] \\ & - E_z^i(-x,y,z)[-E_z^{i*}(x,-y,z) + E_z^{i*}(x,y,z) + E_z^{i*}(-x,-y,z)] \end{aligned} \tag{7.186}$$

In evaluating the expectation of (7.186), the four terms are evaluated by the uniformity property of the field given in (7.15), and the remaining terms involve the transverse correlation function given in (7.63), so that the final result is:

$$\left\langle |E_z^t(x,y,z)|^2 \right\rangle = \frac{E_0^2}{3}\left[1-\rho_t(2y)-\rho_t(2x)+\rho_t\left(2\sqrt{x^2+y^2}\right)\right] \quad (7.187)$$

There are a number of limiting cases of (7.187) that are of interest. For either x or y equal to 0, we have $\left\langle |E_z^t|^2 \right\rangle = 0$, so that the expectation of the square of the z component is zero at the wall surface. For large kx and ky, we have $\left\langle |E_z^t(x,y,z)|^2 \right\rangle \to E_0^2/3$ which is the expected uniform field far from the walls, as in (7.15). For large kx, (7.187) reduces to the single-wall result in (7.166). On the diagonal ($x = y$), (7.187) reduces to:

$$\left\langle |E_z^t(x,x,z)|^2 \right\rangle = \frac{E_0^2}{3}\left[1-2\rho_t(2x)+\rho_t(2\sqrt{2}x)\right] \quad (7.188)$$

This result on the diagonal evolves from 0 at the corner to $E_0^2/3$ at large distances. For determining the useful test volume of a reverberation chamber, (39) is useful because it shows how rapidly the field reaches its uniform asymptotic value. To reach that value, it is necessary that $2kx \geq 1$. This is achieved if x is greater than approximately $\lambda/2$.

The behaviors of E_x^t and E_y^t are somewhat different from that of E_z^t because they are tangential to one wall and normal to the other. We consider only E_x^t because the behavior of E_y^t is the same with an interchange of x and y. The squared magnitude of E_x^t can be written:

$$\begin{aligned}
|E_x^t(xyz)|^2 = {} & |E_x^i(x,y,z)|^2 + |E_{xi}(x,-y,z)|^2 + |E_x^i(-x,-y,z)|^2 + |E_x^i(-x,-y,z)|^2 \\
& + E_x^i(x,y,z)[-E_x^{i*}(x,-y,z) - E_x^{i*}(-x,-y,z) + E_x^{i*}(x,-y,z)] \\
& - E_x^i(x,-y,z)[E_x^{i*}(x,y,z) - E_x^{i*}(-x,-y,z) + E_x^{i*}(-x,y,z)] \\
& - E_x^i(-x,-y,z)[E_x^{i*}(x,y,z) - E_x^{i*}(x,-y,z) + E_x^{i*}(-x,-y,z)] \\
& + E_x^i(-x,y,z)[E_x^{i*}(x,y,z) - E_x^{i*}(x,-y,z) + E_x^{i*}(-x,y,z)]
\end{aligned} \quad (7.189)$$

In evaluating the expectation of (7.189), the first four terms are again evaluated by the uniformity of the field given in (7.15), and the remaining terms involve both the transverse and longitudinal correlation functions in (7.63) and (7.58), so that the final result is:

$$\begin{aligned}
\left\langle |E_x^t(x,y,z)|^2 \right\rangle = \frac{E_0^2}{3}\Bigg[& 1-\rho_t(2y)+\rho_l(2x) - \frac{y^2}{x^2+y^2}\rho_t\left(2\sqrt{x^2+y^2}\right) \\
& - \frac{x^2}{x^2+y^2}\rho_l\left(2\sqrt{x^2+y^2}\right)\Bigg]
\end{aligned} \quad (7.190)$$

There are a number of special cases of (7.190) that are of interest. For $y = 0$, we have $\left\langle |E_x^t(x,0,z)|^2 \right\rangle = 0$, so that the expectation of the square of the tangential electric field is zero at the wall surface. For $x = 0$, we have:

$$\left\langle |E_x^t(0,y,z)|^2 \right\rangle = \frac{2E_0^2}{3}[1-\rho_t(2y)] \tag{7.191}$$

This is twice the result of that in (7.166) for a single wall. For large kx and ky, we have $\left\langle |E_x^t(x,y,z)|^2 \right\rangle \to E_0^2/3$, which is the expected field far from the walls. For large kx, we have:

$$\left\langle |E_x^t(x,y,z)|^2 \right\rangle = \frac{E_0^2}{3}[1-\rho_t(2y)], \tag{7.192}$$

which is the same as (7.166) for a single wall. For large ky, we have:

$$\left\langle |E_y^t(x,y,z)|^2 \right\rangle = \frac{E_0^2}{3}[1+\rho_l(2y)], \tag{7.193}$$

which is analogous to (7.162) for the electric field normal to a single wall.

The analysis for the expectation of the squares of the magnetic field components is similar to that of the electric field components. So we shall skip some intermediate steps and proceed directly to the final results. Consider first the z (tangential) component H_z^t of the magnetic field. The expectation of its squared magnitude is:

$$\left\langle |H_z^t(x,y,z)|^2 \right\rangle = \frac{E_0^2}{3\eta^2}\left[1+\rho_t(2x)+\rho_t(2y)+\rho_t\left(2\sqrt{x^2+y^2}\right)\right] \tag{7.194}$$

A number of limiting cases of (7.194) are of interest. For $x = 0$, we have:

$$\left\langle |H_z^t(0,y,z)|^2 \right\rangle = \frac{2E_0^2}{3\eta^2}[1+\rho_t(2y)], \tag{7.195}$$

which is twice the result in (7.174). For large ky, (7.195) reduces to $\left\langle |H_z^t(0,y,z)|^2 \right\rangle \to 2E_0^2/3\eta^2$, which is the same result as (7.176) for the tangential magnetic field at a single wall. For $y = 0$, we have:

$$\left\langle |H_z^t(x,0,z)|^2 \right\rangle = \frac{2E_0^2}{3\eta^2}[1+\rho_t(2x)], \tag{7.196}$$

which is analogous to (7.195). For both $x = y = 0$, both (7.195) and (7.196) yield:

$$\left\langle |H_z^t(0,0,z)|^2 \right\rangle = \frac{4E_0^2}{3\eta^2} \tag{7.197}$$

For large kx, we have:

$$\left\langle |H_z^t(x,y,z)|^2 \right\rangle \to \frac{E_0^2}{3\eta^2}[1+\rho_t(2y)], \tag{7.198}$$

which is the same as the single-wall result in (7.174). For large ky, we have:

$$\left\langle |H_x^t(x,y,z)|^2 \right\rangle \to \frac{E_0^2}{3\eta^2}[1+\rho_t(2x)], \tag{7.199}$$

which is analogous to (7.198).

On the diagonal, $x = y$, we have:

$$\left\langle |H_z^t(x,x,z)|^2 \right\rangle = \frac{E_0^2}{3\eta^2}\left[1+2\rho_t(2x)+\rho_t(2\sqrt{2}x)\right] \tag{7.200}$$

The result on the diagonal evolves from $4E_0^2/(3\eta^2)$ at the corner to $E_0^2/(3\eta^2)$ at large distances. As with (7.188), (7.200) is useful for determining the useful test volume of a reverberation chamber because it shows how rapidly the magnetic field reaches its asymptotic value. As with the electric field, it is necessary that $2kx \gg 1$.

The behaviors of H_x^t and H_y^t are somewhat different from that of H_z^t because they are tangential to one wall and normal to the other. We consider only H_x^t because the behavior of H_y^t is the same with an interchange of x and y. The expectation of the squared magnitude is:

$$\begin{aligned}\left\langle |H_x^t(x,y,z)|^2 \right\rangle = \frac{E_0^2}{3\eta^2}\Bigg[1+\rho_t(2y)-\rho_t(2x) &- \frac{y^2}{x^2+y^2}\rho_t\left(2\sqrt{x^2+y^2}\right) \\ &- \frac{x^2}{x^2+y^2}\rho_l\left(2\sqrt{x^2+y^2}\right)\Bigg]\end{aligned} \tag{7.201}$$

There are a number of special cases of (7.201) that are of interest. For $x = 0$, we have $\left\langle |H_x^t(0,y,z)|^2 \right\rangle = 0$ so that the expectation of the square of the normal magnetic field is zero at the wall surface. For $y = 0$, we have:

$$\left\langle |H_x^t(x,0,z)|^2 \right\rangle = \frac{2E_0^2}{3\eta^2}[1-\rho_l(2x)], \tag{7.202}$$

which is twice the analogous result for the normal magnetic field in (7.170). For large kx and ky, we have $\left\langle |H_x^t(x,y,z)|^2 \right\rangle \to E_0^2/(3\eta^2)$, which is the expected result far from the walls. For large kx, we have:

$$\left\langle |H_x^t(x,y,z)|^2 \right\rangle \to \frac{E_0^2}{3\eta^2}[1+\rho_t(2y)], \tag{7.203}$$

which is equal to the result for the magnetic field tangential to a single wall in (7.174). For large ky, we have:

$$\left\langle |H_x^t(x,y,z)|^2 \right\rangle \to \frac{E_0^2}{3\eta^2}[1-\rho_l(2x)], \tag{7.204}$$

which is equal to the result for the magnetic field normal to a single wall in (7.170).

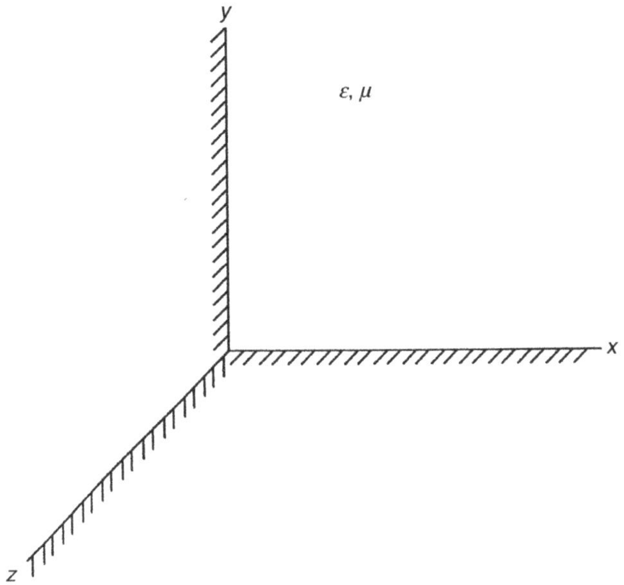

FIGURE 7.21 Junction of three planar walls (right-angle corner) in a reverberation chamber [97].

7.8.3 Right-Angle Corner

The geometry of a right-angle corner in Figure 7.21 applies to the case where the field point is close to all three walls that make up a corner. The expression for the incident electric field is similar to that in (7.1) except that the solid angle integration is now performed over only $\pi/2$ steradians:

$$\vec{E}^i(\vec{r}) = \iint_{\pi/2} \vec{F}(\Omega)\exp(i\vec{k}^i \bullet \vec{r})\mathrm{d}\Omega \tag{7.205}$$

The integral over solid angle $\pi/2$ steradians in (7.205) actually represents the following double integral:

$$\iint_{\pi/2} [\quad]\mathrm{d}\Omega = \int_{\beta=0}^{\pi/2} \int_{\alpha=0}^{\pi/2} [\quad]\sin\alpha\,\mathrm{d}\alpha\,\mathrm{d}\beta \tag{7.206}$$

The ranges of α and β are both 0 to $\pi/2$ because the incident field includes only plane waves propagating toward all three walls of the right-angle corner.

The incident field is again written in rectangular coordinates as in (7.154). The reflected field is more complicated yet, because seven images are needed to satisfy the boundary conditions on all three walls ($x = 0$, $y = 0$, and $z = 0$). Hence, each rectangular component of the total (incident plus reflected) electric field has eight terms. Since each field component is normal to one wall and tangential to two walls, all three components have this behavior. So we will analyze only one electric field component E_z^t, which can be written:

$$\begin{aligned} E_z^t(x,y,z) = \; & E_z^i(x,y,z) - E_z^i(x,-y,z) + E_z^i(-x,-y,z) \\ & - E_z^i(-x,y,z) + E_z^i(x,y,-z) - E_z^i(x,-y,-z) \\ & + E_z^i(-x,-y,-z) - E_z^i(-x,y,-z) \end{aligned} \quad (7.207)$$

The expressions in this section are valid for $x, y, z \geq 0$. At the interface $x = 0$, we have $E_z^t(0,y,z) = 0$, as expected for a tangential component. Similarly, at the interface $y = 0$ we have $E_z^t(x,0,z) = 0$. The z component of the electric field is normal to the interface $z = 0$, and we have:

$$E_z^t(x,y,0) = 2[E_z^i(x,y,0) + E_z^i(-x,-y,0) - E_z^i(x,-y,0) - E_z^i(-x,y,0)], \quad (7.208)$$

which is similar to (7.181) and (7.182).

For the magnetic field we again analyze only one component H_z^t, which can be written:

$$\begin{aligned} H_z^t(x,y,z) = \; & H_z^i(x,y,z) + H_z^i(x,-y,z) + H_z^i(-x,-y,z) \\ & + H_z^i(-x,y,z) - H_z^i(x,y,-z) - H_z^i(x,-y,-z) \\ & - H_z^i(-x-y-z) - H_z^i(-x,y,-z) \end{aligned} \quad (7.209)$$

At the interface $z = 0$, we have $H_z^t(x,y,0) = 0$, as expected for the normal component of the magnetic field. At the interface $x = 0$, we have:

$$H_z^t(0,y,z) = 2[H_z^i(0,y,z) + H_z^i(0,-y,z) - H_z^i(0,y,-z) - H_z^i(0,-y,-z)], \quad (7.210)$$

which is a combination of four doubled terms. At the interface $y = 0$, we have:

$$H_z^t(x,0,z) = 2[H_z^t(x,0,z) + H_z^t(-x,0,z) - H_z^t(x,0,-z) - H_z^t(-x,0,-z)], \quad (7.211)$$

which is similar to (7.210). It can be shown that (7.208), (7.210), and (7.211) agree with the earlier results in the previous section on right-angle bends if one of the other coordinates is set to zero.

As in the two previous cases (planar interface and right-angle bend), the average values of each scalar field component is zero because the average value of the angular spectrum $\langle \vec{F} \rangle$ is zero. We follow the previous method of determining the average values of the squares of the z components of the electric and magnetic fields. Because there are so many terms in (7.207) and (7.209), the squares have many more terms. For brevity, we skip the expressions for the squares of the field components and give just the results for the averages of the squared magnitude. For the expectation of the squared magnitude of E_z^t, we have:

$$\begin{aligned}\left\langle |E_z^t(x,y,z)|^2 \right\rangle = \frac{E_0^2}{3}\Bigg[&1-\rho_t(2x)-\rho_t(2y)+\rho_t\left(2\sqrt{x^2+y^2}\right)+\rho_l(2z) \\ &-\frac{x^2}{x^2+z^2}\rho_t\left(2\sqrt{x^2+z^2}\right)-\frac{z^2}{x^2+z^2}\rho_l\left(2\sqrt{x^2+z^2}\right) \\ &-\frac{y^2}{y^2+z^2}\rho_t\left(2\sqrt{y^2+z^2}\right)-\frac{z^2}{y^2+z^2}\rho_l\left(2\sqrt{y^2+z^2}\right) \\ &+\frac{x^2+y^2}{x^2+y^2+z^2}\rho_t\left(2\sqrt{x^2+y^2+z^2}\right) \\ &+\frac{z^2}{x^2+y^2+z^2}\rho_l\left(2\sqrt{x^2+y^2+z^2}\right)\Bigg] \end{aligned} \tag{7.212}$$

Because (7.212) is so complex, we can again usefully take various limits for both checks and insight. For either x or y equal 0, we have $\left\langle |E_z^t|^2 \right\rangle = 0$, so that the expectation of the square of the tangential electric field is zero at the wall surface. For $z = 0$, we have:

$$\left\langle |E_z^t(x,y,0)|^2 \right\rangle = \frac{2E_0^2}{3}\left[1-\rho_t(2y)-\rho_t(2x)+\rho_t\left(2\sqrt{x^2+y^2}\right)\right], \tag{7.213}$$

which is twice the value for the right-angle bend in (7.187). For large kx, ky, and kz, we have $\left\langle |E_z^t(x,y,z)|^2 \right\rangle \to E_0^2/3$, which is the expected uniform result far from the walls. For large kz, we have:

$$\left\langle |E_z^t(x,y,\infty)|^2 \right\rangle = \frac{E_0^2}{3}\left[1-\rho_t(2y)-\rho_t(2x)+\rho_t\left(2\sqrt{x^2+y^2}\right)\right], \tag{7.214}$$

which is the same as the right-angle result in (7.187). For large kx, we have:

$$\begin{aligned}\left\langle |E_z^t(\infty,y,z)|^2 \right\rangle = \frac{E_0^2}{3}\Bigg[&1-\rho_t(2y)+\rho_l(2z)-\frac{y^2}{z^2+y^2}\rho_t\left(2\sqrt{x^2+y^2}\right) \\ &-\frac{z^2}{z^2+y^2}\rho_l\left(2\sqrt{z^2+y^2}\right)\Bigg], \end{aligned} \tag{7.215}$$

which is analogous to the right-angle bend result in (7.190).

On the diagonal ($x = y = z = r/\sqrt{3}$) from the corner, (7.212) reduces to:

$$\left\langle \left| E_z^t\left(\frac{r}{\sqrt{3}}, \frac{r}{\sqrt{3}}, \frac{r}{\sqrt{3}}\right)^2 \right| \right\rangle = \frac{E_0^2}{3}\left[1 - 2\rho_t\left(\frac{2r}{\sqrt{3}}\right) + \rho_t\left(\frac{2\sqrt{2}r}{\sqrt{3}}\right) + \rho_l\left(\frac{2r}{\sqrt{3}}\right)\right.$$
$$\left. -2\rho_t\left(\frac{2\sqrt{2}r}{\sqrt{3}}\right) - 2\rho_l\left(\frac{2\sqrt{2}r}{\sqrt{3}}\right) + \frac{2}{3}\rho_t(2r) + \frac{1}{3}\rho_t(2r)\right] \tag{7.216}$$

All of the terms in the square bracket in (7.216) involving either ρ_t or ρ_l decay to zero for large kr. The slowest decay is of order $(2kr)^{-1}$ in terms involving ρ_t. So (7.216) reaches its large kr limit of $E_0^2/3$ when r is approximately $\lambda/2$. This is similar to the results for the right-angle bend in this chapter and with Dunn's results [87] for the planar wall. The same result is obtained for the x and y components of the electric field.

We deal now with the magnetic field. Starting with (7.209), we obtain the following for the expectation of squared magnitude of H_z^t:

$$\begin{aligned}\langle |H_z^t(x,y,z)|^2\rangle = \frac{E_0^2}{3\eta^2}\Bigg[& 1 + \rho_t(2x) + \rho_t(2y) + \rho_t\left(2\sqrt{x^2+y^2}\right) - \rho_l(2z) \\ & - \frac{x^2}{x^2+z^2}\rho_t\left(2\sqrt{x^2+z^2}\right) - \frac{z^2}{x^2+z^2}\rho_l\left(2\sqrt{x^2+z^2}\right) \\ & - \frac{y^2}{y^2+z^2}\rho_t\left(2\sqrt{y^2+z^2}\right) - \frac{z^2}{y^2+z^2}\rho_l\left(2\sqrt{y^2+z^2}\right) \\ & - \frac{x^2+y^2}{x^2+y^2+z^2}\rho_t\left(2\sqrt{x^2+y^2+z^2}\right) \\ & - \frac{z^2}{x^2+y^2+z^2}\rho_l\left(2\sqrt{x^2+y^2+z^2}\right)\Bigg]\end{aligned} \tag{7.217}$$

As with (7.212), we can take various limits of (7.217) for both checks and insight. For $z = 0$, we have $\left\langle |H_z^t|^2 \right\rangle = 0$, so that the expectation of the square of the normal magnetic field at the wall surface is zero. For $x = 0$, we have:

$$\begin{aligned}\langle |H_z^t(0,y,z)|^2\rangle = \frac{2E_0^2}{3\eta^2}\Bigg[& 1 + \rho_t(2y) - \rho_l(2z) - \frac{y^2}{y^2+z^2}\rho_t\left(2\sqrt{y^2+z^2}\right) \\ & - \frac{z^2}{y^2+z^2}\rho_l\left(2\sqrt{y^2+z^2}\right)\Bigg],\end{aligned} \tag{7.218}$$

which is twice the analogue of (7.201) for the right-angle bend. For $y = 0$, we obtain a similar result. For large kz, we have:

$$\langle |H_z^t(x, y, \infty)|^2 \rangle = \frac{E_0^2}{3\eta^2}\left[1 + \rho_t(2x) + \rho_t(2y) + \rho_t\left(2\sqrt{x^2+y^2}\right)\right], \tag{7.219}$$

which is the same as (7.194) for the right-angle bend. For large kx, we have:

$$\begin{aligned}\langle |H_z^t(\infty, y, z)|^2 \rangle = \frac{E_0^2}{3\eta^2}\Bigg[&1 + \rho_t(2y) - \rho_l(2z) - \frac{y^2}{y^2+z^2}\rho_t\left(2\sqrt{y^2+z^2}\right)\\ &- \frac{z^2}{y^2+z^2}\rho_l\left(2\sqrt{y^2+z^2}\right)\Bigg],\end{aligned} \tag{7.220}$$

which is analogous to (7.201) for the right-angle bend. The result for large ky is similar. The results for the magnetic field on the diagonal ($x = y = z = r/\sqrt{3}$) from the corner are similar to that for the electric field in (7.216) so that $\left\langle |H_z^t(r/\sqrt{3}, r/\sqrt{3}, r/\sqrt{3})|^2 \right\rangle$ reaches its large kr limit of $E_0^2/3\eta^2$ when r is approximately $\lambda/2$. The same result is obtained for the x and y components of the magnetic field.

7.8.4 Probability Density Functions

In the previous parts of this section, we have used the statistical properties of the angular spectrum [69] and the boundary conditions at walls, bends, or corners to derive a number of useful ensemble averages. These results have not required a knowledge of the particular forms of the probability density functions. However, such knowledge would be very useful for analysis of measured data, which is always based on some limited number of samples (stirrer positions).

We choose to treat only the z component of the electric field, but the same methods are applicable to the other components of the electric and magnetic fields. The starting point for deriving the probability density functions of interest is to write E_z^t in terms of real and imaginary parts:

$$E_z^t(x, y, z) = E_{zr}^t(x, y, z) + iE_{zi}^t(x, y, z) \tag{7.221}$$

Because the average value of the angular spectrum $\langle \vec{F} \rangle$ is zero [69], the average values of both the real and imaginary parts of (7.221) are zero:

$$\langle E_{zr}^t(x, y, z) \rangle = \langle E_{zi}^t(x, y, z) \rangle = 0 \tag{7.222}$$

The variances of the real and imaginary parts are equal and are given by one half the values given for the three geometries earlier in this section:

$$\langle E_{zr}^{t2}(x, y, z) \rangle = \langle E_{zr}^{t2}(x, y, z) \rangle \equiv \sigma^2, \tag{7.223}$$

where, for convenience, we omit the dependence of σ^2 on x, y, and z in the equations to follow. The mean and variance of the real and imaginary parts in (7.222) and (7.223) are all the information that can be derived from the assumptions of the properties of $\vec{F}$ and the wall boundary conditions. From the maximum-entropy method [64,65], we can show that the most probable probability density function f of both the real and imaginary parts of E_z^t is Gaussian:

$$
\begin{aligned}
f[E_{zr}^t(x,y,z)] &= \frac{1}{\sqrt{2\pi}\sigma}\exp\left[-\frac{E_{zr}^{t2}(x,y,z)}{2\sigma^2}\right],\\
f[E_{zi}^t(x,y,z)] &= \frac{1}{\sqrt{2\pi}\sigma}\exp\left[-\frac{E_{zr}^{t2}(x,y,z)}{2\sigma^2}\right]
\end{aligned}
\tag{7.224}
$$

We have shown in (7.35) that the real and imaginary parts of the components of $\vec{E}^t$ are uncorrelated. Since they are Gaussian, they are also independent [57]. Since the real and imaginary parts of the z component of the electric field are normally distributed with zero mean and equal variances and are independent, the probability density functions of the magnitude or squared magnitude of E_z^t is χ or χ-squared distributed with two degrees of freedom. Consequently, the magnitude of E_z^t has a Rayleigh distribution [57]:

$$
f(|E_z^t(x,y,z)|) = \frac{|E_z^t(x,y,z)|}{\sigma^2}\exp\left[-\frac{|E_z^t(x,y,z)|^2}{2\sigma^2}\right] \tag{7.225}
$$

The squared magnitude of E_z^t has an exponential distribution [57]:

$$
f(|E_z^t(x,y,z)|^2) = \frac{1}{2\sigma^2}\exp\left[-\frac{|E_z^t(x,y,z)|^2}{2\sigma^2}\right] \tag{7.226}
$$

The probability density functions in (7.225) and (7.226) agree with [69] and Kostas and Boverie [72]. The magnitude and squared magnitude of the electric and magnetic field components have Rayleigh and exponential probability density functions, but the variances are different and are functions of position. Thus we cannot write the probability density functions of the magnitude and squared magnitude of the total electric and magnetic fields as χ and χ-squared with six degrees of freedom, as was done in [69]. However, the variances do become equal at large distances from the walls, so that the limiting probability density functions do agree with those in [69].

7.9 ENHANCED BACKSCATTER AT THE TRANSMITTING ANTENNA

Transmission between a pair of antennas in a reverberation chamber was covered in Section 7.5. When the receiving antenna (which we will identify as antenna 2) is located at a sufficient distance from the transmitting antenna (which we will identify as antenna 1) and the chamber walls and stirrer, the ensemble average of the received power is independent of the location and orientation of the receiving antenna, as shown in (7.103) and (7.110). The receiving antenna is frequently called the reference antenna because its average received power $\langle P_2 \rangle$ can be used to determine the field strength in the chamber, as shown by (7.103). The transmitting antenna also has power scattered back to its location and receives power P_1. In order to eliminate the need for a reference antenna, it is necessary to understand how P_1 relates to P_2 or the scattered field strength in the chamber. This relationship has been studied theoretically and experimentally via scattering parameters [98]. The square of the absolute value of the scattering parameter S_{21} is proportional to P_2:

$$|S_{21}|^2 \propto P_2, \tag{7.227}$$

and the same proportionality applies to the ensemble averages:

$$\langle |S_{21}|^2 \rangle \propto \langle P_2 \rangle \tag{7.228}$$

The constant of proportionality is not required for this analysis. In general, it depends on the characteristics of the receiving antenna and the chamber.

7.9.1 Geometrical Optics Formulation

The simplest way to compare the scattering parameters S_{11} (whose square is proportional to the power scattered back to the transmitting antenna) and S_{12} is via geometrical optics. This generally provides a good approximation because the relevant dimensions (chamber size, stirrer size, and antenna separation) are electrically large. The scattering parameter S_{21} can be approximated by a large, but finite, number N of rays:

$$S_{21} = \sum_{p=1}^{N} A_p \frac{\exp(ikr_p)}{r_p}, \tag{7.229}$$

where r_p is the length of the pth ray and A_p is a complex coefficient that takes into account the antenna patterns and the reflection characteristics of the chamber walls and stirrer. A typical ray from antenna 1 to antenna 2 is shown in Figure 7.22. The number of rays is finite because the imperfect conductivity of the chamber walls and stirrer is taken into account [99]. Because we assume a well stirred field, the average value of S_{21} is zero:

$$\langle S_{21} \rangle = 0 \tag{7.230}$$

From the central limit theorem [57] or from maximum entropy [69], we can determine that the real and imaginary parts of S_{21} are Gaussian distributed.

Using (7.229), we write the square of the absolute value of S_{21} as:

$$|S_{21}|^2 = S_{21}S_{21}^* = \sum_{p=1}^{N} A_p \frac{\exp(ikr_p)}{r_p} \sum_{q=1}^{N} A_q^* \frac{\exp(-ikr_q)}{r_q} \tag{7.231}$$

Because of the randomness in the ray paths, the rays for $p \neq q$ are uncorrelated. Hence the average value of (7.231) is:

$$\langle |S_{21}|^2 \rangle = \left\langle \sum_{p=1}^{N} \frac{|A_p|^2}{r_p^2} \right\rangle = N \left\langle \frac{|A_p|^2}{r_p^2} \right\rangle \tag{7.232}$$

The second average $\langle\ \rangle$ in (7.232) is actually over both ensemble (stirrer position) and ray number p. As with the received power, the probability density function of $|S_{21}|^2$ is exponential.

For the scattering parameter S_{11}, the transmitting and receiving locations are identical. Hence, reciprocity [94] requires that every ray has a companion ray traveling the same path in the opposite direction, as indicated in Figure 7.22. So S_{11} has half as many separate rays as S_{21}, but each ray contribution is doubled because the two reciprocal rays add in phase:

$$S_{11} = \sum_{p=1}^{N/2} 2A_p \frac{\exp(ikr_p)}{r_p} \tag{7.233}$$

As with S_{21}, the average value of S_{11} is zero:

$$\langle S_{11} \rangle = 0 \tag{7.234}$$

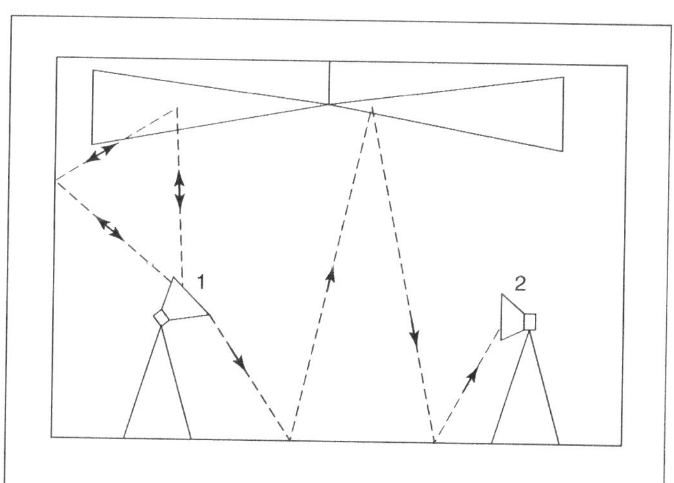

FIGURE 7.22 A typical ray propagating from antenna 1 to antenna 2, and typical back-scattered and reciprocal rays for antenna 1 in a reverberation chamber [98].

From the central limit theorem [57] or from maximum entropy [69], we can determine that the real and imaginary parts of S_{11} are also Gaussian.

Using (7.233), we can write the square of the absolute value of S_{11} as:

$$|S_{11}|^2 = S_{11}S_{11}^* = 4\sum_{p=1}^{N/2} A_p \frac{\exp(ikr_p)}{r_p} \sum_{q=1}^{N/2} A_q^* \frac{\exp(-ikr_q)}{r_q} \tag{7.235}$$

Because of the randomness in the ray paths, the rays for $p \neq q$ are again uncorrelated. Hence, the average of (7.235) is:

$$\langle |S_{11}|^2 \rangle = 4\left\langle \sum_{p=1}^{N/2} \frac{|A_p|^2}{r_p^2} \right\rangle = 2N\left\langle \frac{|A_p|^2}{r_p^2} \right\rangle \tag{7.236}$$

Comparing (2.232) with (2.236), we achieve the desired result:

$$\langle |S_{11}|^2 \rangle = 2\langle |S_{21}|^2 \rangle \tag{7.237}$$

The result in (7.237) is completely analogous to enhanced backscatter [100–102] that has been analyzed for scattering by a random medium, also yielding a factor of 2 for the increase in the backscattered intensity. The physical mechanism, coherent addition of reciprocal rays in the backscatter direction, is the same in both reverberation chambers and in scattering by random media.

An experimental verification of (7.237) is shown in Figure 7.23. The data [66] were taken in the NASA chamber (14 × 7 × 3 m). The agreement with the factor of 2 is good above about 200 MHz where the mode density of the chamber is sufficiently high. The number of samples at each frequency was 225.

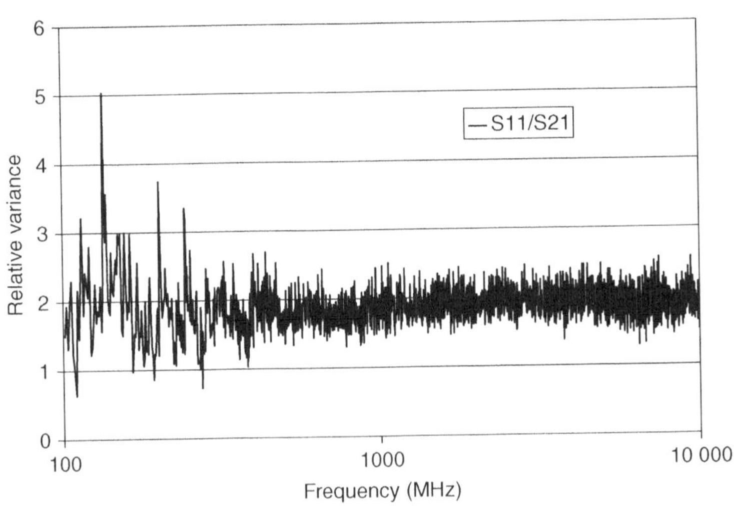

FIGURE 7.23 Ratio of the variances of S_{11} and S_{21} from 100 to 10,000 MHz.

7.9.2 Plane-Wave Integral Formulation

The geometrical optics formulation of enhanced backscatter does well in explaining the factor of 2 in (7.237), but it cannot tell us the size of the region over which enhanced backscatter occurs. To obtain this, we return to the plane-wave integral representation that has been used to describe the spatial and statistical properties of fields in reverberation chambers [69]:

$$\vec{E}(\vec{r}) = \iint_{4\pi} \vec{F}(\Omega)\exp(i\vec{k} \bullet \vec{r})\mathrm{d}\Omega \tag{7.238}$$

The statistical properties of $\vec{F}$ were covered in Section 7.1.

The field representation in (2.238) is valid for a source-free region and requires modification to represent enhanced backscatter at the source. To represent enhanced backscatter for a source at the origin, we replace $\vec{E}$ and $\vec{F}$ in (2.238) by $\vec{E}_e$ and $\vec{F}_e$:

$$\vec{E}_e(\vec{r}) = \iint_{2\pi} \vec{F}_e(\Omega)\exp(i\vec{k} \bullet \vec{r})\mathrm{d}\Omega, \tag{7.239}$$

where $\vec{F}_e(\alpha, \beta) = \vec{F}(\alpha, \beta) + \vec{F}(\alpha', \beta')$, $\alpha' = \pi - \alpha$, and $\beta' = \beta + \pi$. The ranges of α and β in (2.237) are $0 \leq \alpha < \pi/2$ and $0 \leq \beta < 2\pi$. Hence the integration range in (2.237) is reduced to 2π steradians. The geometry for the plane-wave representation is shown in Figure 7.24. Each plane wave propagating in the $\vec{k}$ direction is accompanied by a

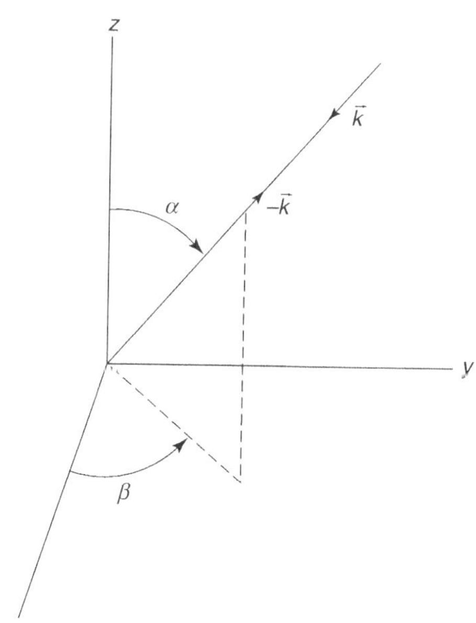

FIGURE 7.24 Geometry for the plane-wave representation of enhanced backscatter.

reciprocal plane wave propagating in the $-\vec{k}$ direction. The average value $\langle \vec{E}_e \rangle$ is zero because the average value $\langle \vec{F}_e \rangle$ is zero.

The square of the magnitude of $\vec{E}_e$ can be written:

$$|\vec{E}_e(\vec{r})|^2 = \iint\limits_{2\pi} \iint\limits_{2\pi} \vec{F}_e(\Omega_1) \bullet \vec{F}_e^*(\Omega_2) \exp[i(\vec{k}_1 - \vec{k}_2)] d\Omega_1 d\Omega_2 \tag{7.240}$$

The ensemble average of (7.240) is:

$$\langle |E_e(\vec{r})|^2 \rangle = \iint\limits_{2\pi} \iint\limits_{2\pi} \langle \vec{F}_e(\Omega_1) \bullet \vec{F}_e^*(\Omega_2) \rangle \exp[i(\vec{k}_1 - \vec{k}_2) \bullet \vec{r}] d\Omega_1 d\Omega_2 \tag{7.241}$$

The mathematics for evaluating the double integral in (7.241) was covered in Sections 7.2 and 7.4 and will not be repeated here. The resulting expression for (7.241) is:

$$\langle |\vec{E}_e(\vec{r})|^2 \rangle = E_0^2 \left[1 + \frac{\sin(2kr)}{2kr} \right] \tag{7.242}$$

At large kr, the mean-square electric field reduces to E_0^2, which is consistent with the uniform-field result in (7.14). For $kr = 0$, (7.242) reduces to:

$$\langle |\vec{E}(0)|^2 \rangle = 2E_0^2 \tag{7.243}$$

Because the average power received by an antenna is proportional to the mean-square electric field as shown in (7.103), the factor of 2 in (7.243) is consistent with the factor of 2 in (7.237).

We can arbitrarily define the region of enhanced backscatter as the distance r_e from the origin for which the value of (7.242) drops to E_0^2:

$$2kr_e = \pi \quad \text{or} \quad r_e = \pi/(2k) = \lambda/4 \tag{7.244}$$

Hence, the region of enhanced backscatter is fairly small (a sphere of radius $\lambda/4$). Beyond that, the mean-square field rapidly approaches its uniform-field value of E_0^2. Thus a receiving antenna will typically be in the statistically uniform field region and will not see an enhanced backscatter effect.

PROBLEMS

7-1 Derive (7.9) and (7.10) from (7.6)–(7.8).

7-2 Derive (7.15). Is this consistent with (7.14)?

7-3 Derive (7.20). Show that a single deterministic plane wave satisfies the same relationship regardless of the propagation direction.

7-4 Following the general approach in (7.33)–(7.35), show that the following correlations are also zero:

$$\langle E_{yr}(\vec{r})E_{yi}(\vec{r})\rangle = \langle E_{zr}(\vec{r})E_{zi}(\vec{r})\rangle = \langle E_{xr}(\vec{r})E_{yr}(\vec{r})\rangle = \langle E_{xi}(\vec{r})E_{yi}(\vec{r})\rangle = 0.$$

7-5 Define the reverberation-chamber electric field in the xy-plane as $\vec{E}_p = \hat{x}E_x + \hat{y}E_y$. How many degrees of freedom does $\vec{E}_p$ have? Determine the probability density functions for $|\vec{E}_p|$ (chi PDF) and $|\vec{E}_p|^2$ (chi-square PDF).

7-6 Derive (7.52) from (7.50).

7-7 Derive (7.62) from (7.61).

7-8 Verify that (7.48), (7.58), and (7.63) satisfy (7.67).

7-9 Derive (7.68).

7-10 Derive (7.73) from (7.72). Derive the small argument approximation in (7.75) from (7.74).

7-11 Derive (7.80) from (7.78) and (7.79).

7-12 Derive (7.82) from (7.81) and (7.79).

7-13 Derive (7.84) from (7.83) and (7.79).

7-14 How many degrees of freedom does the energy density W in (7.85) have? What is the probability density function of W?

7-15 Derive (7.91) from (7.90).

7-16 In the derivation for the average power received by an antenna in a reverberation chamber, fill in the steps from (7.99) to (7.103).

7-17 Consider two reverberation chambers of identical size and shape: one with copper walls ($\sigma_W = 5.7 \times 10^7$ S/m, $\mu_r = 1$) and one with steel walls ($\sigma_W = 10^6$ S/m, $\mu_r = 2000$). (Steel properties vary greatly depending on the particular alloy.) From (7.123), determine the ratio of the Q_1 due to wall loss for the two chambers.

7-18 The NIST rectangular reverberation chamber has dimensions 2.74 m × 3.05 m × 4.57 m. For a matched receiving antenna ($m = 1$), compare the value of Q_4(7.132) at frequencies of 200 MHz and 10 GHz.

7-19 Derive (7.146) from (7.145).

7-20 From (7.161), derive (7.162) for the normal electric field at the wall boundary.

7-21 Derive (7.166) from (7.165). From (7.166), derive the first nonzero term in the small argument (ky) expansion of $\left\langle |E_x^t(x, y, z)|^2 \right\rangle$.

7-22 Derive (7.170) from (7.169). From (7.170) derive the first nonzero term in the small argument (ky) expansion of $\left\langle |H_y^t(x, y, z|^2 \right\rangle$.

7-23 Derive (7.174) from (7.173).

7-24 Derive (7.187) from (7.186). Show that $\left\langle |E_z^t(0,y,z)|^2 \right\rangle = \left\langle |E_z^t(x,0,z)|^2 \right\rangle = 0$.

7-25 Derive (7.190) from (7.189).

7-26 From (7.191), derive the first nonzero term in the small argument (ky) expansion of $\left\langle |E_x^t(0,y,z)|^2 \right\rangle$.

7-27 Derive (7.194). From (7.194), show that $\left\langle |H_z^t(0,0,z)|^2 \right\rangle = \frac{4E_0^2}{3\eta^2}$.

7-28 Derive (7.201). From (7.201), show that $\left\langle |H_x^t(0,y,z)|^2 \right\rangle = 0$.

7-29 Derive (7.212). Show that $\left\langle |E_z^t(0,y,z)|^2 \right\rangle = \left\langle |E_z^t(x,0,z)|^2 \right\rangle = 0$.

7-30 Derive (7.217). Show that $\left\langle |H_z^t(x,y,0)|^2 \right\rangle = 0$.

7-31 Derive (7.242) from (7.241). Show that the squared magnetic field satisfies the analogous expression: $\langle |\vec{H}_e(\vec{r})|^2 \rangle = \frac{E_0^2}{\eta^2}\left[1 + \frac{\sin(2kr)}{kr}\right]$.

CHAPTER 8

Aperture Excitation of Electrically Large, Lossy Cavities

In many electromagnetic interference problems, the important electronic systems are located within a metal enclosure with apertures. In such cases, it is important to know the shielding effectiveness (SE) of the enclosure so that we can relate the interior fields to the external incident fields. The purpose of this chapter is to develop a mathematical model [38] for the shielding effectiveness of electrically large enclosures that contain apertures and interior loading. The method that we present uses a power-balance approach, and much of the mathematical formalism follows that of the reverberation chamber from Chapter 7. The main difference is that the source is an external field incident on an aperture rather than an internal antenna.

8.1 APERTURE EXCITATION

Consider a time-harmonic plane wave of power density S_i incident on the shield apertures, as shown in Figure 8.1. (S_i is actually the magnitude of the incident vector power density.) If the total transmission cross section of the apertures is σ_t, the power P_t transmitted into the cavity is:

$$P_t = \sigma_t S_i \tag{8.1}$$

(Of course power will also leak out through the apertures, but we lump that effect under leakage loss P_{d3} in the cavity Q analysis covered in Section 7.6.) For the general case of N apertures, σ_t can be written as a sum:

$$\sigma_t = \sum_{i=1}^{N} \sigma_{ti}, \tag{8.2}$$

where σ_{ti} is the transmission cross section of the ith aperture. In general, σ_{ti} and σ_t depend on the frequency, incidence angle, and polarization of the incident field.

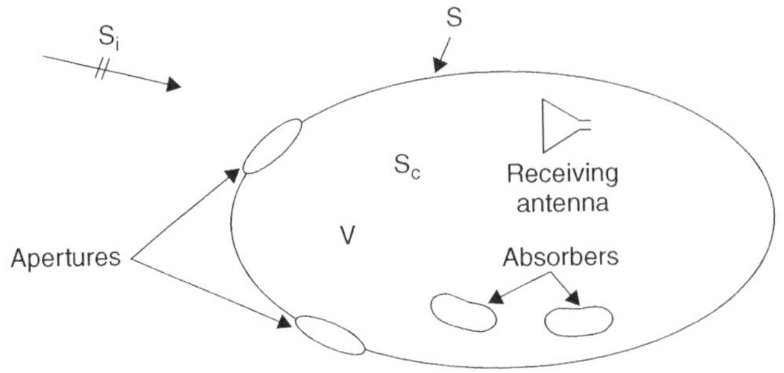

FIGURE 8.1 Aperture excitation of a cavity containing absorbers and a receiving antenna [41].

In many practical applications, the incidence angle and polarization are unknown and are best treated as random. This case is well treated experimentally by illuminating the cavity in a reverberation chamber [38]. Then the transmitted power can be written:

$$P_t = \langle \sigma_t \rangle S_i / 2 \tag{8.3}$$

The factor $\frac{1}{2}$ in (8.3) results from shadowing of the incident field by the electrically large enclosure and is a good approximation for convex shields. The average $\langle\,\rangle$ is over stirrer position for reverberation chamber measurements or over incidence angle and polarization for calculations. The average value of the transmission cross section for N apertures is obtained directly from (8.2):

$$\langle \sigma_t \rangle = \sum_{i=1}^{N} \langle \sigma_{ti} \rangle \tag{8.4}$$

8.1.1 Apertures of Arbitrary Shape

Consider a plane wave incident on an aperture in a perfectly conducting sheet, as shown in Figure 8.2. For convenience, we drop the subscript i that identifies the i_{th} aperture in the shield. Aperture theory has been developed primarily for apertures in flat, perfectly conducting screens of infinite extent and zero thickness [103]. Here we assume that the shield is locally planar and that the shield thickness is small. Aperture theory can be subdivided into three cases, where the aperture dimensions are either small, comparable, or large compared to the wavelength.

For electrically large apertures, the geometrical optics approximation yields:

$$\sigma_t = A \cos \theta^i, \tag{8.5}$$

where A is the aperture area and θ^i is the incident elevation angle. Thus σ_t is independent of frequency, polarization, and azimuth angle of the incident field.

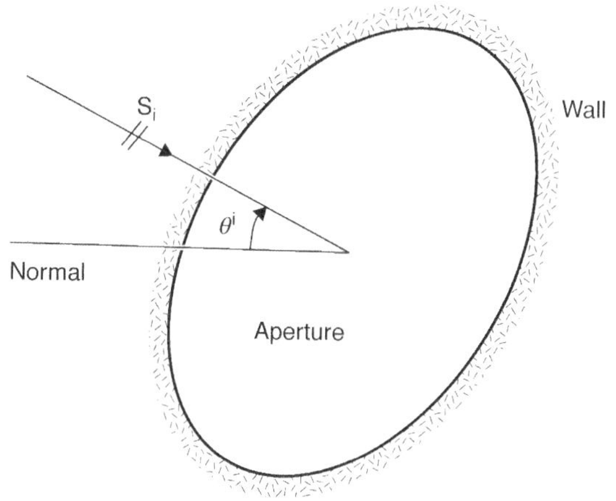

FIGURE 8.2 External field incident on an aperture of arbitrary shape [41].

For this case the average transmission cross section can by written:

$$\langle \sigma_t \rangle = \frac{1}{2\pi} \int_0^{2\pi} \mathrm{d}\phi^i \int_0^{\pi/2} A\cos\theta^i \sin\theta^i \mathrm{d}\theta^i = A/2, \tag{8.6}$$

where we restrict θ^i to angles less than $\pi/2$ because the field is incident from only one side of the screen.

For electrically small apertures, polarizability theory states that the transmitted fields are those of induced electric and magnetic dipole moments [103, 104]. This theory yields a transmission cross section that is proportional to frequency to the fourth power:

$$\sigma_t = Ck^4, \tag{8.7}$$

where C depends on incidence angle and polarization and aperture size and shape, but is independent of frequency. The wavenumber $k = \omega/c$. The specific form of C for a circular aperture will be given in the following section.

In the resonance region, the aperture dimensions are comparable to a wavelength, and the frequency dependence of σ_t depends on the aperture shape. Numerical methods [89] can be used to compute σ_t for such cases, but we will not pursue such methods here.

8.1.2 Circular Aperture

The circular aperture is of particular interest because it has an analytical solution and is easy to work with experimentally. The geometry for a circular aperture of radius a is shown in Figure 8.3. An exact solution for the transmission coefficient is available

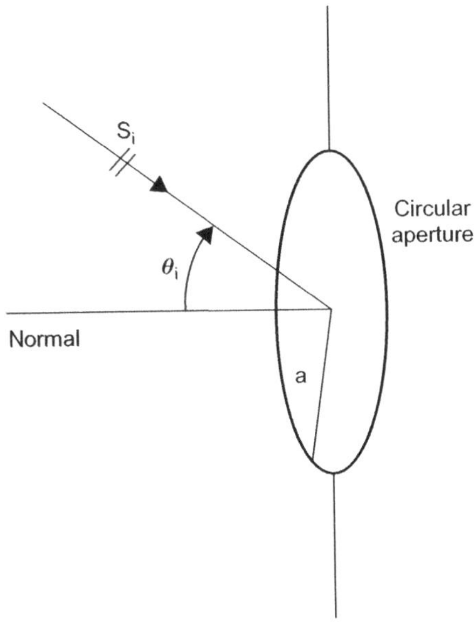

FIGURE 8.3 External field incident on a circular aperture of radius a [41].

in terms of spheroidal functions [104], but we choose to construct a simpler solution in terms of the approximations that are available for electrically large and small circular apertures.

For electrically large circular apertures, the geometrical optics approximations in (8.5) and (8.6) yield the following expressions for the transmission cross section and the averaged transmission cross section:

$$\sigma_t = \pi a^2 \cos\theta^i \quad \text{and} \quad \langle \sigma_t \rangle = \pi a^2/2 \tag{8.8}$$

For electrically small circular apertures, polarizability theory [103] can be used to determine the effective electric and magnetic dipole moments and the resultant transmission cross section. The details are given in Appendix I. The transmission cross section depends on the polarization and the elevation angle of the incident field. For the electric field polarized parallel to the incidence plane—defined by the incident wave vector and the normal to the aperture—we write the transmission cross section as σ_{tpar}:

$$\sigma_{tpar} = \frac{64}{27\pi} k^4 a^6 \left(1 + \frac{1}{4}\sin^2\theta^i\right) \tag{8.9}$$

For perpendicular polarization, we write the transmission cross section as σ_{tperp}:

$$\sigma_{tperp} = \frac{64}{27\pi} k^4 a^6 \cos^2\theta^i \tag{8.10}$$

Both θ_{tpar} and θ_{tperp} have the k^4 dependence given by (8.7), and they are equal for normal incidence ($\theta^i = 0$). We assume that an incident random field will have equal power densities in the parallel and perpendicular waves. Thus the averaged transmission cross section can be written:

$$\langle \sigma_t \rangle = \frac{1}{2} \int_0^{\pi/2} (\sigma_{tpar} + \sigma_{tperp}) \sin\theta^i \mathrm{d}\theta^i, \tag{8.11}$$

where we have used the fact that the transmission cross sections are independent of the incident azimuth angle. If we substitute (8.9) and (8.10) into (8.11) and carry out the integration over θ^i, we obtain.

$$\langle \sigma_t \rangle = \frac{16}{9\pi} k^4 a^6 \tag{8.12}$$

We do not have a simple expression for the transmission cross section that is valid in the resonance region, but the circular aperture does not have strong resonances [105]. Thus, we choose to cover the entire frequency range by using only the electrically small and electrically large approximations. The crossover wavenumber k_c, where we switch from (8.12) to (8.8) for the average transmission cross section, is given by equating (8.8) and (8.12):

$$\pi a^2/2 = \frac{16}{9\pi} k^4 a^6 \tag{8.13}$$

The solution to (8.13) is:

$$k_c a = (9\pi^2/32)^{1/4} \approx 1.29 \tag{8.14}$$

This technique is not valid for long, narrow apertures, which typically have strong resonances.

8.2 POWER BALANCE

In this section we use the technique of power balance to determine the shielding effectiveness and the decay time of a cavity with apertures. The technique is approximate because it assumes that the scalar power density S_c within the cavity is independent of position. This is consistent with the reverberation chamber analysis in Section 7.2 and will use the expression for scalar power density in (7.28).

8.2.1 Shielding Effectiveness

Consider again the geometry in Figure 8.1, where an incident wave is incident on a shielded cavity with apertures. We wish to determine the scalar power density S_c inside the cavity. For steady-state conditions, we require that the power P_t transmitted through the apertures is equal to the power P_d dissipated in the four loss mechanisms considered earlier in Section 7.6:

$$P_t = P_d \tag{8.15}$$

If we substitute (7.107), (7.108), (7.28), and (8.1) into (8.15), we can solve for the scalar power density S_c in the cavity:

$$S_c = \frac{\sigma_t \lambda Q}{2\pi V} S_i \tag{8.16}$$

Since we have assumed that the scalar power density S_c is uniform throughout the cavity, we can define shielding effectiveness (SE) in terms of the ratio of the incident and cavity power densities:

$$SE = 10\log_{10}(S_i/S_c) = 10\log_{10}\left(\frac{2\pi V}{\sigma_t \lambda Q}\right) \text{dB} \tag{8.17}$$

The results in (8.16) and (8.17) are consistent with a related treatment of this problem [106]. We have defined SE to be greater than one (or positive in dB) when the cavity power density is less than the incident power density. The result for SE in (8.17) depends on the cavity volume and Q in addition to the transmission cross section σ_t. A computer code to evaluate SE and Q is included in [41].

The results in (8.16) and (8.17) apply to a single incident plane wave where σ_t depends on the incident direction and polarization. For the case of uniformly random incidence (as in a reverberation chamber), we need to replace σ_t in (8.16) and (8.17) by one-half the averaged value, $\langle \sigma_t \rangle / 2$.

The Q enhancement of the cavity power density is clear in (8.16) and (8.17), and we can see that a lossy cavity (low Q) has a greater shielding effectiveness than a high-Q cavity. The significance of loss is seen if we consider the special case where the cavity is lossless ($P_{d1} = P_{d2} = P_{d4} = 0$), except for leakage. In this case, Q is given by:

$$Q = Q_3 = \frac{4\pi V}{\lambda \langle \sigma_l \rangle} \tag{8.18}$$

If we substitute (8.18) into (8.16), we obtain:

$$S_c = S_i \frac{2\sigma_t}{\langle \sigma_l \rangle} \tag{8.19}$$

For the case of uniformly random excitation, the transmission cross section is replaced by one half the averaged cross section. However, the averaged transmission cross section is equal to the averaged leakage cross section ($\langle \sigma_t \rangle = \langle \sigma_l \rangle$), and so (8.19) reduces to:

$$S_c = S_i \quad \text{or} \quad \text{SE} = 0\,\text{dB} \tag{8.20}$$

Thus the leakage loss equals the transmitted power, and the cavity has zero shielding. This result is independent of the aperture size and shape. Physically, this case corresponds to an apertured (but otherwise lossless) cavity inside a reverberation chamber. We expect real cavities to have additional losses (such as wall loss) and hence positive values of SE (in dB).

8.2.2 Time Constant

Up to this point we have considered only steady-state, single-frequency excitation. Since pulses are important in some applications (for example a radar beam incident on an aircraft), we also need to consider transient effects. In general, this is a complex problem that is best handled with Fourier integral techniques. However, we can analyze the special case of a turned-on or turned-off sinusoid in a simpler manner.

We consider first the case of field decay where the source (the incident power density) is instantaneously turned off. By equating the change in the cavity energy U to the negative of the dissipated power over a time increment $\mathrm{d}t$, we obtain the differential equation:

$$\mathrm{d}U = -P_d \mathrm{d}t \tag{8.21}$$

We can use (7.108) to replace P_d in (8.21):

$$\mathrm{d}U = -(\omega U/Q)\mathrm{d}t = -\frac{U}{\tau}\mathrm{d}t, \tag{8.22}$$

where the time constant $\tau = Q/\omega$. The initial condition is $U = U_s$ at $t = 0$. The solution of (8.22) with this initial condition is:

$$U = U_s \exp(-t/\tau), \quad t > 0 \tag{8.23}$$

The time constant τ has been measured [38, 41] by fitting the decay curve in (8.23) to experimental data. Once τ has been determined, the frequency dependent Q is determined from:

$$Q = \omega\langle\tau\rangle, \tag{8.24}$$

where the average time constant $\langle\tau\rangle$ is used to measure Q. Equation (8.24) was used to measure Q, and comparisons with theory were shown in Figures 7.13 and 7.14.

The closely related case of a turned-on (step-modulated) incident power density involves the same exponential function and time constant:

$$U = U_s[1-\exp(-t/\tau)], \quad t > 0 \tag{8.25}$$

The cavity energy density and scalar power density also follow the same exponential variation with the same time constant, and (8.23) and (8.25) agree with [91]. If a radar pulse duration is long compared to τ, then the cavity fields will reach their steady-state values. However, if the pulse length is short compared to τ, the fields will not reach their steady-state values before the incident pulse is turned off. Some common radars and their pulse characteristics are described in [41].

A high-Q (long-τ) cavity might have poor steady-state SE, but would require a long period for the cavity fields to reach their steady-state values. Physically, a high Q (long τ) means that waves make many reflections within the cavity before they decay.

8.3 EXPERIMENTAL RESULTS FOR SE

Measurements were made on two cavities with apertures. Both cavities were rectangular, with walls made of aluminum. Aluminum was chosen because it has a high electrical conductivity and is easy to weld. Because of the uncertainty in handbook values of the electrical conductivity of aluminum, a conductivity measurement was made at NIST using a parallel-plate, dielectric resonator technique. The measured result for the conductivity was 8.83×10^6 S/m. This value was lower than handbook values, but was considered more reliable than handbook values made at dc. It also gave better agreement with theoretical Q values shown in Figures 7.13 and 7.14.

A rectangular cavity of dimensions 0.514 m × 0.629 m × 1.75 m was constructed at NIST [41]. The cavity was selected to have sufficient mode density at frequencies above 1 GHz, but to be light enough to be manageable. The geometry was as shown in Figure 8.4. The circular aperture had a radius of 1.4 cm. The stirrer was made of the same type of aluminum as was used for the cavity walls. Various numbers of salt water spheres of radius 6.6 cm were used for cavity loading. The salt concentration was selected as that of sea water so that the electrical properties as given by Saxton and Lane [107] could be used in the theory.

For SE measurements, the cavity was placed in the NIST reverberation chamber [19]. Both the reverberation chamber and cavity fields were stirred, and the measured SE in dB was taken as the average power received in the reverberation chamber minus the average power received in the cavity. This value was compared with the theoretical result in (8.17).

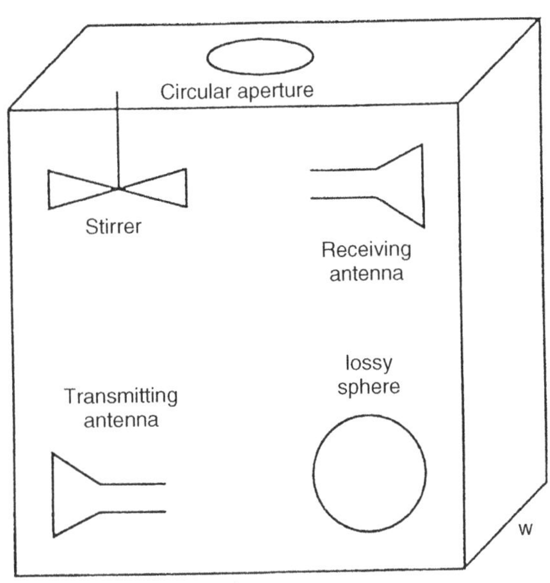

FIGURE 8.4 Rectangular cavity with a circular aperture, a mode stirrer, receiving and transmitting antennas and lossy sphere(s) [38].

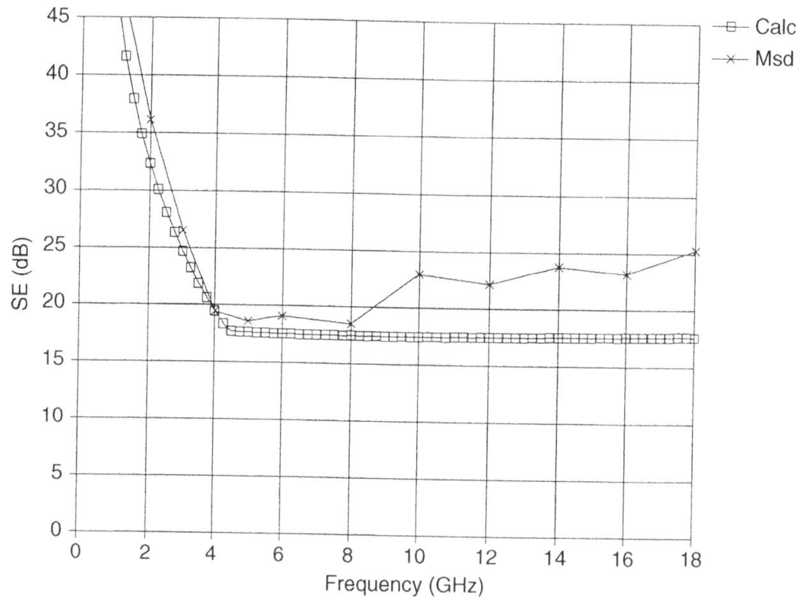

FIGURE 8.5 Calculated and measured values of SE for the rectangular cavity of Figure 8.4 with an aperture radius of 0.014 m, two antennas, and one sphere of radius 0.066 m filled with salt water for absorptive loading [41].

The first comparison of measurement and theory [41] in Figure 8.5 was for the case of a single salt-water sphere for absorptive loading. The theory for the absorption cross section of a lossy sphere is given in Appendix H. Double-ridged horn antennas were used in both the reverberation chamber and the cavity because of their wide bandwidth, 1 to 18 GHz. The agreement between theory and measurement below 8 GHz is typical of that for stirred fields, but the disagreement above 8 GHz is larger than expected.

In Figure 8.6, the cavity was loaded with three salt water spheres. The agreement is slightly better than that of Figure 8.5 at the high frequencies. Also, the SE is larger because of the lower Q, as predicted by (8.17). A practical consequence of this result is that the SE of a cavity can be increased by loading the cavity with lossy material.

A related set of measurements was made with standard-gain, K_u-band horns. These antennas have a high efficiency of about 98 %. This comparison was done because the efficiency of the broadband, double-ridged horns was suspected to be fairly low. A comparison of measured and calculated SE is shown in Figure 8.7. The agreement is improved over the broadband, double-ridged horn results in Figures 8.5 and 8.6. The SE values are also lower because no absorptive loading by salt water spheres was included. Hence the cavity Q was higher.

SE measurements were also performed by Hatfield [41] on a rectangular cavity with a circular aperture placed in the Naval Surface Warfare Center (NSWC) reverberation chamber. The cavity contained a broadband, double ridged receiving

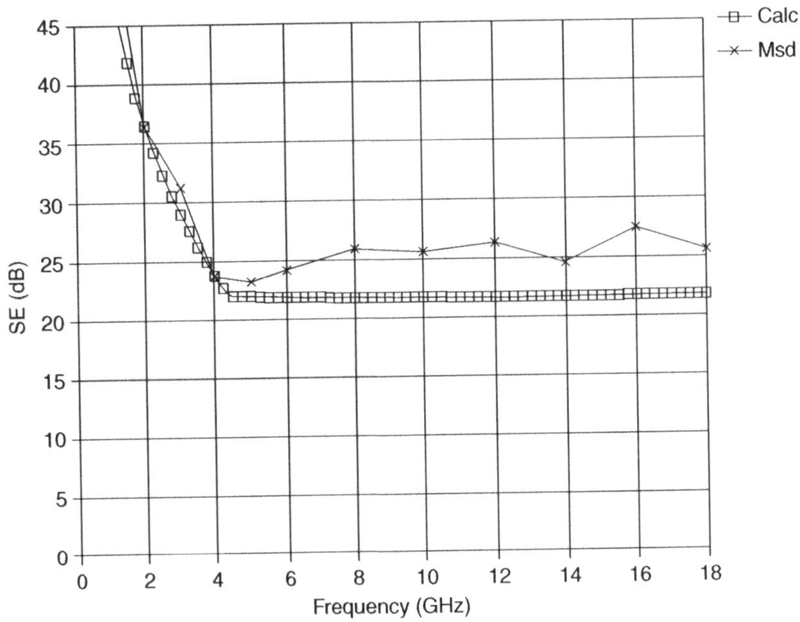

FIGURE 8.6 Calculated and measured values of SE for the rectangular cavity of Figure 8.4 with an aperture radius of 0.014 m, two antennas, and three spheres of radius 0.066 m filed with salt water for increased absorptive loading [41].

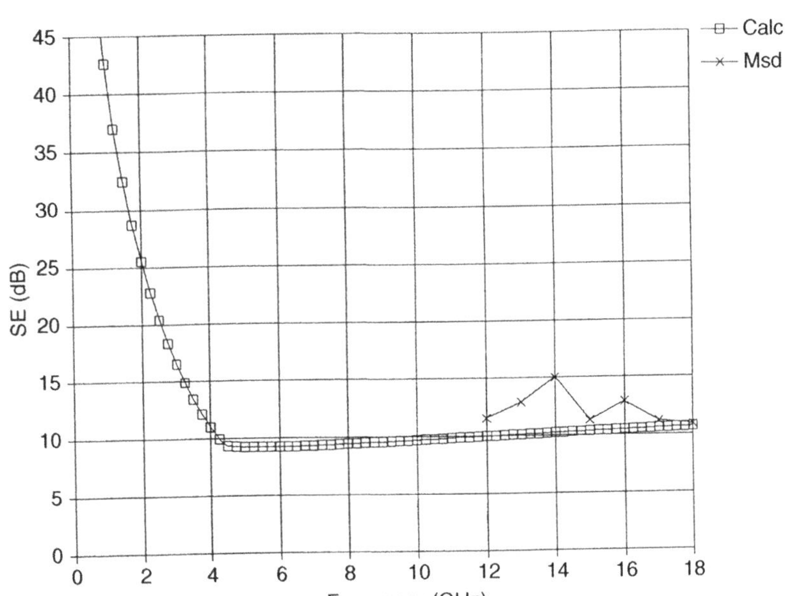

FIGURE 8.7 Calculated and measured values of SE for the rectangular cavity of Figure 8.4 with an aperture radius of 0.014 m, and two K_u-band horn antennas [41].

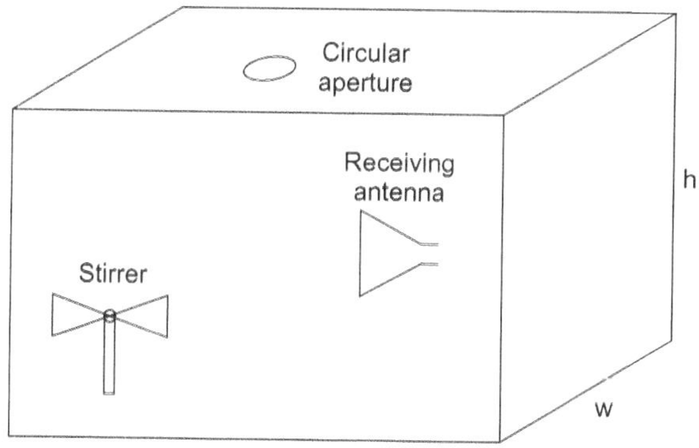

FIGURE 8.8 NSWC rectangular cavity with a circular aperture. A mode stirrer and a receiving antenna are located inside [41].

horn and a mode stirrer as shown in Figure 8.8. No absorptive loading was included. The cavity was made of the same aluminum alloy (conductivity equals 8.83×10^6 S/m) with dimensions: $l = 1.213$ m, $w = 0.603$ m, and $h = 0.937$ m. Two different aperture radii, $a = 2.94$ cm and 3.51 cm, were used.

SE measurements were made from 200 MHz to 18 GHz for both apertures. A comparison with calculated SE values is shown in Figure 8.9. The theory is not

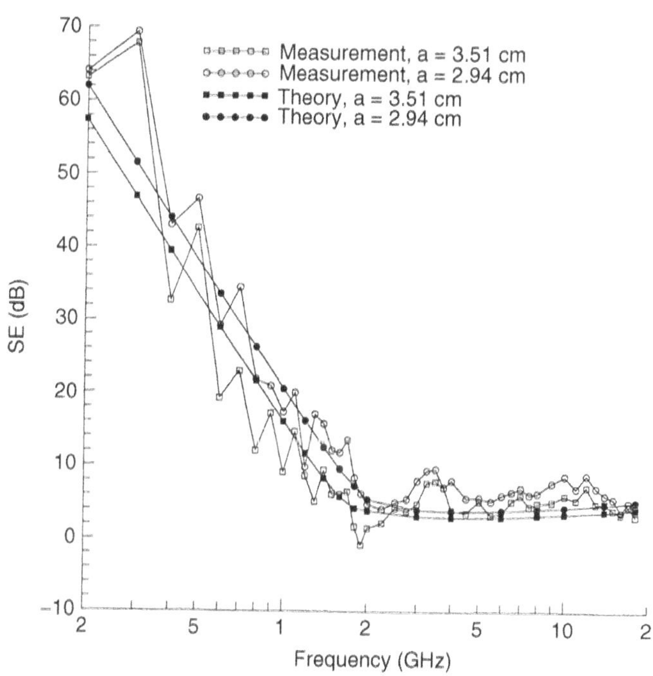

FIGURE 8.9 Comparisons of calculated and measured SE for the NSWC rectangular cavity with two different aperture radii [41].

expected to be valid below 400 MHz because the cavity is not electrically large (the mode density is too low). The measured values show rapid variations with frequency which do not appear in the smooth theory, but rapid variations with frequency are typical of reverberation chamber measurements [19]. The general agreement between theory and measurements is good above 400 MHz for both aperture sizes. The smaller aperture yields greater SE, but the high frequency theoretical SE is low for both apertures. The reason for the decrease in SE with frequency is the increase in both cavity Q and transmission cross section with frequency. Equation (8.17) shows the dependence of SE on both quantities.

A more recent set of data [108] was taken on an aluminum box of dimensions $0.73\,\text{m} \times 0.93\,\text{m} \times 1.03\,\text{m}$ with five 1.6 cm-diameter circular holes punched at random locations in each of the six sides of the box (a total of 30 holes). The box included a paddle (stirrer) and a single receiving antenna. Illumination was with a single approximately plane wave, and the box could be rotated to change the incidence angle. The experimental data generally followed the theory of this section, and, in addition, statistics were also checked to confirm that the interior of the box behaved statistically like a reverberation chamber. This means that the received power probability density function should be exponential [18] as in (7.37). For an exponential PDF, the coefficient of variance (COV), the ratio of the variance to the mean, should equal one. Figure 8.10 shows COV as a function of frequency for a fixed incidence angle and polarization, and the results are centered close to one as expected. Figure 8.11 shows COV as a function of azimuthal incidence angle for a fixed frequency of 3 GHz, and the results are again centered close to one. These results

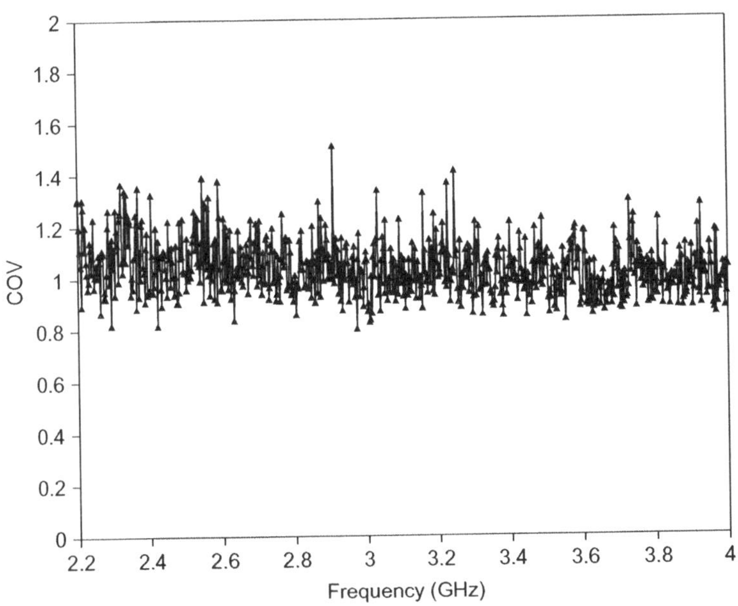

FIGURE 8.10 Coefficient of variance as a function of frequency [108].

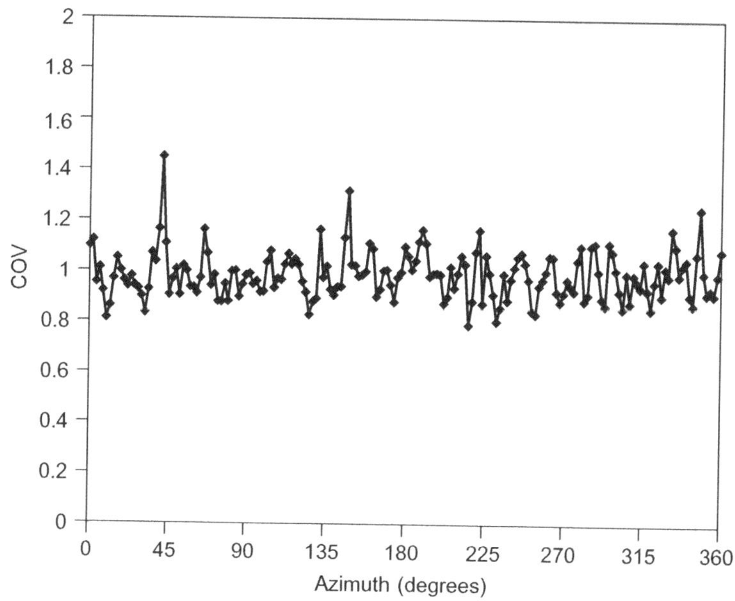

FIGURE 8.11 Coefficient of variance as a function of azimuthal angle [108].

are at least a partial confirmation that reverberation chamber theory is applicable to an apertured cavity illuminated by an external source [108].

The measured results presented in this section could be scaled in size and frequency to match those of practical cavities (such as aircraft). However, the electrical properties of the walls and absorbers would also need to be scaled; the required scaling relationships are discussed in Appendix J.

PROBLEMS

8-1 Derive (8.12) from (8.9)–(8.11).

8-2 Derive (8.16) from (7.107), (7.108), (7.28), (8.1), and (8.16).

8-3 Consider an empty cubic cavity, one meter on a side, with a one-centimeter radius circular aperture, and a Q of 10^4. Calculate the shielding effectiveness (SE) for uniformly random illumination.

8-4 Consider a closed empty cubic cavity so that the Q is determined by wall loss. The cavity is constructed of copper ($\sigma_W = 5.7 \times 10^7$ S/m and $\mu_W = \mu_0$). Calculate the time constant for a turned-off sinusoid of frequency of 10 GHz.

8-5 When the cavity Q is determined by wall loss as in the previous problem, what is the frequency dependence of the time constant?

8-6 Consider the cavity in Problem 8-4. If we wish to increase the cavity dimensions by a factor of 10, what frequency should be used to maintain the electromagnetic behavior of the cavity? (See Appendix J on scaling relations.) What wall conductivity is needed for the large scaled cavity?

8-7 Compare the skin depths of the cavities in Problems 8-4 and 8-6. Does the skin depth also increase by a factor of 10?

8-8 Verify that the Q of the cavity in Problem 8-4 equals the Q of the scaled cavity in Problem 8–6.

8-9 Compare the time constants of the cavities in Problems 8-4 and 8-5. Do they satisfy the equation for τ in (8.22)?

CHAPTER 9

Extensions to the Uniform-Field Model

In the two previous chapters on reverberation chambers and aperture excitation of electrically large, lossy cavities, we dealt with statistically uniform fields for which we could derive probability density functions for the quantities of interest. In this chapter, we examine cases where we do not necessarily have statistically uniform fields.

9.1 FREQUENCY STIRRING

Mechanical mode stirring can be very effective [19,66], but it is fairly slow. In their analysis of mechanical stirring, Wu and Chang [109] pointed out that a rotating mechanical stirrer continuously changes the resonant frequencies of the cavity modes and that mechanical stirring has some equivalence to frequency modulation of the source. Loughry [90] made statistical predictions of the field homogeneity achieved by frequency stirring and performed comparison measurements using a band-limited, white-Gaussian-noise source. Crawford et al. [49] made band-limited, white-Gaussian-noise measurements of radiated immunity of various test objects in a reverberation chamber. In this section, we will study the theory of frequency stirring in an idealized two-dimensional cavity with line-source excitation [48].

9.1.1 Green's Function

The geometry for an electric line source of current I_0 located at (x_0, y_0) in a two-dimensional rectangular cavity $(a \times b)$ is shown in Figure 9.1. The cavity region has permittivity ε and permeability μ (usually the free-space values). Initially, the cavity walls are assumed to be perfect electric conductors so that the tangential electric field is zero at the cavity walls.

The fields are independent of z $(\partial/\partial z = 0)$ and have $\exp(-i\omega t)$time dependence, which is suppressed. (Later we will introduce a nonzero bandwidth which is required for frequency stirring.) For a real, three-dimensional cavity $(a \times b \times c)$, a realistic source will excite additional modes with z variation not included in this analysis.

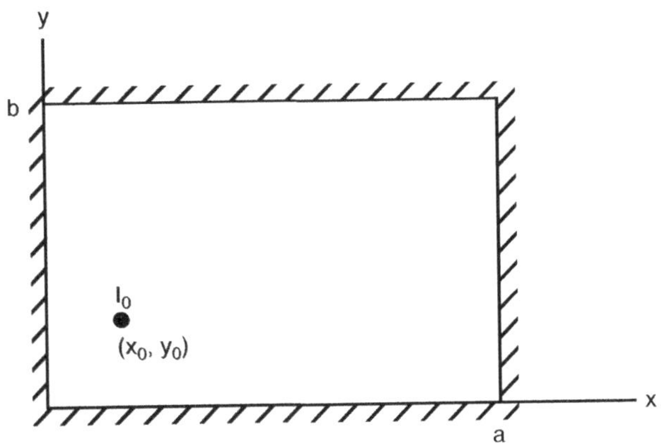

FIGURE 9.1 Geometry for an electric line source in a two-dimensional, rectangular cavity [48].

The nonzero field components are E_z, H_x, and H_y, and the magnetic field components can be derived from the z-directed electric field E_z:

$$H_x = \frac{-1}{i\omega\mu}\frac{\partial E_z}{\partial y} \quad \text{and} \quad H_y = \frac{1}{i\omega\mu}\frac{\partial E_z}{\partial x}. \tag{9.1}$$

The Green's function (for E_z) must satisfy the following scalar equation:

$$\left(\frac{\partial^2}{\partial x^2} + \frac{\partial^2}{\partial y^2} + k^2\right)E_z = -i\omega\mu I_0 \delta(x-x_0)\delta(y-y_0), \tag{9.2}$$

where $k^2 = \omega^2\mu\varepsilon$ and δ is the Dirac delta function. (E_z is used rather than the usual G notation for the Green's function because there is no need to integrate over an extended source region to obtain the electric field E_z.) To make the solution of (9.2) unique, the condition $E_z = 0$ is enforced at the cavity boundaries.

Using standard separation of variables techniques [110], (9.2) can be solved for E_z in the following double summation form:

$$E_z = \frac{-4i\omega\mu I_0}{ab}\sum_{m=1}^{\infty}\sum_{n=1}^{\infty}\frac{\sin(m\pi x_0/a)\sin(m\pi x/a)\sin(n\pi y_0/b)\sin(n\pi y/b)}{k^2-(m\pi/a)^2-(n\pi/b)^2} \tag{9.3}$$

The denominator of (9.3) has zeros at cavity resonance frequencies f_{mn} given by:

$$f_{mn} = (v/2)\sqrt{(m/a)^2 + (n/b)^2}, \tag{9.4}$$

where the velocity $v = 1/(\mu\varepsilon)^{1/2}$ and m and n run over all positive integers. The solution in (9.3) includes only sine terms that satisfy the boundary conditions at the cavity walls on a term-by-term basis.

It is possible to sum the n summation (or the m summation) in (9.3) and obtain the following alternative expression [110] for E_z:

$$E_z = \frac{2i\omega\mu I_0}{a}\sum_{m=1}^{\infty}\frac{\sin(m\pi x_0/a)\sin(m\pi x/a)}{k_m\sin(k_m b)} \times \begin{cases} \sin(k_m y_0)\sin[k_m(b-y)], & y > y_0 \\ \sin[k_m(b-y_0)]\sin(k_m y), & y < y_0 \end{cases}, \tag{9.5}$$

where $k_m = \sqrt{k^2-(m\pi/z)^2}$. Both (9.3) and (9.5) agree with the related scalar Green's function in [110].

Because (9.3) and (9.5) apply to a lossless cavity with perfectly conducting walls, they have singularities at the resonant frequencies given by (9.40). No exact solution exists for the physically realistic case of lossy walls, but (9.3) and (9.5) can be modified in a simple way to obtain a fairly accurate solution for the case of fairly high Q. There are several slightly different forms for the finite Q modification, but for large Q they are approximately equivalent. Here loss is introduced by replacing k in (9.3) and (9.5) with the following complex k_c [3,7,34], and [111]:

$$k_c = k\left(1 + \frac{i}{2Q}\right) \tag{9.6}$$

In the following section, an expression for Q will be derived based on wall loss. However, as shown in Section 7.6, other loss mechanisms can lead to a finite Q; so (9.6) can also represent other types of loss.

For computational efficiency, the expression in (9.5) is preferred because it involves only a single sum. The sum is finite for finite Q because k_m becomes complex with the substitution indicated in (9.6):

$$k_m = \left[k^2\left(1 + \frac{i}{2Q}\right)^2 - \left(\frac{m\pi}{a}\right)^2\right]^{1/2} \tag{9.7}$$

Thus both k_m and $\sin(k_m b)$ are nonzero for all real frequencies so that (9.5) remains finite. An examination of the denominator of (9.3) indicates that the 3 dB bandwidth of any given mode is approximately f_{mn}/Q, where f_{mn} is given by (9.40). Computer programs have been written to evaluate E_z from both (9.3) and (9.5), and they have been shown to agree numerically. However, the program based on (9.5) is much faster, because it is a single sum and because the terms decay exponentially for $m > ka/\pi$. This greater computational speed is important later where repeated calculations are performed for many frequencies and observation points.

9.1.2 Uniform-Field Approximations

Before performing field calculations from the mode theory of the previous section, approximate expressions are developed for the cavity Q and scalar power density based on the statistically uniform field approximation used in room acoustics [51] and in Chapter 7 on reverberation chambers. The first assumption is that the line source radiates the same power in the lossy cavity that it does in a free-space environment. If (9.2) is solved for E_z by use of the radiation condition rather than a cavity wall boundary condition, the expression for E_z is [3]:

$$E_z = -\frac{\omega\mu I_0}{4}\mathrm{H}_0^{(1)}(k\rho), \tag{9.8}$$

where $\rho = \sqrt{x^2+y^2}$ and $\mathrm{H}_0^{(1)}$ is the zero-order Hankel function of the first kind [25]. If the asymptotic expression for $\mathrm{H}_0^{(1)}(k\rho)$ for large $k\rho$ is used, then the radiated power density S_r per unit length is:

$$S_r = \frac{|E_z|^2}{\eta} = \frac{|I_0|^2\eta k}{8\pi\rho}, \tag{9.9}$$

where $\eta = \sqrt{\mu/\varepsilon}$. The power radiated per unit length P_r is obtained by multiplying S_r by the circumference $2\pi\rho$:

$$P_r = 2\pi\rho S_r = |I_0|^2\eta k/4 \tag{9.10}$$

The second assumption is that for a well-stirred cavity the scalar power density S $(= |E_z|^2/\eta)$ and the energy density $W\,(= \eta|E_z|^2)$ are statistically uniform throughout the cavity. By conservation of power, the power radiated must equal the power dissipated in the cavity, and Q can be written:

$$Q = \omega U/P_r, \tag{9.11}$$

where U is the stored energy per unit length in the cavity. U can be written:

$$U = \langle W\rangle A = \varepsilon\langle|E_z|^2\rangle A, \tag{9.12}$$

where the cross-sectional area $A = ab$. In deriving (9.12) the stored electric and magnetic energies are assumed to be equal. This equality holds for a lossless cavity at resonance [3] and holds approximately for a stirred, high-Q cavity. By substituting (9.10) and (9.11) into (9.12), the average of the square of the electric field is found to be:

$$\langle|E_z|^2\rangle = |I_0|^2\eta^2 Q/(4ab) \tag{9.13}$$

The missing piece of information in (9.13) is the cavity Q. In general, it is an involved process to calculate Q because it is difficult to account for all the cavity losses as described in Section 7.6. However, if we consider only wall loss for this idealized two-dimensional cavity, we can use the method in Section 7.6 to obtain the analogous result to (7.124):

$$Q = \frac{2A}{\mu_r\delta L}, \tag{9.14}$$

where μ_r is the relative permeability of the wall, δ is the skin depth of the wall, and $L = 2(a+b)$.

9.1.3 Nonzero Bandwidth

If E_z (or H_x or H_y) is computed from (9.5), rapid variations are found to ocurr with x and y due to standing waves or with frequency due to the mode structure. The mode density (as discussed for various cavities in Part I) is an important quantity in understanding the frequency behavior of fields in cavities. The mode density expressions for three-dimensional cavities are well known (and given in Part I), but here the expression for a two-dimensional cavity is needed. Examination of (9.40) shows that the number N of modes with resonant frequency less than f is approximately:

$$N = \pi abf^2/v^2 \tag{9.15}$$

The mode density is the derivative of the number with respect to frequency:

$$\frac{\mathrm{d}N}{\mathrm{d}f} = 2\pi abf/v^2 \tag{9.16}$$

The specific mode density N_s has been defined as the number of modes within the 3-dB bandwidth f/Q resulting from a finite Q [36]:

$$N_s = \frac{f}{Q}\frac{\mathrm{d}N}{\mathrm{d}f} = \frac{2\pi abf^2}{Qv^2} \tag{9.17}$$

Typically, the bandwidth f/Q is not large enough to bring in a significant number of modes to provide a uniform field through mode mixing. Mechanical mode stirring changes the resonant frequencies sufficiently to provide a well-stirred field [19,109].

If the source has a nonzero bandwidth BW, then the number of modes N_{BW} excited is:

$$N_{BW} = 2\pi abfBW/v^2 \tag{9.18}$$

This assumes that BW is somewhat greater than f/Q, but this is required in order to gain any advantage from the nonzero bandwidth. There is some freedom in the type of signal that is actually used to obtain the bandwidth, and Loughry [90] chose to use band-limited, white, Gaussian noise. Here the source spectrum is assumed flat over the bandwidth BW, and the field contributions from any two unequal frequencies are assumed orthogonal. (These assumptions are consistent with Loughry's source.)

Then the mean square field at any point can be written:

$$|E_z|^2 = \frac{1}{BW}\int_{f-BW/2}^{f+BW/2} |E_z(f')|^2 \mathrm{d}f' \tag{9.19}$$

If perfect field uniformity were achieved and if the line source were to radiate the same power that it would in a free-space environment, then (9.19) would agree with (9.13) at all points within the cavity. This suggests that (9.13) be used to normalize

(9.19) to the ideal case and to compute a normalized field given by:

$$|E_{zn}|^2 = \frac{1}{C_n^2 BW} \int_{f-BW/2}^{f+BW/2} |E_z(f')|^2 df', \tag{9.20}$$

where $C_n^2 = |I_0|^2 \eta^2 Q/(4ab)$. The purpose of the following calculations is to see how closely the ideal case ($|E_{zn}|^2 = 1$) is approached as BW is increased.

In Figures 9.2 through 9.5, the normalized electric field (in decibels) is shown as a function of x for a fixed value of y. For cavity dimensions, two dimensions from the NIST reverberation chamber [19] are chosen: $a = 4.57$ m and $b = 3.05$ m. The source location is fixed at $x_0 = y_0 = 0.5$ m. This is consistent with the practice of locating the transmitting antenna in one of the chamber corners, but not too close to the walls. The remaining parameters for Figure 9.2 are $f = 4$ GHz, $Q = 10^5$, and $y = 1.5$ m. The Q value was selected to match the experimental value for the NIST chamber [19]. Two trends are clear as the bandwidth is increased in Figure 9.2. The field variability as a function of x decreases, and the average field approaches 0 dB. This means that frequency stirring is effective both in improving spatial uniformity of the field and in reducing the interaction between the line source and the chamber walls. The second effect is equivalent to providing a free-space environment for the transmitting antenna, thus reducing impedance mismatch effects. The average value and standard deviation of the normalized field and the number of modes excited are

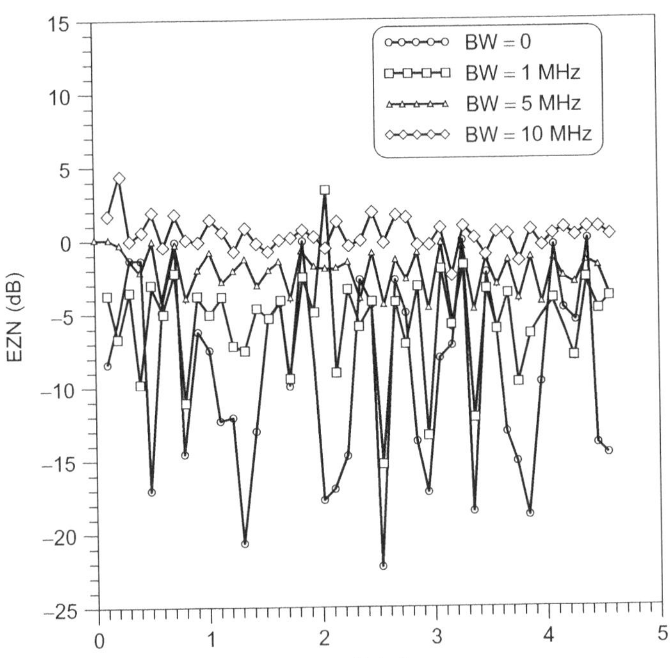

FIGURE 9.2 Normalized electric field magnitude versus x for various bandwidths. Parameters: $f = 4$ GHz, $Q = 10^5$, $y = 1.5$ m, and $x_0 = y_0 = 0.5$ m [48].

TABLE 9.1 Average and Standard Deviation of the Field and the Number of Modes Excited for Various Bandwidths [48].

f (GHz)	*BW* (MHz)	*Q*	*y* (m)	Aver. (dB)	Stand. Dev. (dB)	N_{BW}
4	0.0	1.0×10^5	1.5	−5.81	6.20	0.0
4	1.0	1.0×10^5	1.5	−4.90	3.04	3.9
4	5.0	1.0×10^5	1.5	−1.95	1.54	19.5
4	10.0	1.0×10^5	1.5	0.49	0.88	38.9
4	10.0	1.0×10^5	1.0	0.76	0.72	38.9
4	10.0	1.0×10^5	2.0	0.71	0.89	38.9
4	10.0	5.0×10^4	1.5	0.46	0.98	38.9
4	10.0	2.0×10^5	1.5	0.51	0.85	38.9
8	0.0	1.5×10^5	1.5	−4.83	5.13	0.0
8	1.0	1.5×10^5	1.5	2.04	2.69	7.8
8	5.0	1.5×10^5	1.5	0.30	1.27	38.9

given in Table 9.1 for each curve in Figures. 9.2 through 9.5. The number of modes, as determined from (9.18), is not necessarily an integer because (9.18) is an approximate asymptotic expression. If discrete mode counting had been used, as in [9], then the number of modes would have been an integer.

In Figure 9.3 similar results are shown for a higher frequency of 8 GHz. An increase in Q to 1.5×10^5 reflects the usual increase in chamber Q with frequency [19]. Again,

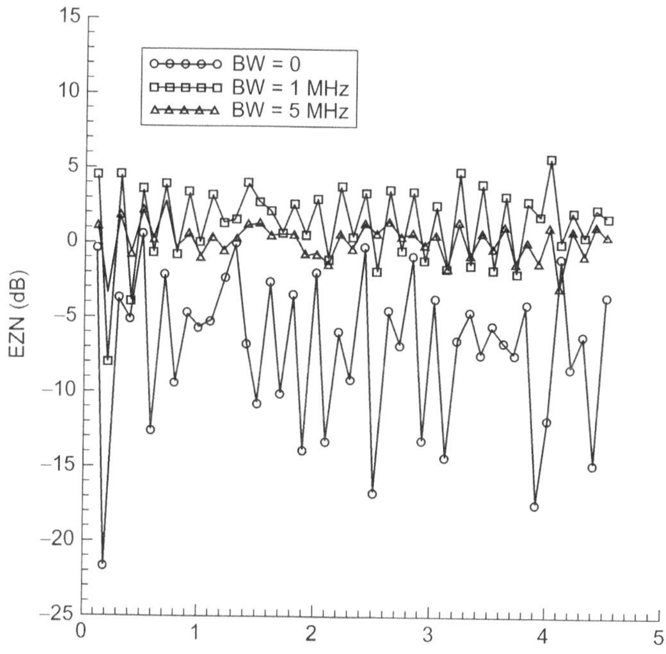

FIGURE 9.3 Normalized electric field magnitude for a higher frequency (8 GHz) and Q (1.5×10^5) [48].

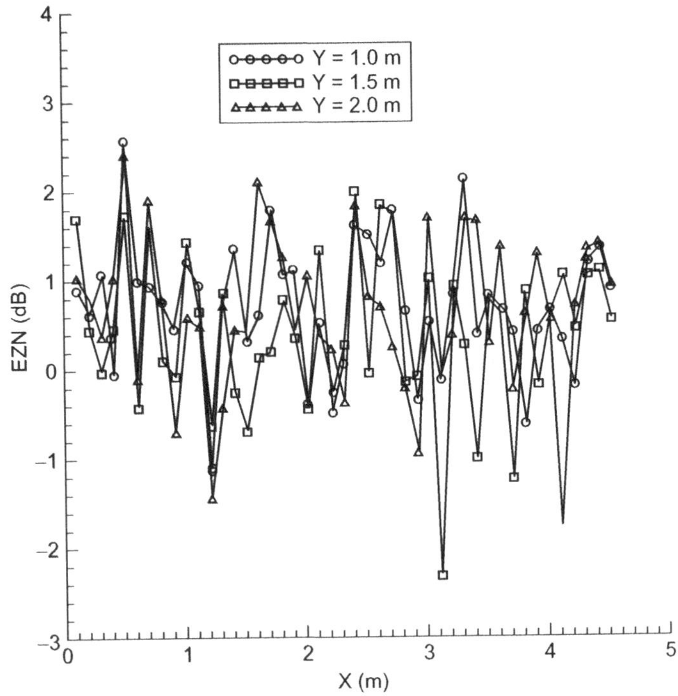

FIGURE 9.4 Normalized electric field magnitude for various values of y. Other parameters: $f = 4\,\text{GHz}$, $Q = 10^5$, and $x_0 = y_0 = 0.5\,\text{m}$ [48].

the field uniformity improves with increasing bandwidth, and the average value approaches 0 dB. Equation (9.18) shows that the number N_{BW} of modes excited is proportional to f_{BW}, so a smaller bandwidth is needed at higher frequencies. Table 9.1 shows that the number of modes N_{BW} is the significant quantity in determining field uniformity, and this is consistent with [90].

In Figure 9.4, results at 4 GHz are shown for three different y values. The three curves are quite distinct, but they have approximately the same statistics, as seen in Table 9.1. All three curves have average values and standard deviations less than 1 dB. This is a good illustration of the type of statistical spatial field uniformity that is to be expected with well-stirred fields in a reverberation chamber.

In Figure 9.5 results are shown for three different Q values. In this case, the actual curves, not just their statistics, are very similar. However, it should be remembered that the normalization in (9.20) involves Q. Thus, the unnormalized field is higher for higher Q. Again, the average values and standard deviations are less than 1 dB for each case.

If the results of Table 9.1 are compared with Loughry's results, fewer modes are required to obtain a given level of field uniformity (for example, 1 dB) for the idealized two-dimensional model. This is to be expected because more modes are required to

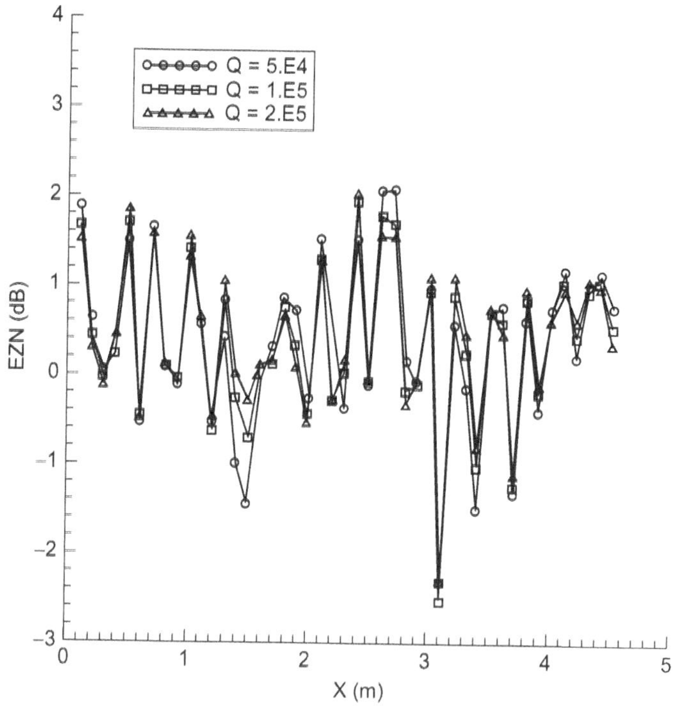

FIGURE 9.5 Normalized electric field magnitude for various values of Q [48].

mix the fields in a fully three-dimensional cavity. If this factor is taken into account, then the results in Table 9.1 are consistent with Loughry's results.

The use of two sources of the same single frequency for exciting the cavity has also been analyzed in [48]. However, this does not provide much improvement in field uniformity even if the sources are incoherent or varied in relative phase. Some additional mechanical or frequency stirring is required to excite additional modes needed for field uniformity.

9.2 UNSTIRRED ENERGY

The term "unstirred energy" has been used to refer to a deterministic field (that does not interact with the rotating stirrer) in a reverberation chamber [111]. A simple analysis of this case has been performed where the unstirred field is assumed to be the direct field of an isotropic antenna, and the usual expression is used for the stirred field [112]. This comparison is useful in determining both how far away from the transmitting antenna the test object should be placed for a valid immunity test and how large the chamber Q should be so that the stirred field dominates the unstirred field throughout most of the chamber.

We first represent the magnitude of the power density S_d from the direct transmission of power P_t by an isotropic antenna in free space:

$$S_d = \frac{P_t}{4\pi r^2}, \tag{9.21}$$

where r is the distance from the antenna. We choose an idealized isotropic antenna because the main beam of the excitation antenna is normally pointed away from the test object (toward a corner or a stirrer). Hence the direct field is coming from the antenna sidelobes, and a directivity of one is a good (conservative) estimate of this field. Also, the isotropic antenna assumption makes the analysis independent of the excitation antenna directivity. The other idealization that is implied by (9.21) is that the unstirred field does not contain any contribution from wall reflections. This assumption simplifies the analysis and is partially justified because reflected paths are longer than the direct path.

Consider now the stirred field. From [38], the mean scalar power density in a reverberation chamber is given by:

$$\langle S_r \rangle = \frac{\lambda Q P_t}{2\pi V}, \tag{9.22}$$

where V is the chamber volume, and λ is the free-space wavelength. At a radius r_e, the power densities in (9.21) and (9.22) become equal. This radius is given by:

$$r_e = \sqrt{\frac{V}{2\lambda Q}} \tag{9.23}$$

For a radius less than r_e, the power density in the chamber is dominated by S_d(direct coupling or unstirred energy), and for a radius greater than r_e, the (stirred) reverberation power density dominates.

The radius r_e corresponds to a spherical volume of:

$$V_{re} = \frac{4}{3}\pi r_e^3 = \frac{4}{3}\pi \left(\frac{V}{2\lambda Q}\right)^{3/2} \tag{9.24}$$

It is worth noting that Q is approximately proportional to the volume (see Sec. 7.6), so the right-hand side of (9.24) is nearly independent of V. For an effective or efficient reverberation chamber, this volume (V_{re}) must be much less than the actual chamber volume V:

$$V \gg V_{re} \tag{9.25}$$

V_{re} can be used as a metric for assessing the chamber performance. If the volume of the chamber is much larger than V_{re}, the chamber can be considered an effective reverberation chamber because throughout most of the chamber the stirred energy exceeds the unstirred energy. Hence, the useful test volume is large.

A threshold of the chamber Q can be obtained by returning to expression (9.24), and realizing that (9.25) also implies:

$$Q \gg Q_{thr}, \tag{9.26}$$

where:

$$Q_{thr} = \left(\frac{4\pi}{3}\right)^{2/3} \frac{V^{1/3}}{2\lambda} \tag{9.27}$$

This is the value that the chamber Q must exceed for the reverberation chamber to be effective. The factor λ^{-1} on the right hand side of (9.27) does not imply that λ should be increased without limit. The value of λ needs to remain small enough compared to the chamber dimensions such that mode density is sufficient [9].

We can examine (9.26) and (9.27) for the case of an aluminum chamber with dimensions $1.213 \times 0.603 \times 0.937$ m [38]. At 12 GHz, (9.27) yields $Q_{thr} \cong 40$. The actual measured Q of the aluminum chamber was found to be approximately 8×10^4. Thus (9.26) was easily satisfied.

Some related measurements were performed in the NIST chamber by loading it with 500-ml bottles filled with lossy liquid [113]. Figure 9.6 shows the decrease in Q as a function of the number of bottles. Figure 9.7 shows the degradation of spatial uniformity as a function of the number bottles (as the Q decreases).

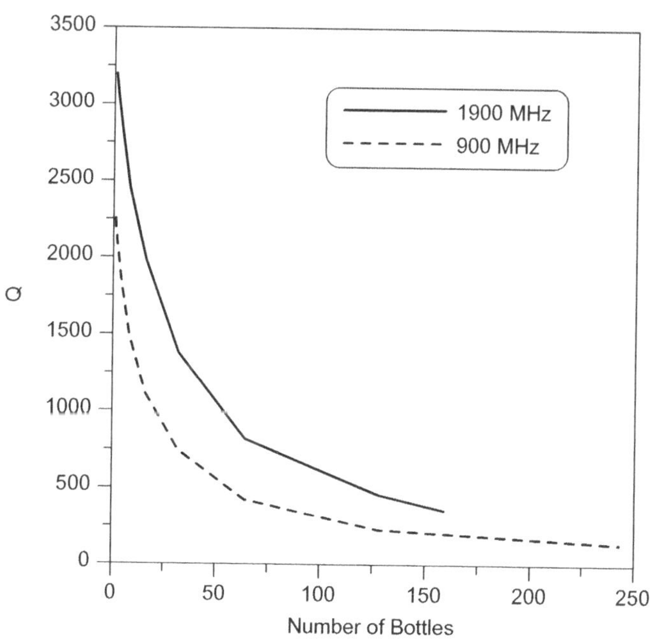

FIGURE 9.6 Effect of loading (500-ml bottles filled with lossy liquid) on the Q of the NIST reverberation chamber [112].

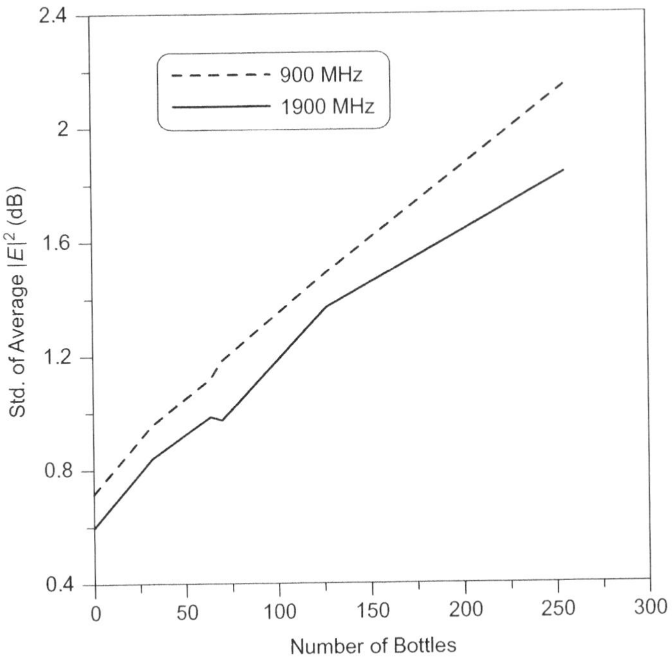

FIGURE 9.7 Effect of loading (500-ml bottles filled with lossy liquid) on the standard deviation of the average squared total electric field in the NIST reverberation chamber [112].

Although the theory in this section was applied to the performance of reverberation chambers [113], the theory is also applicable to the behavior of fields in a large cavity excited through an aperture [38]. If (9.25) and (9.26) are satisfied for an aperture-excited cavity, then the uniform-field theory in [38] is applicable to the aperture penetration problem where the fields throughout most of the cavity (away from the aperture) are uniform and calculated by the theory in Chapter 8.

9.3 ALTERNATIVE PROBABILITY DENSITY FUNCTION

In the previous section, we examined the relationships between the direct (unstirred) and stirred reverberation fields. It is also useful to examine the difference in the probability density function of the field magnitude when the direct (unstirred) field cannot be ignored [113].

For simplicity of analysis, we assume that the direct electric field is linearly polarized in the θ direction and denote that spherical component as $E_{d\theta}$. (The origin of the spherical coordinate system is at the transmitting antenna.) Then the magnitude of the power density can be written as:

$$S_d = \frac{|E_{d\theta}|^2}{\eta} = \frac{P_t}{4\pi r^2}, \tag{9.28}$$

where η is the impedance of free space. (We have again assumed a nondirectional transmitting antenna.) For the stirred field, the scalar power density can be written as

$$\langle S_r \rangle = \frac{\langle |E_s|^2 \rangle}{\eta} = \frac{\lambda Q P_t}{2\pi V} \tag{9.29}$$

If we examine just the θ component of the stirred electric field $E_{s\theta}$, the mean square value in an idealized chamber is $\frac{1}{3}$ of the total value in (9.29):

$$\langle |E_{s\theta}|^2 \rangle = \frac{1}{3} \frac{\eta \lambda Q P_t}{2\pi V} \tag{9.30}$$

The total θ component of the electric field can be written as the sum of the stirred and unstirred (direct) components:

$$E_\theta = E_{s\theta} + E_{d\theta} \tag{9.31}$$

We now write the stirred field as the sum of the real and imaginary parts:

$$E_{s\theta} = E_{s\theta r} + iE_{s\theta i} \tag{9.32}$$

As shown in Section 7.2, the mean values of $E_{s\theta r}$ and $E_{s\theta i}$ are zero, and the variances are:

$$\langle E_{s\theta r}^2 \rangle = \langle E_{s\theta i}^2 \rangle = \frac{\eta \lambda Q P_t}{12\pi V} \equiv \sigma^2 \tag{9.33}$$

Equation (9.33) actually holds for any scalar component of $\vec{E}_s$, but we discuss only the θ component here.

As shown in Section 7.3, both $E_{s\theta r}$ and $E_{s\theta i}$ are Gaussian distributed. Hence, the amplitude of the θ component of electric field has a Rice probability density function [57,111,112]:

$$f(|E_\theta|) = \frac{|E_\theta|}{\sigma^2} \mathrm{I}_0 \left(\frac{|E_{s\theta}||E_{d\theta}|}{2\sigma^2} \right) \exp\left(-\frac{|E_{s\theta}|^2 + |E_{d\theta}|^2}{2\sigma^2} \right) \mathrm{U}(|E_\theta|), \tag{9.34}$$

where I_0 is the modified Bessel function of zero order [25] and U is the unit step function.

In regions where the direct component of the field is insignificant, we expect to have a Rayleigh PDF for the magnitude of a scalar component of the electric field (see Section 7.3). In order for (9.34) to reduce to a Rayleigh PDF, we require:

$$|E_{d\theta}|^2 \ll 2\sigma^2 \tag{9.35}$$

Then (9.35) reduces to a Rayleigh PDF:

$$f(|E_\theta|) = \frac{|E_\theta|}{\sigma^2} \exp\left(-\frac{|E_\theta|^2}{2\sigma^2} \right) \mathrm{U}(|E_\theta|) \tag{9.36}$$

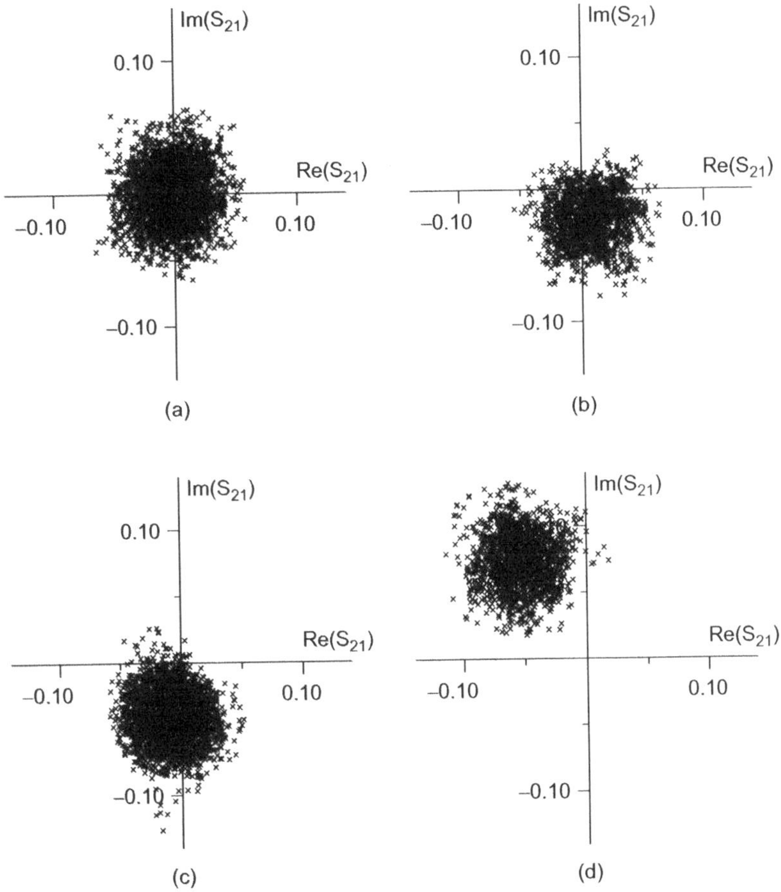

FIGURE 9.8 Scatter plots of measured S_{21} for two antennas in the NIST reverberation chamber at a frequency of 2 GHz [112].

The magnitudes of the φ and r components satisfy a Rayleigh PDF because we have assumed that the direct (unstirred) field has only a θ component. Although we have considered only the electric field E_θ, identical results would be obtained by analyzing the magnetic field component H_ϕ.

The effect of the inequality in (9.35) not being met can be seen in a measurement of the scattering parameter S_{21} for two antennas placed in a reverberation chamber. If the condition in (9.35) is satisfied (the stirred energy dominates the unstirred energy), then a scatter plot of the real and imaginary parts of S_{21} for different stirrer positions results in the data being clustered in a circle and centered about the origin [see Figure 9.8(a)]. As the direct energy (or unstirred energy) becomes comparable to the stirred energy, the cluster of data moves off the origin [as shown in Figure 9.8 (b)–(d)]. For example, the data in Fig. 9.8(d) represent the case where strong direct antenna coupling is present. This is undesirable if a reverberation chamber is to perform well. The data in Figure 9.8 were collected in NIST's reverberation chamber by use of two horn antennas at a frequency of 2 GHz.

Following the procedure of the previous section, (9.35) implies the following volume requirement for an effective chamber:

$$V_{rep} \ll V, \quad \text{where} \quad V_{rep} = \frac{4}{3}\pi\left(\frac{3V}{2\lambda Q}\right)^{3/2} \tag{9.37}$$

This relationship is obtained by determining an effective radius and using this radius to obtain a spherical volume. This effective radius is obtained by substituting (9.28) into the left-hand side of (9.35) and substituting (9.33) into the right-hand side of (9.35), and is expressed as:

$$r_{ep} = \sqrt{\frac{3V}{2\lambda Q}} \tag{9.38}$$

We have added a subscript p to the quantities in (9.37) and (9.38) to indicate that these quantities are based on the probability density function rather than the power density. The only difference between (9.24) and (9.37) is the factor of $3^{3/2}$ in (9.37). This differing factor is not of much significance in this approximate analysis, but we retain it to show that the requirement based on probability density function [i.e., (9.37)] is more stringent than the one given in (9.24). This is partly because we have assumed linear polarization for the direct (unstirred) electric field. This is the most demanding case.

Following the procedure in the previous section, we can also use (9.35) to obtain the following Q requirement for an effective chamber:

$$Q \gg Q_{thrp}, \tag{9.39}$$

where:

$$Q_{thrp} = \left(\frac{4}{3}\pi\right)^{2/3}\frac{3V^{1/3}}{2\lambda} \tag{9.40}$$

We have again added a subscript p to indicate that this result is based on a probability density function rather than power density. The only difference between (9.27) and (9.40) is the factor of 3 in (9.40). Thus, the requirement based on the probability density is again more stringent.

With the use of (9.21), (9.33), and (9.35), it is possible to obtain an alternative requirement for the chamber quality factor Q:

$$Q \gg \frac{6\pi V}{\lambda}\frac{S_d}{P_t} \tag{9.41}$$

Written in this way, it is interesting to note that the requirement for the chamber Q is expressed in terms of the power density of the direct coupling term (unstirred energy).

One way to interpret this expression is that since S_d is inversely proportional to r^2, this expression states that measurements made close to the transmitting antenna require chambers with higher quality factors.

PROBLEMS

9-1 Derive (9.3) from (9.2).

9-2 Derive (9.5) from (9.3).

9-3 Derive (9.8) from (9.2).

9-4 Derive the asymptotic forms (for large $k\rho$) of E_z and H_ϕ from (9.8). Show that these results are consistent with (9.9).

9-5 Derive (9.14).

9-6 Derive (9.15) from (9.40). Hint: use the two-dimensional analogy of the method used in Problem 2-5.

9-7 Derive (9.18).

9-8 Verify that the Rice PDF in (9.34) reduces to the Rayleigh PDF in (9.36) under the condition in (9.35).

9-9 Consider the application of (9.41) to a reverberation chamber of volume of 30 m^3. If a 1 GHz test measurement is made at a distance of 1 m from the test antenna, what is the requirement on chamber Q for the stirred field to dominate the direct field?

9-10 For the test setup in Problem 9, what is the Q requirement at 10 GHz? If wall loss is dominant, what is the frequency dependence of chamber Q?

CHAPTER 10

Further Applications of Reverberation Chambers

Although reverberation chambers have traditionally been used for electromagnetic immunity and emissions testing, they are versatile facilities that have recently been used for several other measurement applications (shielding effectiveness, antenna efficiency, and absorption cross section) that will be covered in this chapter. Reverberation chambers also have many other applications in wireless communications, and those applications will be covered in Chapter 11.

10.1 NESTED CHAMBERS FOR SHIELDING EFFECTIVENESS MEASUREMENTS

Materials used for the shielding of electromagnetic fields range from simple metallic wire meshes to sophisticated composite materials. Composites are very popular because of superior mechanical and chemical properties (low weight, high stiffness and strength, low corrosion, low tooling costs, and ease of fabrication). Despite these benefits, composites have much lower electrical conductivity, and hence lower shielding effectiveness (SE), than metals. Even carbon-fiber-reinforced composites have much lower electrical conductivities than metals. Since most composites are too complicated to allow for calculation of SE, measurement methods must be used.

SE (in dB) is typically used to quantify the shielding properties of materials and can be defined as the ratio of the incident power P_i to the power P_t transmitted through the material:

$$SE = 10\log_{10}\left(\frac{P_i}{P_t}\right) \tag{10.1}$$

Equation (10.1) generally results in a positive value for SE. A coaxial fixture [113] is commonly used to determine the far-field equivalent SE, and other methods are

Electromagnetic Fields in Cavities: Deterministic and Statistical Theories, by David A. Hill

available [114]. However, these methods determine SE for only a limited set of incident field conditions. In most applications, shielding materials are exposed to complex electromagnetic environments where fields are incident on the material with various polarizations and incidence angles. Therefore, a test method that utilizes a complex field environment is useful, and a reverberation-chamber SE test provides such a complex field environment where the incident field is a superposition of incidence angles and polarizations.

In this section we describe initial nested reverberation chamber methods (two reverberation chambers) and introduce a revised approach [115] for determining SE. The revised approach accounts for aperture, cavity size, and chamber loading effects, which are not taken into account in the initial methods.

10.1.1 Initial Test Methods

Figure 10.1 illustrates a typical experimental setup with nested reverberation chambers. Each chamber contains a stirrer and two antennas, and an aperture between the two chambers has a sample whose SE is to be determined. With this setup, one method of determining SE (which we will label SE_1) is based on the following equation [116]:

$$SE_1 = 10 \log_{10}\left(\frac{\langle P_{oc,s}\rangle}{\langle P_{ic,s}\rangle}\right), \tag{10.2}$$

where $\langle P_{ic,s}\rangle$ is the averaged power received inside the inner chamber with a sample in the aperture, $\langle P_{oc,s}\rangle$ is the averaged power received in the outer chamber with a sample in the aperture, and the source is in the outer chamber. A limiting case that any method should satisfy is that with no sample in the chamber, SE should go to zero.

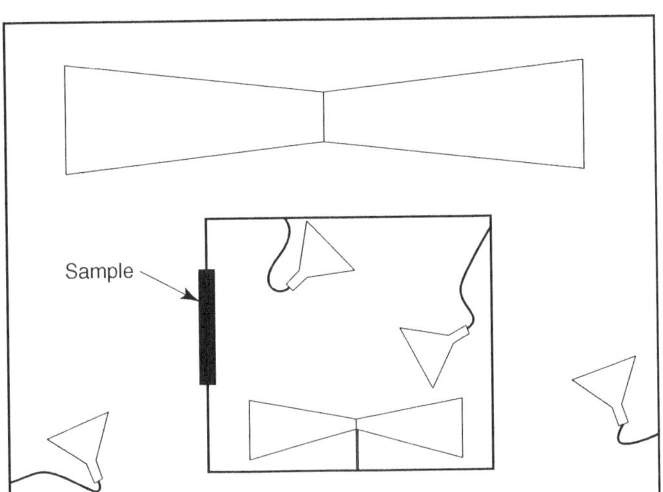

FIGURE 10.1 Nested reverberation chambers with a sample to be evaluated [115].

However, we will see that depending on the chamber and aperture properties, (10.2) will not generally satisfy this condition.

Another approach has been suggested to account for the effects of coupling into the inner chamber [117]:

$$\mathrm{SE}_2 = 10\log_{10}\left(\frac{\langle P_{oc,s}\rangle}{\langle P_{ic,s}\rangle}\right) + \mathrm{CF}, \tag{10.3}$$

where CF is referred to to as either the test-fixture calibration factor or loss factor. It is the ratio of the received power to the input power inside the inner chamber with the sample in the aperture:

$$\mathrm{CF} = 10\log_{10}\left(\frac{\langle P_{rQ,in,s}\rangle}{P_{tx,in,s}}\right), \tag{10.4}$$

where $\langle P_{rQ,in,s}\rangle$ is the averaged measured power in the inner chamber with a sample in the aperture for a transmitting antenna located in the inner chamber with an output power $P_{tx,in,s}$. From (7.112), we see that (10.4) is related to the quality factor Q of the inner chamber. However, as we will see, this method also suffers from not providing a zero value for SE when there is no sample in the aperture.

10.1.2 Revised Method

In deriving a revised method, we start by first defining the shielding effectiveness of a material sample as follows [115]:

$$\mathrm{SE}_3 = \log_{10}\left(\frac{\dfrac{\langle P_{t,ns}\rangle}{\langle S_{ns}^{inc}\rangle}}{\dfrac{\langle P_{t,s}\rangle}{\langle S_{s}^{inc}\rangle}}\right), \tag{10.5}$$

where $\langle P_{t,s}\rangle$ is the averaged power transmitted through the aperture with a sample, $\langle P_{t,ns}\rangle$ is the averaged power transmitted through the same aperture with no sample (open aperture), and $\langle S_{s}^{inc}\rangle$ and $\langle S_{ns}^{inc}\rangle$ are respectively the scalar power densities incident on the aperture with and without the sample. This is approximately equivalent to the IEEE definition of shielding effectiveness [118, p. 831], which compares two measured quantities with and without the shield (sample). Defined in this way, the environmental effects have been removed or normalized out, and only the effects of the material (sample) in the aperture are accounted for.

The averaged powers transmitted through the aperture can be expressed in terms of averaged cross sections:

$$\langle P_{t,s}\rangle = \langle \sigma_{t,s}\rangle\langle S_{s}^{inc}\rangle \quad \text{and} \quad \langle P_{t,ns}\rangle = \langle \sigma_{t,ns}\rangle\langle S_{ns}^{inc}\rangle \tag{10.6}$$

In (10.6), $\langle \sigma_{t,s}\rangle$ and $\langle \sigma_{t,ns}\rangle$ are the respectively averaged transmission cross sections of the aperture with and without the sample. It should be kept in mind that these transmission cross sections are averages over incidence angle and polarization, as in (7.130). Substitution of (10.6) into (10.5) gives the following for SE_3:

$$SE_3 = 10\log_{10}\left(\frac{\langle\sigma_{t,ns}\rangle}{\langle\sigma_{t,s}\rangle}\right) \tag{10.7}$$

This expression states that SE_3 involves just the ratio of the averaged transmission cross sections of the aperture with and without the sample. It is clear that this ratio reduces to one for no sample, and that SE_3 reduces to 0 dB, as it should. This definition is now basically a function of only the material under test.

The next step is to determine how to obtain $\langle\sigma_{t,s}\rangle$ and $\langle\sigma_{t,ns}\rangle$ in a nested reverberation chamber. Using (8.16), the averaged transmission cross sections can be written:

$$\begin{aligned}\langle\sigma_{t,s}\rangle &= \frac{\langle S_{in,s}\rangle}{\langle S_{o,s}\rangle}\frac{2\pi V}{\lambda Q_{in,s}},\\ \langle\sigma_{t,ns}\rangle &= \frac{\langle S_{in,ns}\rangle}{\langle S_{o,ns}\rangle}\frac{2\pi V}{\lambda Q_{in,ns}},\end{aligned} \tag{10.8}$$

where $\langle S_{in,s}\rangle$ and $\langle S_{in,ns}\rangle$ are respectively the averaged scalar power densities in the inner chamber with and without the sample, $\langle S_{o,s}\rangle$and $\langle S_{o,ns}\rangle$ are respectively the averaged scalar power densities in the outer chamber with and without the sample, $Q_{in,s}$ and $Q_{in,ns}$ are respectively the quality factors with and without the sample, V is the volume of the inner chamber, and λ is the wavelength. From (7.104), each of the averaged scalar power densities in (10.8) can be expressed in terms of the average measured power $\langle P\rangle$ through the effective area $\lambda^2/(8\pi)$ of the receiving antenna by:

$$\langle S\rangle = \frac{8\pi}{\lambda^2}\langle P\rangle \tag{10.9}$$

If we substitute (10.8) and (10.9) into (10.7), SE_3 reduces to:

$$SE_3 = 10\log_{10}\left(\frac{\langle P_{r,in,ns}\rangle}{\langle P_{r,in,s}\rangle}\frac{\langle P_{r,o,s}\rangle}{\langle P_{r,o,ns}\rangle}\frac{Q_{in,s}}{Q_{in,ns}}\right), \tag{10.10}$$

where $\langle P_{r,in,s}\rangle$ and $\langle P_{r,in,ns}\rangle$ are respectively the average measured powers in the inner chamber with and without the sample, and $\langle P_{r,o,s}\rangle$ and $\langle P_{r,o,ns}\rangle$ are respectively the averaged measured powers in the outer chamber with and without the sample. These four different received powers are obtained for a source in the outer chamber.

From (10.10) it is shown that the SE is a function of the ratio of the two Qs of the inner chamber (with and without a sample), and not just a function of a single Q of the inner chamber with a sample covering the aperture (as suggested in (10.4)). From

(7.111), the quality factors $Q_{in,s}$ and $Q_{in,ns}$ can be expressed as:

$$Q_{in,s} = \frac{16\pi^2 V}{\lambda^2} \frac{\langle P_{rQ,in,s} \rangle}{P_{tx,in,s}},$$
$$Q_{in,ns} = \frac{16\pi^2 V}{\lambda^2} \frac{\langle P_{rQ,in,ns} \rangle}{P_{tx,in,ns}}, \tag{10.11}$$

where $\langle P_{rQ,in,s} \rangle$ is the average measured power in the inner chamber with a sample in the aperture for a transmitting antenna located in the inner chamber with an output power $P_{tx,in,s}$. Similarly, $\langle P_{rQ,in,ns} \rangle$ is the averaged measured power in the inner chamber without a sample in the aperture for a transmitting antenna located in the inner chamber with an output power $P_{tx,in,ns}$. The SE can now be expressed as:

$$\text{SE}_3 = 10 \log \left(\frac{\langle P_{r,in,ns} \rangle}{\langle P_{r,in,s} \rangle} \frac{\langle P_{r,o,s} \rangle}{\langle P_{r,o,ns} \rangle} \frac{\langle P_{rQ,in,s} \rangle}{\langle P_{rQ,in,s} \rangle} \frac{P_{tx,in,ns}}{P_{tx,in,s}} \right) \tag{10.12}$$

It is readily seen in (10.12) that all four power ratios are equal to one with no sample in the aperture and that SE_3 reduces to 0 dB. Figure 10.2 shows measured results for SE_1, SE_2, and SE_3 with no sample in the aperture, and it is seen that only SE_3 is equal to 0 dB.

Equation (10.12) can be thought of as a first-order measurement of the shielding effectiveness. A zero-order shielding effectiveness can be obtained by assuming that

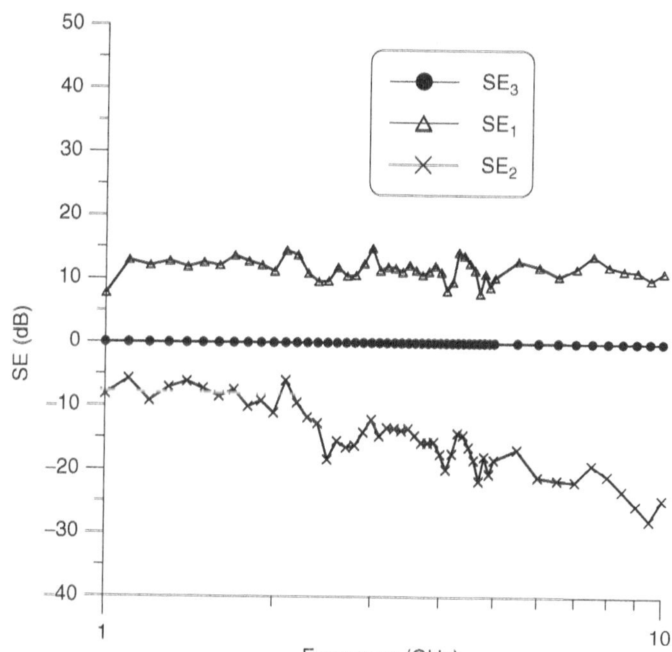

FIGURE 10.2 SE obtained from the three approaches with no sample in the aperture [115].

the wall loss is dominant in both cavities. Under this condition, we have:

$$\frac{\langle P_{r,o,s}\rangle}{\langle P_{r,o,ns}\rangle} \approx 1 \quad \text{and} \quad \frac{Q_{in,s}}{Q_{in,ns}} \approx 1 \tag{10.13}$$

By substituting (10.13) into (10.10), we obtain a zero-order shielding effectiveness SE_4 given by:

$$SE_4 = 10 \log_{10}\left(\frac{\langle P_{r,in,ns}\rangle}{\langle P_{r,in,s}\rangle}\right) \tag{10.14}$$

The result in (10.14) matches the IEEE definition of shielding effectiveness [118, p. 831], but neglects changes in chamber loading and chamber Q. The first-order result in (10.12) includes such effects, but does not include the possible effects of multiple interactions between the two chambers. (These effects are expected to be negligible.)

10.1.3 Measured Results

A series of SE measurements [115] was performed for various types of composite materials as samples in the nested chamber geometry of Figure 10.1. The outer chamber has dimensions of 2.76 × 3.05 × 4.57 m, the inner chamber has dimensions of 1.46 × 1.17 × 1.41 m, and the aperture dimensions are 0.25 × 0.25 m. Ridged horns were used as the transmitting and receiving antennas, and the inner chamber was placed on the center of the floor of the outer chamber.

Table 10.1 describes the composite materials used in the study. Figures 10.3 through 10.6 show SE determined by the three methods, (10.2), (10.3), and (10.12), for the four materials in Table 10.1. It is interesting that SE_1 and SE_3 have similar results, while SE_2 tends to give results that have 20 dB less shielding at frequencies above a few gigahertz.

Figure 10.7 shows SE_3 for the four different materials. This comparison shows that Material 3 offers the best shielding, while Material 2 has the worst shielding.

If SE_3 in (10.12) correctly accounts for cavity and aperture size effects, then the same SE results if one or both of the cavity size or aperture size is varied. To confirm that this is the case, the SE for the four different materials were measured in two different chambers. Only one inner chamber with a fixed aperture size was available. Therefore, to simulate a different inner chamber, electromagnetic absorbing material was placed in the inner chamber. This had the effect of altering the inner chamber by lowering its Q. Figure 10.8 shows the ratio of the Q of the inner chamber without

TABLE 10.1 Descriptions of Composite Materials Used in [115].

Material #	Type	Thickness
Material 1	Carbon fiber	1 mm
Material 2	Sandwich: external fiber glass with inside carbon fiber	4 mm
Material 3	Carbon fiber	1.5 mm
Material 4	Carbon fabric with external rubber coating	0.5 mm

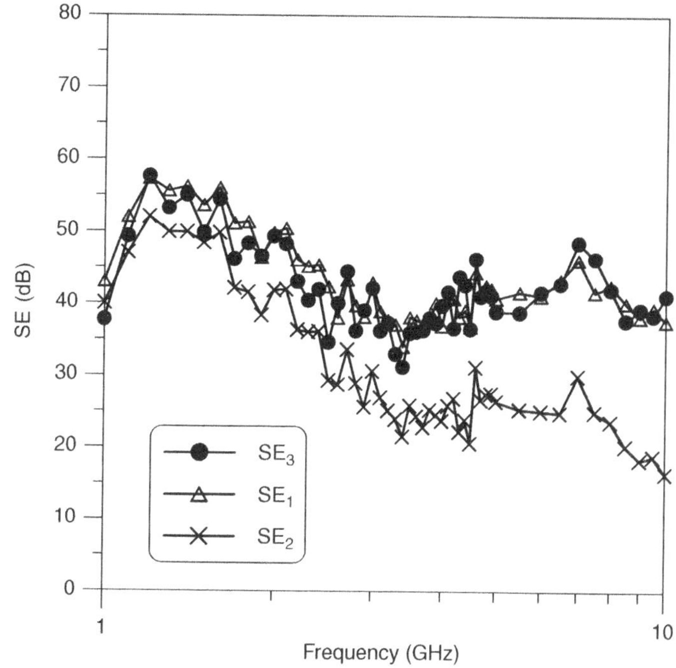

FIGURE 10.3 SE obtained from the three approaches with Material 1 in the aperture [115].

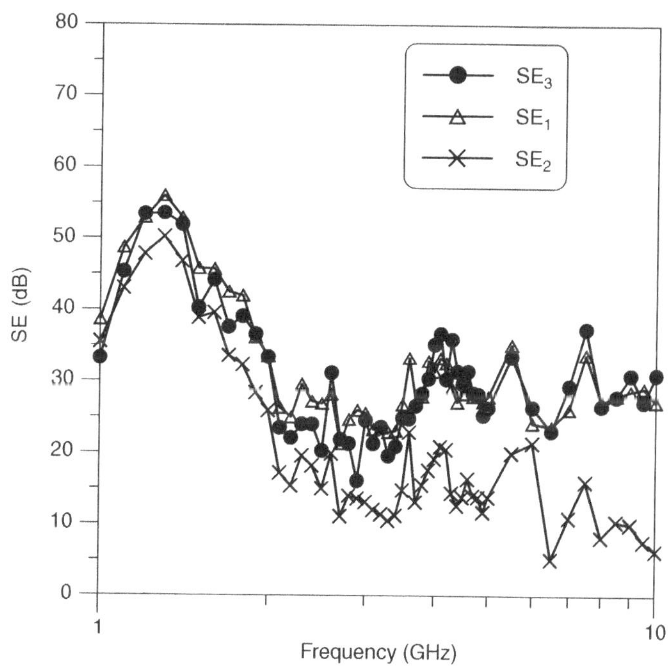

FIGURE 10.4 SE obtained from the three approaches with Material 2 in the aperture [115].

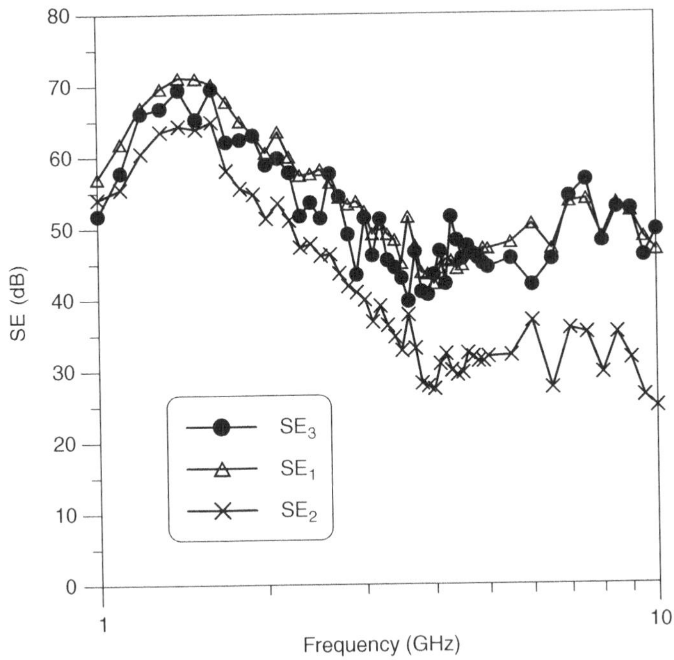

FIGURE 10.5 SE obtained from the three approaches with Material 3 in the aperture [115].

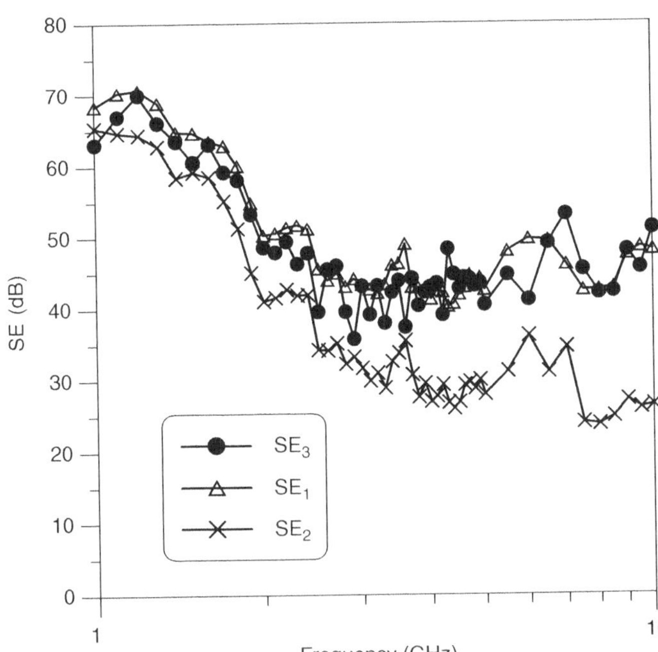

FIGURE 10.6 SE obtained from the three approaches with Material 4 in the aperture [115].

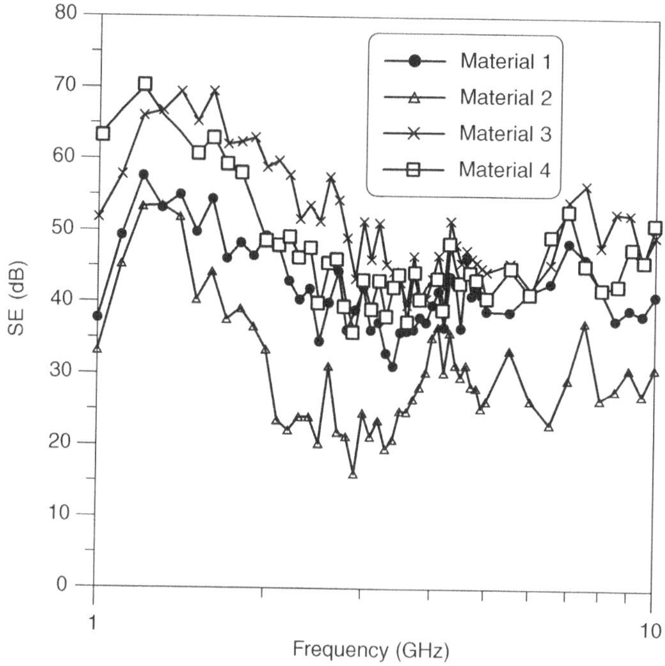

FIGURE 10.7 Comparison of the SE for the four different materials obtained using SE_3 [115].

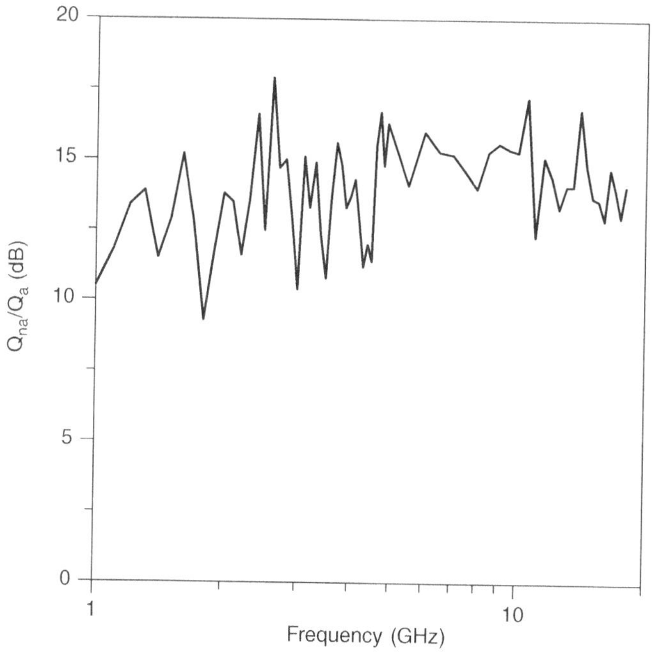

FIGURE 10.8 Ratio of the Q of the inner chamber without the absorber installed (Q_{na}) to the Q with the absorber installed (Q_a) [115].

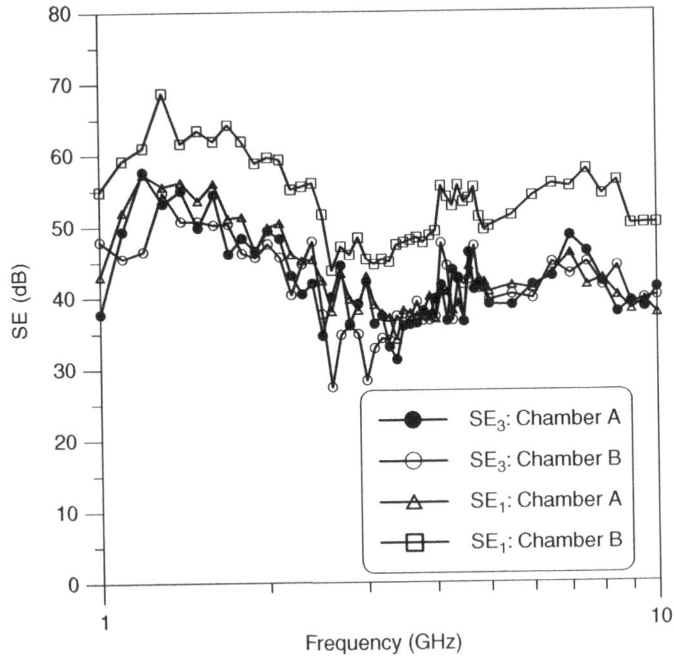

FIGURE 10.9 Comparison of SE for the two different chambers with Material 1 in the aperture [115].

the absorber installed (Q_{na}) to the Q of the inner chamber with the absorber installed (Q_a). Notice that the ratio has changed by 10 to 15 dB over the frequency range. Figures 10.9 through 10.12 show a comparison of the measured SE_3 of the original chamber to the measured SE_3 for the loaded inner chamber for all four materials. In these figures, Chamber A corresponds to no absorber in the inner chamber, and Chamber B corresponds to absorber placed in the inner chamber. Also shown in these figures are the results for SE_1. Note that the SE changes by about 10 dB for the two different chambers when obtained with SE_1. On the other hand, the results obtained using SE_3 are consistent for the two different chambers.

There is, however, some remaining variability in the results obtained for SE_3. It is believed that this is due to the fact that the received powers were obtained from a measurement of the peak values and not from a measurement of the averaged power. Although the ratios should, in theory, be equal for peak and average values, it has been demonstrated that use of peak measurements results in more variability than use of average power measurements [66]. Results based on measurements of maximum received power generally have larger associated uncertainties. Typical measurement uncertainties reported in [66] are ± 2 dB (standard deviations of ± 1 dB) for each measurement of maximum received power. Since each SE value is based on multiple measurements of maximum received power, the resulting uncertainty for the estimated SE will be larger. Further discussion of uncertainties in reverberation chamber measurements can be found elsewhere [119].

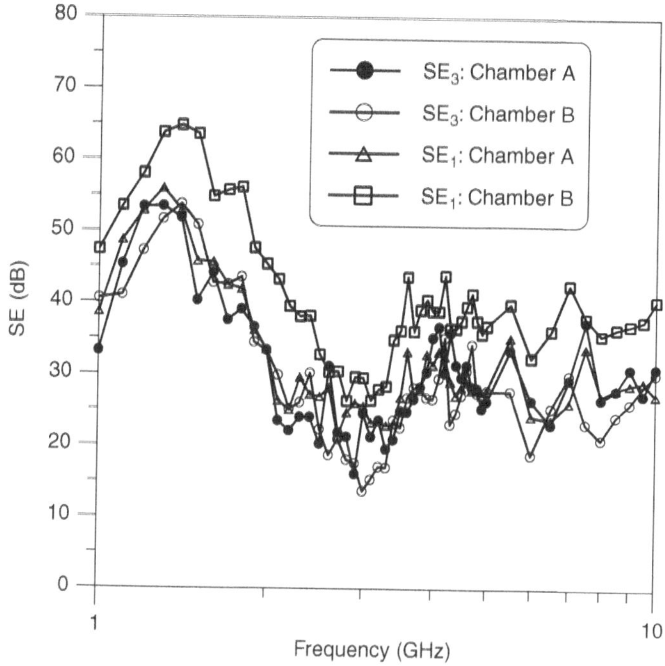

FIGURE 10.10 Comparison of SE for the two different chambers with Material 2 in the aperture [115].

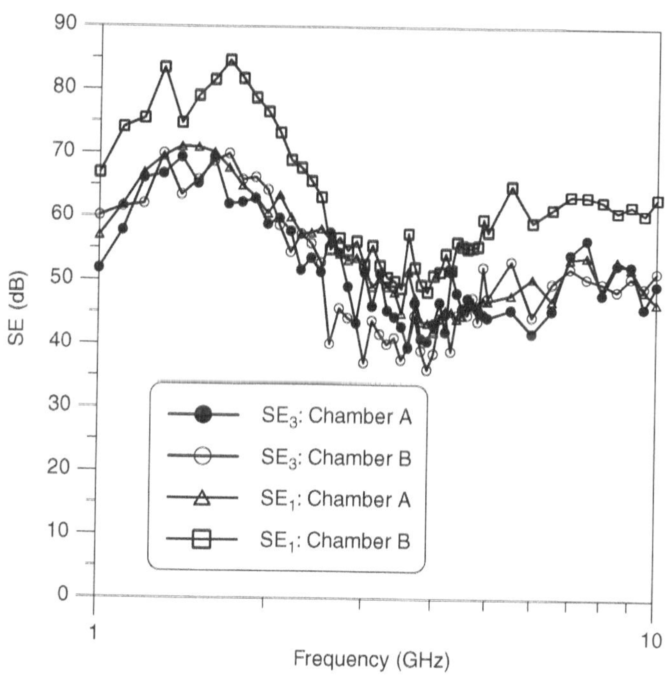

FIGURE 10.11 Comparison of SE for the two different chambers with Material 3 in the aperture [115].

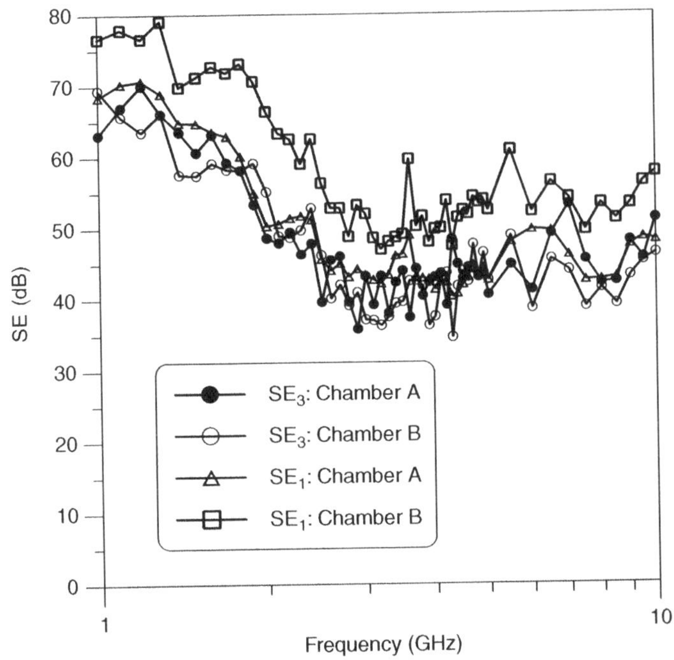

FIGURE 10.12 Comparison of SE for the two different chambers with Material 4 in the aperture [115].

10.2 EVALUATION OF SHIELDED ENCLOSURES

In many applications, shielded enclosures are used to control either immunity or emissions from electronic devices. One way to evaluate the shielding effectiveness (SE) of a shielded enclosure is to place it in a reverberation chamber so that it is illuminated from all incidence angles with all polarizations. In measuring or defining the SE of shielded enclosures, it is necessary to deal with the issues of internal cavity resonances and standing waves. One way to deal with these issues is to sample the field level at several locations inside the enclosure and to perform some sort of spatial averaging. This method would require many field probes (receiving antennas) and is typically not very practical. This is particularly true when it is difficult to place many probes inside the enclosure or to move one around.

10.2.1 Nested Reverberation Chamber Approach

In this method, the shielded enclosure (interior chamber) is treated as a reverberation chamber, and the fields are stirred [120]. In this case, the shielding effectiveness (in dB) can be written as:

$$\mathrm{SE} = 10\log_{10}\left(\frac{\langle S_{out}\rangle}{\langle S_{in}\rangle}\right), \tag{10.15}$$

where $\langle S_{out}\rangle$ is the average scalar power density outside the enclosure and $\langle S_{in}\rangle$ is the average power density inside the enclosure. With this definition, SE is normally positive. Since the average received power is proportional to the average scalar power density (see Sec. 7.5), (10.15) can be rewritten in terms of the average power received by antennas:

$$\mathrm{SE} = 10\log_{10}\left(\frac{\langle P_{out}\rangle}{\langle P_{in}\rangle}\right), \tag{10.16}$$

where $\langle P_{out}\rangle$ is the average power received by an antenna located outside the enclosure and $\langle P_{in}\rangle$ is the average power received by an antenna located inside the enclosure.

10.2.2 Experimental Setup and Results

For a sufficiently large shielded enclosure, the conventional approach for evaluating (10.16) is to use mechanical stirrers in both the (outer) reverberation chamber and the (inner) shielded enclosure. Then the power measurements are performed with receiving horns in both the (outer) reverberation chamber and the (inner) shielded enclosure. When the shielded enclosure is too small to conveniently house a stirrer and a receiving horn inside, alternative methods can be used [120].

For example, a small monopole antenna can be located on one of the chamber walls (but not near a corner). In this case, the average value of the square of the normal component of the electric field E_n is twice that of rectangular components $E_{x,y,z}$ far from the wall [97]:

$$\langle |E_n|^2\rangle = 2\langle |E_{x,y,z}|^2\rangle \tag{10.17}$$

This value is the same as that normal to a ground plane for a monopole located far from the wall. Hence, the average power received by a monopole antenna located at the chamber wall is the same as that for a monopole antenna far from the chamber walls. So (10.16) is still applicable for determining SE when a monopole antenna is located at the chamber wall. This receiving antenna has the advantage that it is easy to feed (through the chamber wall) and takes up less space in the enclosure.

When the enclosure is too small to hold a mechanical stirrer, frequency stirring can be used [48,120]. The receiving antenna in the enclosure can still be either a horn or a wall-mounted monopole. The combination of a wall-mounted monopole with frequency stirring is the most space efficient for measuring SE for small enclosures [120].

To verify that the four combinations (two stirring methods and two types of receiving antennas) discussed give equivalent results for SE, the measurement setup in Figure 10.13 was implemented. All measurements were performed with a multichannel Vector Network Analyzer (VNA) with port 1 connected to the transmitting horn in the outer chamber, port 2 connected to a receiving horn in the outer chamber, port 3 connected to a receiving horn in the enclosure (inner chamber), and port 4 connected to

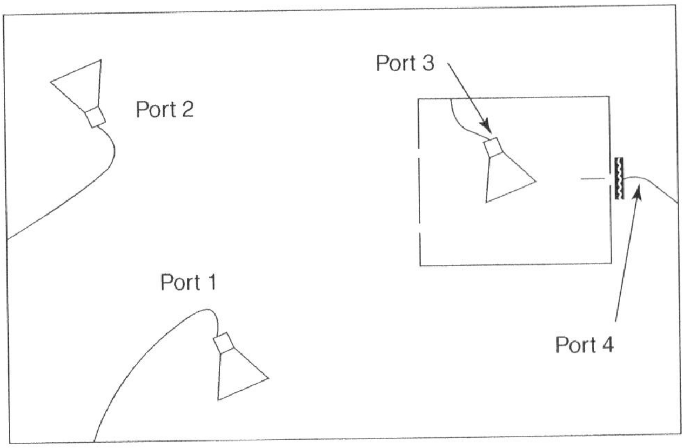

FIGURE 10.13 Experimental set-up for SE measurement [120].

a wall-mounted monopole in the enclosure. The VNA was used as three separate two-port VNAs, with calibrations between ports 1 and 2, 1 and 3, and 1 and 4. With the different S parameters, SE as defined in (10.16) can be measured directly for the four different reverberation chamber approaches. Impedance mismatch, which is particularly significant for the monopole, was taken into account [120].

The reverberation chamber (outer chamber) has dimensions of 4.60 m by 3.04 m by 2.76 m. The enclosure (inner chamber) has dimensions of 1.49 m by 1.45 m by 1.16 m. The enclosure has a square aperture of side 25.3 cm. Four different panels with different aperture sizes and shapes and different values of SE were used in the square aperture. The results in the following SE Figures 10.14 through 10.17, are labeled as

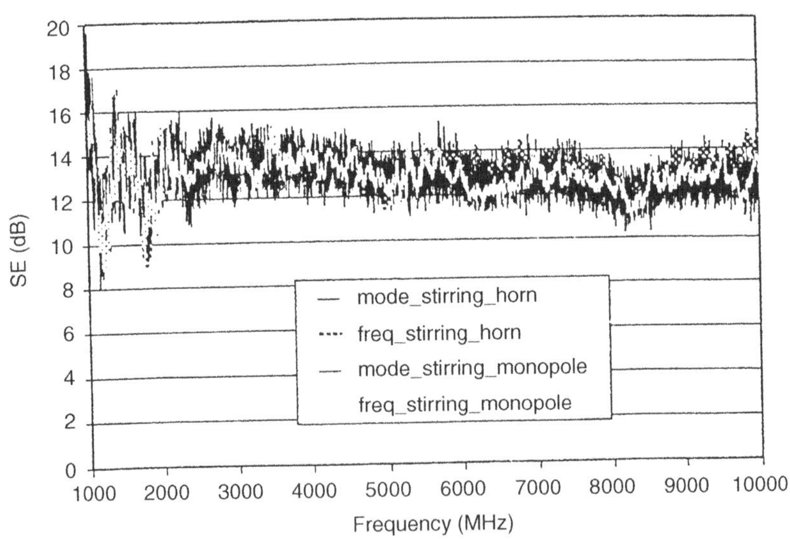

FIGURE 10.14 SE for the four different reverberation chamber approaches for the narrow-slot aperture [120].

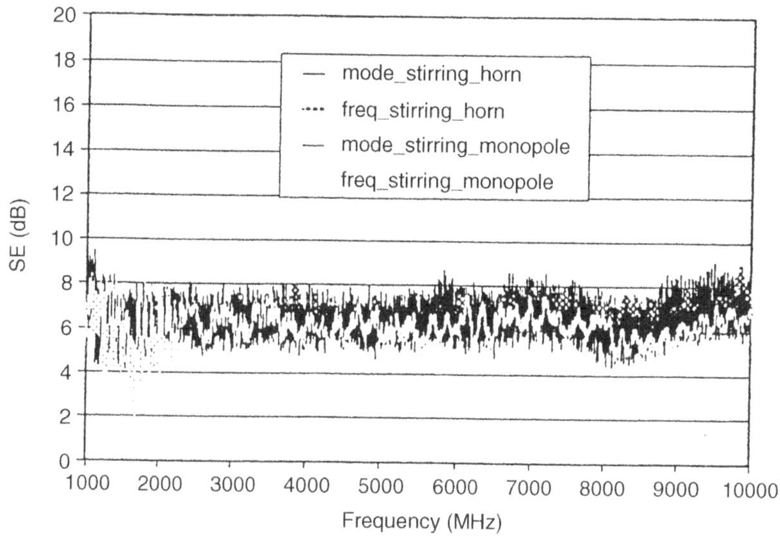

FIGURE 10.15 SE for the four different reverberation chamber approaches for the half-filled aperture [120].

follows: (1) mechanical stirring with the horn antenna in the enclosure is labeled "mode_stirring_horn", (2) mechanical stirring with the monopole antenna in the enclosure is labeled "mode_stirring_monopole", (3) frequency stirring with the horn antenna in the enclosure is labeled "freq_stirring_horn", and (4) frequency stirring with the monopole antenna is labeled as "freq_stirring_monopole".

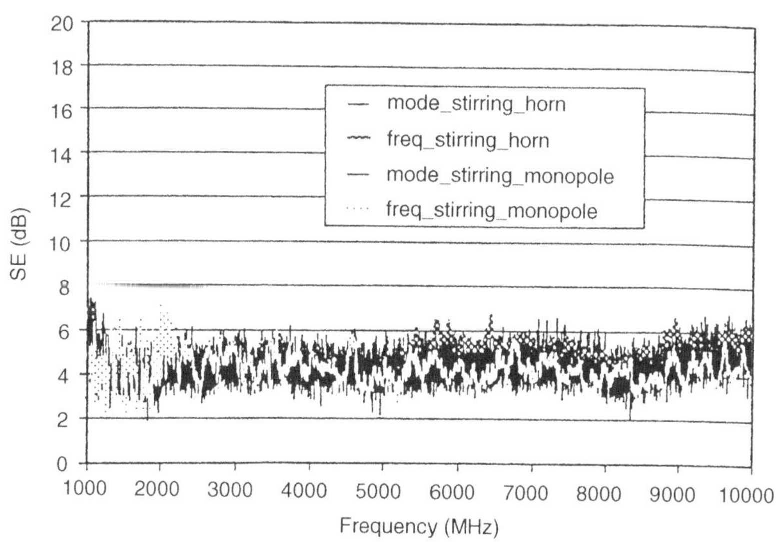

FIGURE 10.16 SE for the four different reverberation chamber approaches for the open aperture [120].

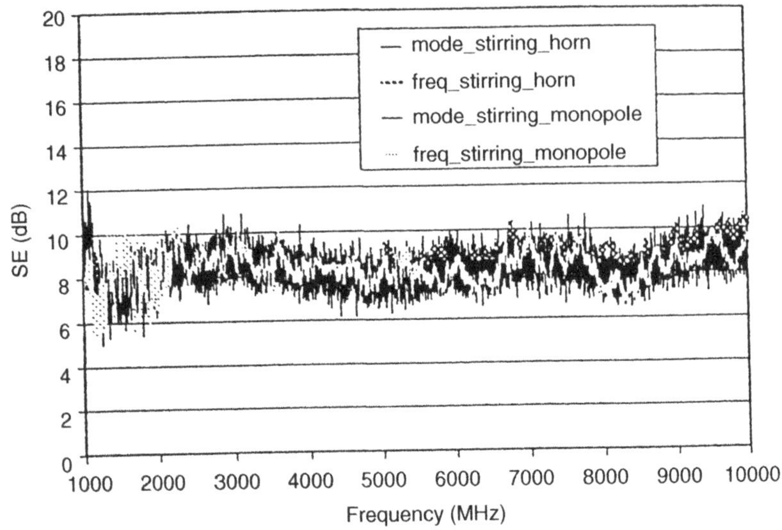

FIGURE 10.17 SE for the four different reverberation chamber approaches for the generic aperture [120].

Figure 10.14 shows SE for a narrow slot aperture obtained from all four of the reverberation-chamber approaches. From this comparison, it is seen that all four approaches give approximately the same result (approximately 13 dB). This agreement shows that frequency stirring with a wall-mounted monopole gives approximately the same results as the other reverberation-chamber approaches. This is an important practical result because frequency stirring with the wall-mounted monopole (which can be very short) takes up the least amount of space, a desirable feature when evaluating small enclosures.

Figures 10.15 through 10.17 show SE for three other apertures for all four approaches. Figure 10.15 shows SE results for the half-covered aperture (25.3 cm by 12.65 cm). The measured SE is approximately 6.5 dB, but the important point is that all four methods give approximately the same SE result. Figure 10.16 shows the SE results for an open square aperture (23.5 cm by 23.5 cm), and the four methods give approximately the same SE result (4 dB). Figure 10.17 shows the SE results for a generic aperture with a combination of circular holes and rectangular slots [120], and the four methods again show good agreement for SE (approximately 8.5 dB).

Thus the method of using frequency stirring and a small, wall-mounted monopole antenna (most convenient for small enclosures) is well verified. However, it should be kept in mind that even though the method is useful for physically small enclosures, the frequency-stirring method still requires that the enclosure be electrically large.

10.3 MEASUREMENT OF ANTENNA EFFICIENCY

Because reverberation chambers typically involve received power measurements, they are well suited for measurement of antenna efficiency. The results in this section are closely related to those in Section 7.7, but are applied specifically to antennas rather than general test objects. Because we do not need to assume reciprocal antennas, we will treat the receiving and transmitting antennas separately.

10.3.1 Receiving Antenna Efficiency

A measurement setup for receiving antenna efficiency is shown in Figure 10.18. The reverberation chamber includes a transmitting antenna, a reference receiving antenna, and a receiving antenna under test (RAUT). The two antennas receive simultaneously. The reference receiving antenna is selected to have a high efficiency and low impedance mismatch (both factors assumed to be one). As in (7.104), the average power received $\langle P_{rref} \rangle$ by the reference antenna can be written:

$$\langle P_{rref} \rangle = \frac{E_0^2}{\eta} \frac{\lambda^2}{8\pi}, \tag{10.18}$$

where E_0^2 is the mean-square electric field in the chamber. It will turn out that the value of E_0^2 will be unimportant because it will cancel (both the reference antenna and the antenna under test will be in the same statistical environment).

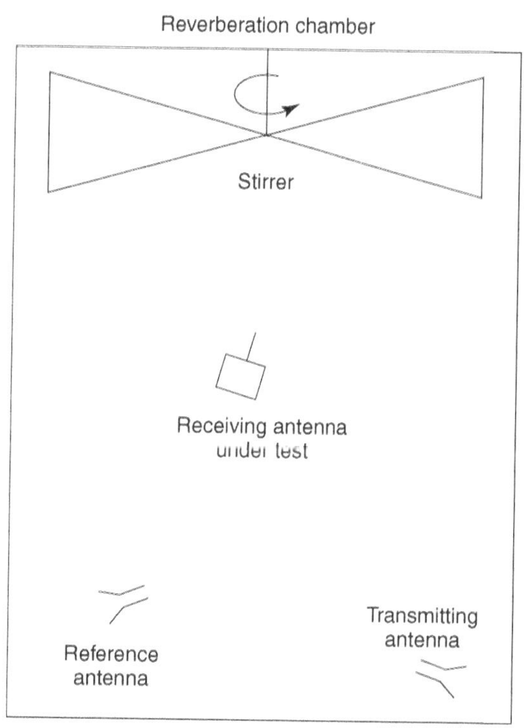

FIGURE 10.18 Measurement setup for receiving antenna efficiency.

Drawing on (7.105), the average power $\langle P_{RAUT}\rangle$ received by the antenna under test can be written:

$$\langle P_{RAUT}\rangle = \frac{E_0^2}{\eta}\frac{\lambda^2}{8\pi} m_{RAUT}\eta_{RAUT}, \tag{10.19}$$

where m_{RAUT} is the impedance mismatch of the RAUT, and η_{RAUT} is the efficiency of the RAUT. Equations (10.18) and (10.19) can be solved for the efficiency:

$$\eta_{RAUT} = \frac{\langle P_{RAUT}\rangle}{\langle P_{rref}\rangle m_{RAUT}} \tag{10.20}$$

In (10.20), $\langle P_{RAUT}\rangle$ and $\langle P_{rref}\rangle$ are the measured averaged powers. The impedance mismatch factor m_{RAUR} is close to one for a well designed antenna, but it can be measured with a network analyzer, as shown in the previous section. Related efficiency measurements for more complex array antennas have also been measured in reverberation chambers [121].

10.3.2 Transmitting Antenna Efficiency

A measurement setup for transmitting antenna efficiency is shown in Figure 10.19. The reverberation chamber contains an efficient (reference) receiving antenna, a

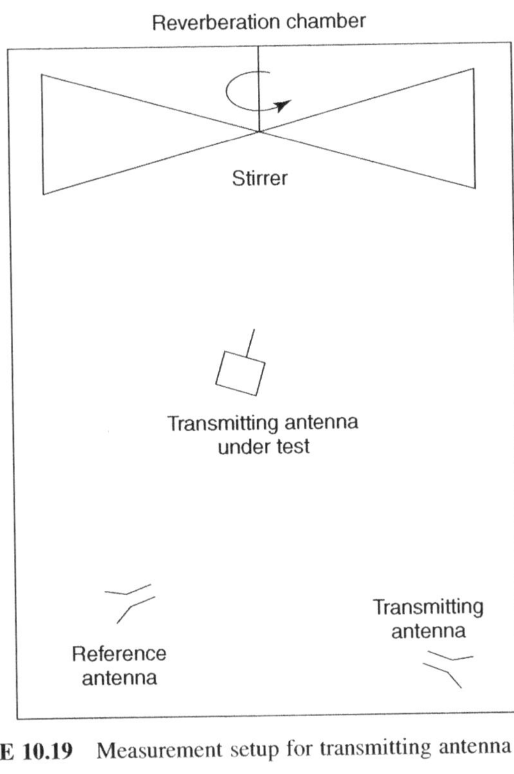

FIGURE 10.19 Measurement setup for transmitting antenna efficiency.

reference transmitting antenna, and a transmitting antenna under test (TAUT). The reference transmitting antenna is chosen to have both the efficiency and the impedance mismatch factor close to one. In this case, two transmission measurements are made with equal power fed to the reference and test antennas. The result for the efficiency of the transmitting antenna under test η_{TAUT} is analogous to that in (10.20):

$$\eta_{TAUT} = \frac{\langle P_{TAUT} \rangle}{\langle P_{tref} \rangle m_{TAUT}}, \tag{10.21}$$

where $\langle P_{TAUT} \rangle$ is the power received by the reference receiving antenna when the antenna under test is transmitting, and $\langle P_{tref} \rangle$ is the power received by the reference receiving antenna when the reference antenna is transmitting. The impedance mismatch factor m_{TAUT} is close to one for a well designed antenna, but again it can be measured with a network analyzer, as shown in the previous section.

For a reciprocal antenna, the receiving and transmitting antenna efficiencies are equal:

$$\eta_{RAUT} = \eta_{TAUT} \tag{10.22}$$

This result is analogous to (7.150) for reciprocal test objects.

10.4 MEASUREMENT OF ABSORPTION CROSS SECTION

In Section 7.6, the Q of reverberation chambers was analyzed for the general case of four loss mechanisms (wall loss, absorption loss, leakage, and extraction due to receiving antennas). If we wish to know the averaged absorption cross section of an absorbing object $\langle \sigma_a \rangle_\Omega$, we can determine it from its contribution to chamber Q from (7.127):

$$\langle \sigma_a \rangle_\Omega = \frac{2\pi V}{\lambda} Q_2^{-1}, \tag{10.23}$$

where the subscript Ω indicates average with respect to incidence angle and polarization. When there is no absorber in the chamber (the unloaded case), we can write the unloaded quality factor Q_u in the following manner by setting Q_2^{-1} equal to zero in (7.113):

$$Q_u^{-1} = Q_1^{-1} + Q_3^{-1} + Q_4^{-1} \tag{10.24}$$

With the absorbing object in the chamber (the loaded case), we can derive the loaded quality factor Q_l by again using (7.113):

$$Q_l^{-1} = Q_u^{-1} + Q_2^{-1} \tag{10.25}$$

From (10.23) and (10.25), the absorption cross section can be written in terms of measurements of loaded and unloaded chamber Q:

$$\langle \sigma_a \rangle_\Omega = \frac{2\pi V}{\lambda}\left(Q_l^{-1} - Q_u^{-1}\right) \tag{10.26}$$

From (7.111), the loaded and unloaded Q can be written:

$$Q_l = \frac{16\pi^2 V}{\lambda^3}\frac{\langle P_{rl}\rangle}{P_t} \quad \text{and} \quad Q_u = \frac{16\pi^2 V}{\lambda^3}\frac{\langle P_{ru}\rangle}{P_t}, \tag{10.27}$$

where P_t is the transmitted power, $\langle P_{rl}\rangle$ is the average received power for the loaded case, and $\langle P_{ru}\rangle$ is the average received power for the unloaded case. From (10.26) and (10.27), we can write the average absorption cross section in the following form:

$$\langle \sigma_a \rangle_\Omega = \frac{\lambda^2 P_t}{8\pi}\left(\frac{1}{\langle P_{rl}\rangle} - \frac{1}{\langle P_{ru}\rangle}\right) \tag{10.28}$$

Note that the result in (10.28) is independent of chamber volume V. From (7.127), we see that (10.28) also applies to the sum of average absorption cross sections if more than one absorbing object is involved.

A form equivalent to (10.28) has been used to determine the absorption cross-section of a lossy cylinder [122]. The experimental result was compared with numerical calculations for a lossy cylinder, and the agreement as shown in Figure 10.20

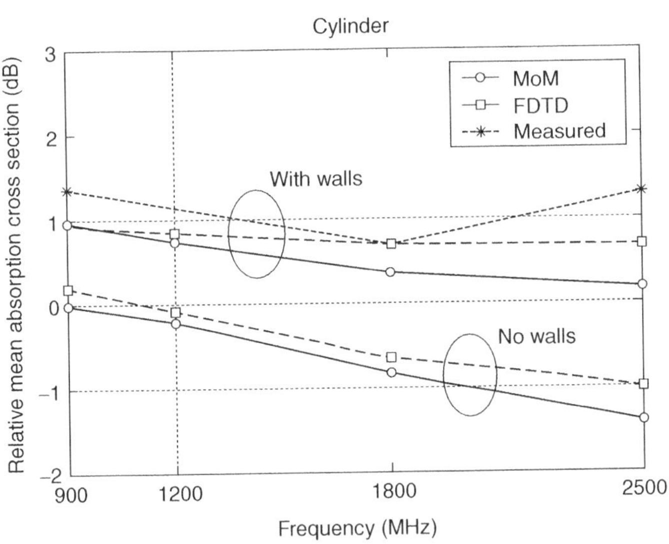

FIGURE 10.20 Mean absorption cross section of a lossy cylinder as a function of frequency. The reference value is 271.13 cm^2 [122].

was good. It has also been pointed out that (10.28) can be used to determine the electrical properties of electrically large dielectric objects [123].

Although (10.26) is the fundamental equation for determining absorption cross section, the loaded and unloaded chamber Q can also be determined from chamber time constant. From (8.24) the loaded and unloaded Q can be written:

$$Q_l = \omega\langle\tau_l\rangle \quad \text{and} \quad Q_u = \omega\langle\tau_u\rangle, \tag{10.29}$$

where $\langle\tau_l\rangle$ is the loaded chamber time constant and $\langle\tau_u\rangle$ is the unloaded chamber time constant. If we substitute (10.29) into (10.26) and use $\omega = 2\pi c/\lambda$, we can also write the absorption cross section in the following form:

$$\langle\sigma_a\rangle_\Omega = \frac{V}{c}\left(\frac{1}{\langle\tau_l\rangle} - \frac{1}{\langle\tau_u\rangle}\right), \tag{10.30}$$

where c is the free-space speed of light.

PROBLEMS

10-1 Derive both results in (10.8).

10-2 Show that if wall losses are dominant in both chambers, then SE_3 in (10.12) reduces to SE_4.

10-3 Compare the short monopole antenna in Figure 10.13 with the short dipole in Figure E1. Show that if the monopole antenna is half the length of the dipole and is impedance matched, then the received power is equal to that of the short dipole in (E4). Hint: make use of (10.17) and the fact that the radiation resistance of a monopole is half that of a dipole.

10-4 Verify that antenna transmitting and receiving efficiencies are equal as in (10.22) if the transmitting and receiving mismatch factors are equal ($m_{TAUT} = m_{RAUT}$).

10-5 Verify that the expressions for the absorption cross-sections in (10.28) and (10.30) are equivalent. Why does (10.30) require the chamber volume V whereas (10.28) does not?

CHAPTER 11

Indoor Wireless Propagation

This chapter represents a departure from the rest of Part II. Commercial and residential buildings and rooms come in many varieties [124], but they generally have fairly low Q values because of windows, penetrable walls, absorbing materials, etc. However, there are some exceptions to this—metal-walled manufacturing plants, airplane hangars, etc. In any case, buildings and rooms are cavities in the sense that they exhibit internal multipath propagation. Since indoor communication is important to the very large wireless communication industry, it is useful to summarize some of the propagation models and to compare their similarities and differences with the statistical techniques discussed in the rest of Part II.

11.1 GENERAL CONSIDERATIONS

The interiors of buildings are typically complicated environments because of the complex construction—walls, doors, windows, scatterers, absorbers, etc. Also, the environment changes: doors and windows are opened and closed, furniture and other objects are moved, and people move around. Even though ray tracing [125] and other computational methods have recently been applied to such complex environments, these methods require a very large amount of site-specific information for a deterministic calculation. Hence, we will continue the philosophy of statistical methods based on partial information as described in the rest of Part II.

Two thorough literature surveys on indoor propagation [124,126] are available. This chapter will concentrate on the case where both the transmitting and receiving antennas are located inside the building, but the case of an external antenna is also of some interest. Penetration loss (or building attenuation) has been defined by Rice [127] as the difference between the received signal inside a building and the average of the received signal around the perimeter of the building. This is not a very precise definition, but it is probably adequate for most cases when one considers the variation of field strength likely to occur within and around the perimeter of most buildings. It has been found that penetration loss is dependent on the construction materials of a building, internal layout, floor height, number and size of windows, incident field

Electromagnetic Fields in Cavities: Deterministic and Statistical Theories, by David A. Hill

angle of arrival and polarization, and frequency. For example, building attenuation for houses of various construction has been found to vary from −2 to 24 dB [128–130] and to increase with frequency [128].

Indoor propagation has been more thoroughly studied for the case where both the transmitting and receiving antennas are located within a building, and the rest of this chapter will deal with this case.

11.2 PATH LOSS MODELS

Path loss is defined as the ratio of transmitted to received power in dB. Hence it is a positive real number. Path loss models for indoor propagation [42,131] tend to be empirical because they are based on experimental data. Consequently, it is difficult to attach much physical meaning to the models and their various adjustable parameters. However, the models can still be useful, and in some limiting cases they do have physical interpretations. Here, we will discuss a few of the more popular models.

Many researchers have shown indoor path loss to obey the following distance power law [131]:

$$PL(\text{dB}) = PL(d_0) + 10n\log\left(\frac{d}{d_0}\right) + X_\sigma \tag{11.1}$$

where $PL(\text{dB})$ is the path loss in dB for an antenna separation d, $PL(d_0)$ is the path loss at some small reference distance d_0, the value of n depends on the building characteristics, and X_σ is a normal random variable in dB with a standard deviation of σ dB. The term $PL(d_0)$ is separated from the rest of the right side of (11.1) so that it includes primarily the effects of the transmitting and receiving antennas, and in some cases d_0 is chosen to be 1 m [42]. The term $10n\log(\frac{d}{d_0})$ represents propagation as a power law d^{-n} for the power density. (In this section, log is taken to the base 10 because we are expressing quantities in terms of dB.) If propagation is dominated by line of sight with spherical spreading, then $n = 2$. If both the transmitting and receiving antennas are located near a flat interface (such as a floor), then the direct and reflected rays cancel and n tends toward 4 [33]. The value of $n = 4$ represents lateral wave propagation along the flat interface. For the special case of $X_\sigma = 0$, (11.1) becomes a deterministic equation for the mean value of the path loss. For propagation in complex buildings, the values of n and σ are fit to experimental data and have no simple physical interpretation. A table of values for n and σ, as measured in different buildings, is given in [132].

A model similar to (11.1) has been shown to be successful for cases where the transmitting and receiving antennas are located on different floors [131,133]:

$$PL(\text{dB}) = PL(d_0) + 10n_{SF}\log\left(\frac{d}{d_0}\right) + FAF \tag{11.2}$$

where n_{SF} represents the exponent for propagation between antennas located on the same floor, and FAF is the floor attenuation factor in dB. A table for measured values of FAF and its standard deviation is given in [131] for propagation through one up to as many as four floors. A variation of (11.2) has been obtained by eliminating FAF and changing the exponent to account for propagation through the appropriate number of floors [131]:

$$PL(\text{dB}) = PL(d_0) + 10n_{MF}\log\left(\frac{d}{d_0}\right), \tag{11.3}$$

where n_{MF} indicates a path loss exponent based on measurements through multiple floors. A table of measured values of n_{MF} for various numbers of floors and numbers of receiver locations is given in [131].

Devasirvatham, et al., [46] found that path loss in some buildings could be fit by free-space path loss plus exponential attenuation:

$$PL(\text{dB}) = PL(d_0) + 20\log\left(\frac{d}{d_0}\right) + \alpha d \tag{11.4}$$

where α is the attenuation rate in dB/m. The term $20\log(\frac{d}{d_0})$ represents spherical spreading loss ($n = 2$ in the previous models), and the term αd could be physically interpreted as attenuation in a lossy medium. Propagation in inhomogeneous, random media is a topic with a large literature (see [53] plus references), and attenuation in such media is due to both absorption and scattering. For simple models, α can be calculated, but for propagation in buildings it must be fit to measurements. Measurements made in a large commercial metropolitan building at frequencies of 850 MHz, 1.9 GHz, 4.0 GHz, and 5.8 GHz [46] yielded α values of 0.54, 0.49, 0.62, and 0.55 dB/m, respectively. These values exhibit remarkably little frequency dependence.

11.3 TEMPORAL CHARACTERISTICS

It is important to have a quantitative knowledge of the temporal characteristics of indoor propagation channels in order to determine limits on data rates due to intersymbol interference. Because of the variety and complexity of indoor propagation conditions, several types of propagation models have been proposed and compared to measurements. Typically the models yield the RMS delay spread, which is a limiting factor in data rates for wideband communications. In this section, we will discuss several models that have been found useful for determining temporal characteristics of indoor channels.

11.3.1 Reverberation Model

For buildings that support many internal reflections, such as metal-wall factories, the fields and energy density follow the characteristics of reverberation chambers described in Chapter 7. The fields are statistically uniform in space, and the Q is fairly high. The average received power $\langle P_r \rangle$ follows the same time decay dependence as the cavity energy given in (8.23):

$$\langle P_r(t) \rangle = P_0 \exp(-t/\tau)\mathrm{U}(t), \tag{11.5}$$

where P_0 is a constant depending on the transmitted power and $\tau = Q/(2\pi f)$. Here we assume that f is the carrier frequency of a short pulse that is turned off at $t = 0$.

For this simple time dependence, we can calculate the RMS delay spread as follows. We first calculate the mean time delay $\langle \tau \rangle$ from [134]:

$$\langle \tau \rangle = \frac{\int_0^\infty t \exp(-t/\tau)\mathrm{d}t}{\int_0^\infty \exp(-t/\tau)\mathrm{d}t} = \tau \tag{11.6}$$

We have set $P_0 = 1$ in (11.6) because the result is independent of P_0. The RMS delay spread τ_{rms} is then determined from [134]:

$$\tau_{rms} = \sqrt{\frac{\int_0^\infty (t-\langle \tau \rangle)^2 \exp(-t/\tau)\mathrm{d}t}{\int_0^\infty \exp(-t/\tau)\mathrm{d}t}} = \tau \tag{11.7}$$

The outcome of (11.6) and (11.7) that $\langle \tau \rangle = \tau_{rms}$ is specific to the exponential time dependence in (11.5) and is not a general result.

Measurements of τ and Q have been made in the main cabins of small airplanes and compared with theory [135]. Not enough information was available to calculate all the losses in the main cabin of the hanger queen airplane, but the approximate volume was $V = 7.25\ \mathrm{m}^3$ and the approximate window area was $A = 2.61\ \mathrm{m}^2$. If we assume that the windows are electrically large, then the theoretical value of Q_3 due to leakage in (7.129) reduces to [38]:

$$Q_3 = \frac{8\pi V}{\lambda A}, \tag{11.8}$$

where we have neglected any effects of window glass. Because Q_3 accounts only for leakage loss, it can be considered a loose upper bound for Q. The Q of the main cabin was measured using both the power-ratio and the time-constant methods. Transverse electromagnetic (TEM) horns were used for the time-constant measurements, and broadband ridged horns were used for the power-ratio measurements. The theoretical value of Q_3 is compared with measured values of Q in Figure 11.1 for frequencies from 4 to 18 GHz [135]. The calculated curve for Q_3 exceeds the measured Q values, as expected, because it is only an upper bound for Q. The scatter in the measured Q values is probably due to the smaller-than-ideal stirrer that was used in the measurements due to lack of space.

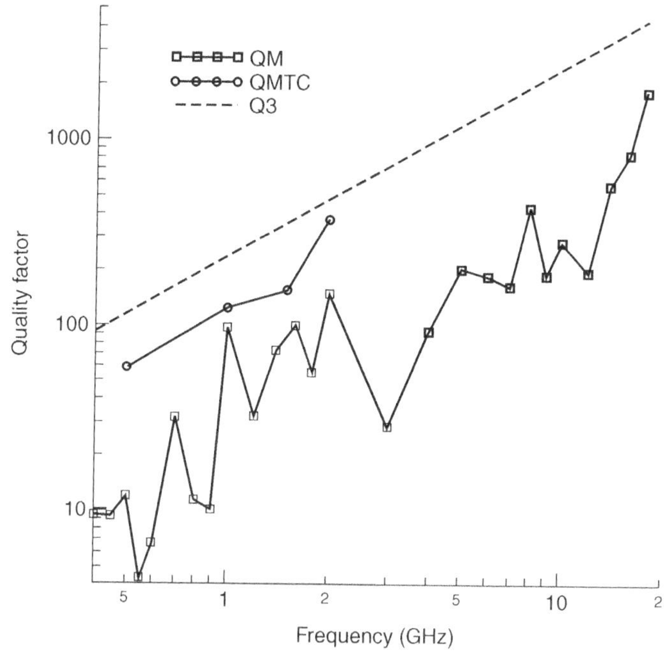

FIGURE 11.1 Quality factor Q of the main cabin of the hangar queen airplane determined by cw measurement (QM), time-domain measurement (QMTC), and leakage calculation (Q3) [135].

It is also possible to calculate a theoretical time constant τ_3 from (11.8) and the relationship between quality factor and decay time:

$$\tau_3 = Q_3/(2\pi f) = \frac{4V}{cA}, \tag{11.9}$$

where c is the free-space speed of light. This decay time is independent of frequency and can be considered an upper bound because Q_3 is an upper bound for Q. If we substitute the volume V and the window area A for the hangar queen into (11.9), the result is $\tau_3 = 37.0$ ns. As expected, this value is higher than the measured values in Table 11.1.

TABLE 11.1 Measured Time Constant for the Main Cabin of the Hangar Queen [135].

Frequency (GHz)	τ (ns)
0.5	18.63
1.0	19.49
1.5	16.35
2.0	29.72

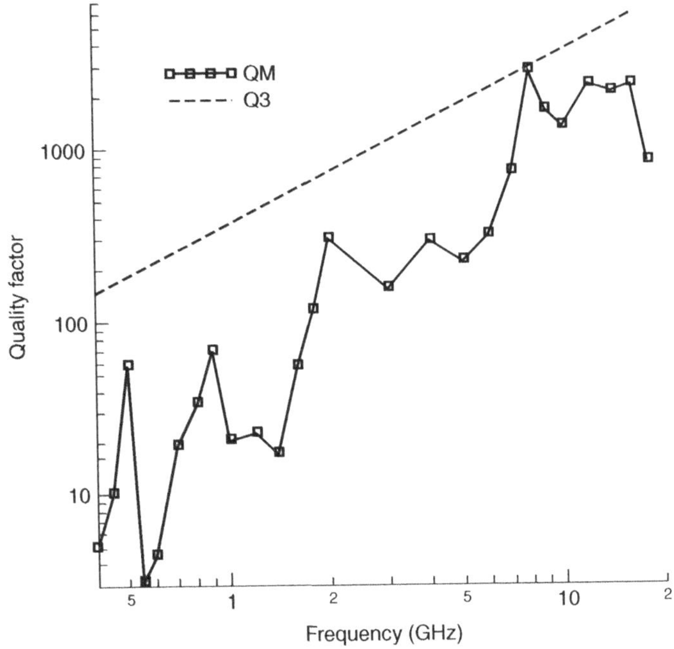

FIGURE 11.2 Quality factor Q of the main cabin of Airplane 1 determined by cw measurement (QM) and leakage calculation (Q3) [135].

Airplane 1 is a twin-engine, six-passenger plane. The estimated volume V of the main cabin is $9.46\,\mathrm{m}^3$, and its estimated window area A is $2.15\,\mathrm{m}^2$. In Figure 11.2, we show the Q measured by the power-ratio method and the calculated Q_3 from (11.8) for frequencies from 4 to 18 GHz. Q_3 again serves as an upper bound for the measured Q because it accounts only for window leakage loss.

Wireless propagation measurements have been performed in a large ($500\,\mathrm{m} \times 250\,\mathrm{m} \times 15\,\mathrm{m}$) assembly plant constructed out of metal [136]. The time-decay characteristics at three frequencies (950 MHz, 2450 MHz, and 5200 MHz) obtained with a 200 MHz averaging bandwidth are shown in Figure 11.3. The large values of Q (greater than 1000) indicate a reverberant environment. The large decay times (greater than 100 ns), which are approximately equal to the rms delay spreads, might make reliable wireless technology difficult. No comparison of the results in Figure 11.3 with theory was made because insufficient information was available for Q calculations.

11.3.2 Discrete Multipath Model

A scalar multipath model that treats individual reflections separately has been developed and used to analyze measured data from factories [137]. This model has increased generality in that it does not assume exponential decay as in (11.5), but the parameters need to be determined experimentally. Let $x(t)$ represent the transmitted

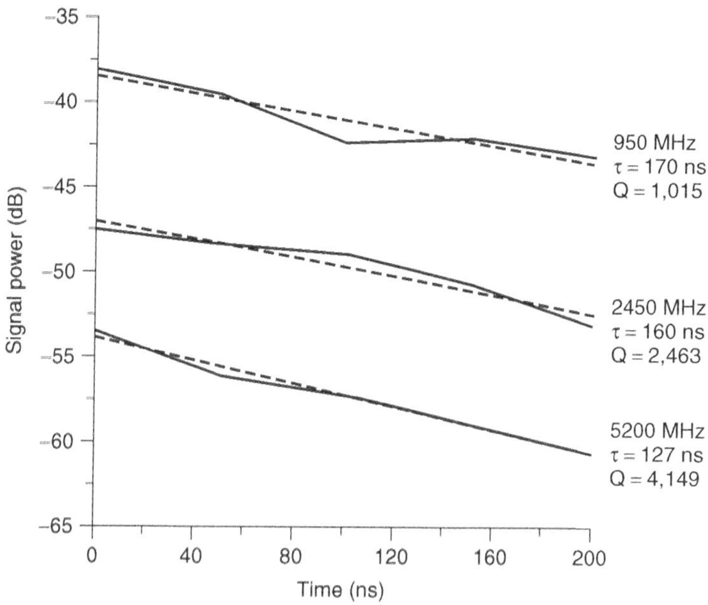

FIGURE 11.3 Measured (solid) and fit (dashed) time decay characteristics at three wireless frequencies with a 200 MHz averaging bandwidth [136].

waveform and $y(t)$ represent the received waveform. For a discrete channel model, $y(t)$ can be written as [138]:

$$y(t) = \sum_k \alpha_k(t)x[t-\tau_k(t)] \tag{11.10}$$

Typically, α_k and τ_k are essentially independent of time. Then the impulse response $h(t)$ of the channel can be written as [137]:

$$h(t) = \sum_{k=0}^{N-1} \alpha_k \delta(t-\tau_k), \tag{11.11}$$

where τ_0 is the arrival time of the first observable pulse and N is the number of observable pulses.

Consider a transmitted signal of the form:

$$x(t) = \mathrm{Re}[p(t)\exp(-i2\pi f_c t)], \text{ where } \quad p(t) = \begin{cases} 1, & \text{for } 0 \le t \le t_p \\ 0, & \text{elsewhere} \end{cases}, \tag{11.12}$$

and f_c is the carrier frequency. The channel output is obtained by convolution:

$$y(t) = \int_{-\infty}^{\infty} x(\zeta)h(t-\zeta)\mathrm{d}\zeta = \mathrm{Re}[r(t)\exp(-i2\pi f_c t)], \tag{11.13}$$

where:

$$r(t) = \sum_{k=0}^{N-1} \alpha_k \exp(i2\pi f_c \tau_k) p(t-\tau_k) \tag{11.14}$$

To simplify the model, the channel may be equivalently described by the baseband impulse response $h_b(t)$, having an output $r(t)$ that is the complex envelope of $y(t)$. The low-pass characterization removes the high frequency variations caused by the carrier. Thus, the low-pass equivalent channel impulse response $h_b(t)$ is given by [137]:

$$h_b(t) = \sum_{k=0}^{N-1} \alpha_k \exp(i2\pi f_c \tau_k) \delta(t-\tau_k), \tag{11.15}$$

where α_k represents a real attenuation factor, $\exp(i2\pi f_c \tau_k)$ represents a linear phase shift due to propagation, and τ_k is the time delay of the kth path in the channel.

In general, the appropriate pulse width t_p is chosen according to the carrier frequency and the desired path resolution. For example, in [137] the pulse width t_p was chosen to be 10 ns so that the output of the low-pass channel closely approximates the impulse response $h_b(t)$. As in [139], instead of measuring the output $r(t)$, the squared magnitude $|r(t)|^2$ is measured. If $|\tau_j - \tau_k| > 10$ ns for all $j \neq k$, then:

$$|r(t)|^2 = \sum_{k=0}^{N-1} \alpha_k^2 p^2(t-\tau_k), \tag{11.16}$$

and the power profile measurement has a path resolution of 10 ns. For $|\tau_j - \tau_k| < 10$ ns, there is pulse overlap, and there are unresolvable subpaths that combine to form one observable path.

Wide-band multipath channels are grossly quantified by their mean excess delay $\langle\tau\rangle$ and RMS delay spread τ_{rms} [138,139]. The discrete analogy to the integral form for the mean time delay in (11.6) is [137]:

$$\langle\tau\rangle = \frac{\sum_{k=0}^{N-1} \alpha_k^2 \tau_k}{\sum_{k=0}^{N-1} \alpha_k^2} \tag{11.17}$$

The discrete analogy to the integral form for the RMS delay spread in (11.7)—the second central moment of the profile—is [137]:

$$\tau_{rms} = \sqrt{\langle\tau^2\rangle - (\langle\tau\rangle)^2}, \quad \text{where } \langle\tau^2\rangle = \frac{\sum_{k=0}^{N-1} \alpha_k^2 \tau_k^2}{\sum_{k=0}^{N-1} \alpha_k^2} \tag{11.18}$$

TABLE 11.2 RMS Delay Spread Data (10–25 m Paths) [137].

RMS delay spread as a function of factory topography (ns) T-R separation of 10–25 m					
Topography	Site B	Site C	Site D	Site E	Site F
LOS light clutter	87.6	118.8	51.1	—	—
LOS heavy clutter	45.6	46.9	106.7	48.7	124.3
LOS along wall	—	122.4	—	—	—
Obstructed light clutter	27.7	102.6	103.2	—	—
Obstructed heavy clutter	70.9	101.5	52.0	79.3	49.6

An advantage of this model is that it can be used regardless of whether or not a strong line-of-sight (LOS) path exists. Thus the model is much more general than the reverberation model, which assumes that the LOS contribution to the total received signal is small. A disadvantage is that it is measurement intensive, in that all values of α_k^2 and τ_k must be determined experimentally in order to characterize the channel. As seen in (11.17) and (11.18), this is true even for the gross channel properties, mean time delay and RMS delay spread.

Measurements made at multiple locations in five factories have been used to determine RMS delay spread [137]. The results are shown in Table 11.2 for short paths (10 to 25 m in length) and in Table 11.3 for longer paths (40 to 75 m in length). Both tables include numerous cases of large delay spread (greater than 100 ns). The values of delay spread are not correlated with path length or topography (LOS, clutter, etc.). These findings agree with some measurements in office buildings [139,140], but disagree with measurements made in a much larger office building [141].

11.3.3 Low-Q Rooms

As indicated in (11.5), the received power in a high-Q (reverberating) room decays exponentially with a decay time of $\tau = Q/\omega$. In this section, we consider a room where the walls are not highly reflecting. In this case, wall loss is dominant, and we can approximate Q by (7.116):

$$Q = Q_1 = \frac{2kV}{A\langle(1-|\Gamma|^2)\cos\theta\rangle_\Omega} \tag{11.19}$$

TABLE 11.3 RMS Delay Spread Data (40–75 m Paths) [137].

RMS delay spread as a function of factory topography (ns) T-R separation of 40–75 m					
Topography	Site B	Site C	Site D	Site E	Site F
LOS light clutter	33.9	43.2	118.5	—	—
LOS heavy clutter	39.5	201.5	33.3	93.6	44.3
LOS along wall	—	92.7	—	—	—
Obstructed light clutter	—	118.5	108.9	—	—
Obstructed heavy clutter	77.2	114.7	106.8	52.5	129.6

If we divide (11.19) by ω, we obtain the following for the decay time:

$$\tau = \frac{2V}{cA\langle(1-|\Gamma|)\cos\theta\rangle} \tag{11.20}$$

To cast the decay time in the form used in the acoustics community [142], we can rewrite (11.20) as:

$$\tau = \frac{4V}{cA\alpha}, \tag{11.21}$$

where the absorption coefficient α is [134]:

$$\alpha = 2\int_0^{\pi/2}\left[1-\frac{1}{2}\left(|\Gamma_{TE}|^2+|\Gamma_{TM}|^2\right)\right]\cos\theta\sin\theta\,\mathrm{d}\theta \tag{11.22}$$

For a homogeneous half space, the reflection coefficient Γ_{TE} for TE (perpendicular) polarization is given in (7.117), and the reflection coefficient Γ_{TM} for TM (parallel) polarization is given in (7.118). For layered media (more applicable to room walls), the reflection coefficients are given in [143] and [144]. For the acoustic case [142], c in (11.21) is replaced by the speed of sound.

The exponential decay time in (11.21) is valid for highly reflecting walls ($\alpha \ll 1$). However, for poorly reflecting walls, the exponential decay model with the decay time given by (11.21) is not valid. To illustrate this failure, consider the case where the reflection coefficients are zero. In this case, (11.22) reduces to:

$$\alpha_{nr} = 2\int_0^{\pi/2}\cos\theta\sin\theta\,\mathrm{d}\theta = 1, \tag{11.23}$$

where the α subscript nr refers to nonreflecting walls. Then, the decay time in (11.21) reduces to:

$$\tau_{nr} = \frac{4V}{cA} \tag{11.24}$$

Hence, the decay time τ_{nr} for nonreflecting walls approaches a constant rather than the expected value of zero. This same dilemma of nonzero decay time for rooms with nonreflecting walls has been noted in the analogous acoustic problem [145]. A solution to this dilemma for acoustic problems was given by Eyring [145] where he approximated the characteristic decay time of so-called "dead" rooms as:

$$\tau = \frac{l_c}{-c\ln(1-\alpha)} \tag{11.25}$$

The length l_c is defined as the mean-free path between wall reflections, and for a rectangular room is given by [146]:

$$l_c = \frac{4V}{S} \tag{11.26}$$

For $\alpha = 1$, (11.25) gives the expected value of $\tau = 0$. For small α, (11.25) and (11.26) agree with (11.21). The "dead" room formula in (11.25) has been used to analyze electromagnetic anechoic test chambers [146].

Dunens and Lambert [147] define reverberation as occurring when several wall reflections are present, or equivalently reverberation occurs after approximately $10l_c/c$. For indoor wireless communications in rooms with walls that are not highly reflecting, a large amount of energy is lost through the walls, and few reflections occur. So before the time $10l_c/c$ elapses, only a small amount of energy remains in the room. This case where few wall reflections occur can be referred to as the nonreverberating regime. Holloway et al. [134], have developed a power delay profile (PDP) model to cover this nonreverberating case.

Their model separates received power according to time intervals depending on the number of reflections that have occurred. The characteristic time t_c of a room that is required before a given set of rays makes one reflection is given by a function of the mean-free path l_c and by utilizing (11.26) can be expressed as [134]:

$$t_c = 2\frac{l_c}{c} = \frac{8V}{cA} \tag{11.27}$$

Equation (11.27) has been justified in [134] by using a ray tracing model for rays making n wall bounces for integer values of n from 1 through 10. The authors [134] demonstrated that by $t = nt_c$, the majority of the rays making n bounces have reached the receiver.

By using the characteristic parameters of a room, it is possible to approximate the power levels at different times. The average power level of the bundle of rays that corresponds to rays after n reflections is approximated by:

$$P_n = A\frac{\gamma^n}{d_n^2} \tag{11.28}$$

In (11.28), A is a constant that is a function of the transmitting and receiving antennas and transmitted power, and d_n is the characteristic distance that a bundle of rays making n reflections travels and is determined by the time it takes these rays to reach the receiver (nt_c). Using the definition of t_c in (11.27), d_n is expressed in terms of the mean-free path l_c as:

$$d_n = nt_cc = 2nl_c \tag{11.29}$$

The average power reflection γ is defined as:

$$\gamma = 1-\alpha, \tag{11.30}$$

where α is given by (11.22).

The direct ray arrives at the receiver at a time delay determined by the transmitter and receiver separation d_0. The power level of the direct ray at the receiver is given by:

$$P_0 = \frac{A}{d_0^2} \tag{11.31}$$

The antenna separation is known for a specific configuration, but the goal of the analysis in [134] is to determine the PDP of the room in an average sense; that is, to determine the global behavior without knowing the exact location of the transmitter and receiver. Thus, it is assumed that the direct path equals an average distance equal to one characteristic length of the room $d_0 = l_c$ and the direct ray arrives at the receiver at $t = t_0 = l_c/c$.

With the power level and delay times of the direct and reflected rays determined, the PDP can be modeled. By initializing the delay time of the direct ray to zero and normalizing the power to P_0, the power levels at different delay are approximated by:

$$\begin{aligned} \mathrm{PDP}_0 &= 1, \quad \tau = 0, \quad \text{for} \quad n = 0 \\ \mathrm{PDP}_n &= \frac{1}{4}\frac{\gamma^n}{n^2}, \quad \tau_n = \frac{t_c}{2}(2n-1), \quad \text{for} \quad n \neq 0. \end{aligned} \tag{11.32}$$

The normalized PDP is shown in Figure 11.4. By connecting the arrows in this figure, an approximation to the PDP is obtained. One needs to keep in mind that this PDP is not for a particular location in a room; it corresponds to the average room behavior.

The average reflected power given by (11.30) assumes that all the reflecting surfaces are identical. When different reflecting surfaces are present in a room the average power reflection coefficient is calculated as a weighted average of all the surfaces. The effective average absorption and the resultant average power reflection

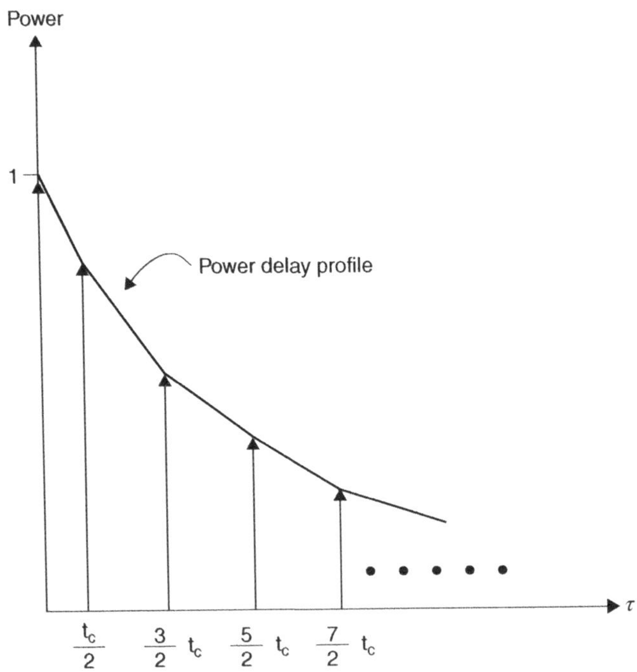

FIGURE 11.4 Normalized PDP model for an in-room wireless radio propagation channel [134].

coefficient in a room with different reflecting surfaces are given by:

$$\alpha_{eff} = \frac{\sum_n A_n \alpha_n}{A} \quad \text{and} \quad \gamma_{eff} = 1 - \alpha_{eff}, \tag{11.33}$$

where A is the total surface area of the room, A_n is the area of surface n, and α_n is the average absorption of surface n.

This model has been compared to measurements made at a carrier frequency of 1.5 GHz with a bandwidth of 500 MHz in two different rooms. The measurement system is described in [148] and [149]. The first room is a small office with a height of 3.20 m, a width of 2.31 m, and a length of 5.26 m. The second room is a laboratory with a height of 5.0 m, a width of 7.18 m, and a length of 9.35 m. The walls in the office and laboratory were composed of concrete slabs and concrete blocks of thickness 14.5 cm with $\varepsilon_r = 6.0$ and $\sigma = 1.95 \times 10^{-3}$ S/m [150]. Figures 11.5 and 11.6 show comparisons of the PDP model to measured data for the two rooms. The measured data in both rooms were obtained with the transmitter located near a corner of the room at a height of 1.8 m, and the receiver was placed on a cart with an antenna height of 1.8 m. The impulse responses for several locations distributed throughout the rooms were obtained. The magnitude of all the

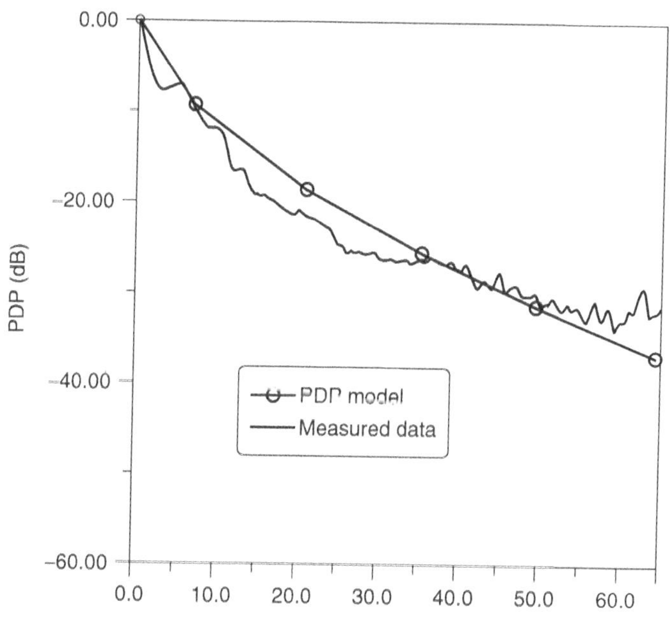

FIGURE 11.5 Comparison of the PDP model to the measured data obtained by averaging several locations through the office with a length of 5.26 m, a width of 2.31 m, and a height of 3.20 m [134].

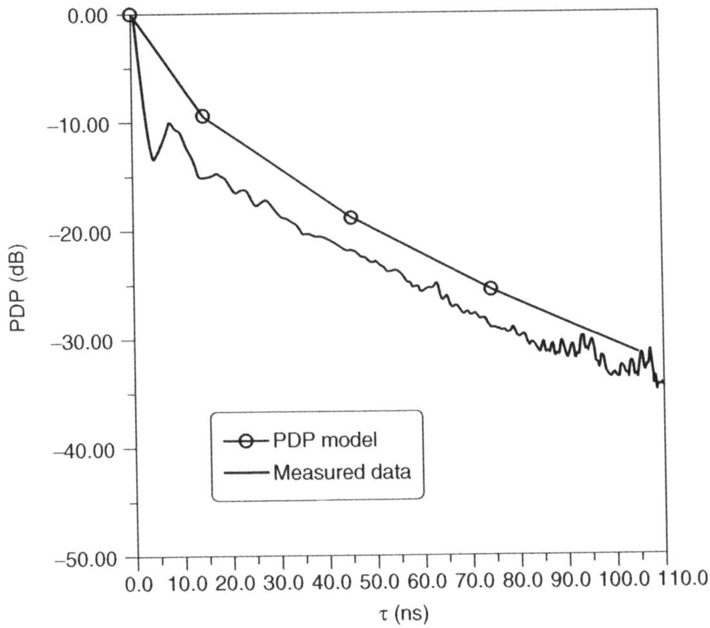

FIGURE 11.6 Comparison of the PDP model to the measured data obtained by averaging several locations through the laboratory with a length of 9.35 m, a width of 7.18 m, and a height of 5.00 m [134].

impulse responses in each of the two rooms were averaged together to obtain an effective average PDP of each room. The comparisons in these two figures illustrate that the PDP model predicts the same decay characteristics in PDP as seen in the measurements.

The PDP model can be used to estimate the rms delay spread for the two rooms [134]. The rms delay spread is calculated by the following expression, which is analogous to (11.7):

$$\tau_{rms} = \sqrt{\frac{\int_0^{\infty} (t-\langle\tau\rangle)^2 \mathrm{PDP}(t)\mathrm{d}t}{\int_0^{\infty} \mathrm{PDP}(t)\mathrm{d}t}}, \tag{11.34}$$

where:

$$\langle\tau\rangle = \frac{\int_0^{\infty} t\,\mathrm{PDP}(t)\mathrm{d}t}{\int_0^{\infty} \mathrm{PDP}(t)\mathrm{d}t} \tag{11.35}$$

For the office room, the calculated RMS delay spread was 6.1 ns, and the measured average PDP yielded an rms delay spread of 7.5 ns. For the laboratory, the calculated RMS delay spread was 13.2 ns, and the measured average PDP yielded an RMS delay spread of 15.6 ns. Considering the uncertainties in the measured values and the approximations in the PDP model, these agreements are reasonably good. Also, these RMS delay spreads are sufficiently small that it is clear that the rooms are not reverberating (low Q).

In summary, the PDP model is useful for analyzing nonreverberating rooms, which includes most office and residential rooms. It has the advantage that the PDP profile and the rms delay spread can be calculated, but the room parameters (dimensions and wall properties) must be known, at least approximately. The PDP model is not likely to be useful for metal-wall rooms, such as factories, where the Q can be fairly high and reverberation can occur.

11.4 ANGLE OF ARRIVAL

Indoor propagation channels are characterized by multipath, as discussed previously in this chapter. Although most indoor propagation research has dealt with path loss and temporal characteristics (such as time of arrival and RMS delay spread), as discussed in the previous two sections, less attention has been paid to angle of arrival. Yet the angle of arrival of the multipath signals is important in predicting the performance of adaptive array systems. In this section, we will discuss the ideal reverberation chamber environment and an empirical statistical [47] model based primarily on experimental results.

11.4.1 Reverberation Model

For buildings that support many internal reflections, such as metal-wall factories, the fields and the angular spectrum approximately follow the theory of reverberation chambers described in Chapter 7. As in (7.1), we can write the electric field $\vec{E}$ as:

$$\vec{E}(\vec{r}) = \iint_{4\pi} \vec{F}(\Omega)\exp(i\vec{k} \bullet \vec{r})\mathrm{d}\Omega, \tag{11.36}$$

where the angular spectrum $\vec{F}(\Omega)$ provides the information on angle of arrival. The angular spectrum contains both α (elevation) and β (azimuthal) components that have zero ensemble averages as indicated in (7.6). Equations (7.10) and (7.14) yield the following useful statistical properties of the components of $\vec{F}$:

$$\langle F_\alpha(\Omega_1)F_\alpha^*(\Omega_2)\rangle = \langle F_\beta(\Omega_1)F_\beta^*(\Omega_2)\rangle = \frac{E_0^2}{16\pi}\delta(\Omega_1 - \Omega_2), \tag{11.37}$$

where E_0^2 is the mean-square electric field and δ is the Dirac delta function. Since the argument of the delta function depends only on the angular difference, $\Omega_1 - \Omega_2$, the expectations of the squares of both angular spectrum components, $\langle|F_\alpha|^2\rangle$ and $\langle|F_\beta|^2\rangle$, contain delta functions which peak (at zero argument) for any values of Ω (shorthand for α and β).

Strictly speaking, the angular spectrum property in (11.37) has no physical meaning because the delta function is a distribution or generalized function. However, if we think of the delta function as a limit of sequence of ordinary (but highly peaked) functions [151], then we can picture (11.37) as representing plane waves propagating in all directions with both orthogonal polarizations. Hence a highly reverberant cavity produces all possible angles of arrival uniformly distributed.

A better way to justify the previous statement is to examine the expectation of the power received by a lossless, impedance-matched antenna, as given previously in (7.103):

$$\langle P_r\rangle = \frac{E_0^2}{\eta}\frac{\lambda^2}{8\pi} \tag{11.38}$$

Equation (11.38) is independent of the antenna pattern and the antenna orientation. So it is valid for the test case of a highly peaked antenna pattern with the antenna pointed in any arbitrary direction. Hence, we again conclude that a highly reverberant cavity generates all possible angles of arrival uniformly distributed.

Important properties of the field in a highly reverberant cavity are statistical spatial uniformity and isotropy, as shown previously in (7.15):

$$\langle|E_x|^2\rangle = \langle|E_y|^2\rangle = \langle|E_z|^2\rangle = \frac{E_0^2}{3} \tag{11.39}$$

The magnetic field has the same statistical spatial uniformity and isotropy properties as shown previously in (7.21):

$$\langle|H_x|^2\rangle = \langle|H_y|^2\rangle = \langle|H_x|^2\rangle = \frac{E_0^2}{3\eta^2} \tag{11.40}$$

Results for the spatial correlation functions of the electric and magnetic fields have been given in Section 7.4. The spatial correlation functions of antenna response in a highly reverberant cavity are similar, but in contrast to (11.38), they are dependent on the antenna receiving pattern [152]. Spatial correlation functions are important in cases where multiple receiving antennas are used to provide diversity in multipath environments. This type of wireless communication system is commonly called multiple-input, multiple-output (MIMO) [42].

11.4.2 Results for Realistic Buildings

Most indoor propagation experiments have concentrated on the time of arrival of multipath reflections rather than angle of arrival. However, because some indoor wireless systems use multiple antennas to combat multipath interference, some indoor angle of arrival measurements have been made. Angle of arrival measurements have been made at 950 MHz in a simple building with concrete walls at ranges of about 20 m [153]. The measurements were made only in the horizontal plane, but strong multipath lobes were measured. Another set of horizontal-plane measurements was made at 60.5 GHz in an office room (6 m $\times$ 4.65 m $\times$ 3 m) with and without furniture [154]. The presence of furniture made a significant difference in the angle of arrival results, particularly when it blocked the line-of-sight path. A planar array was used to scan in azimuth and elevation at 1 GHz in a large convention hall [155]. This type of scanning is useful because multipath lobes were detected at elevation angles out of the horizontal plane. None of these measurements was compared with any theoretical model.

One attempt at a comparison of theoretical and experimental results has been made in the frequency band from 6.75 to 7.25 GHz [156,157]. The model was based on a clustering phenomenon in which the multipath arrivals came in clusters in time. Within a given cluster, the multipath arrivals decayed with time. These effects had been noted by Saleh and Valenzuela [139], but they did not study angle of arrival. The conclusion from 65 sets of data taken in two buildings was that temporal and angular effects were statistically independent [157]. If there had been a correlation, then it would have been expected that a longer time delay would correspond to a larger angular variance from the mean of a cluster. That effect was not observed in the data; so an assumption of independence was made. (However, further study of this issue is probably warranted.)

The consequence of independence is that the impulse response with respect to time and angle $h(t, \theta)$ can be approximately written as a product [156]:

$$h(t, \theta) \approx h(t)h(\theta) \tag{11.41}$$

As a result, we will address only $h(\theta)$ because temporal effects were addressed previously in Section 11.3.

The proposed model for $h(\theta)$ is [156]:

$$h(\theta) = \sum_{l=0}^{\infty} \sum_{k=0}^{\infty} \beta_{kl} \delta(\theta - \Theta_l - \omega_{kl}), \tag{11.42}$$

where β_{kl} is the multipath amplitude for the kth arrival in the lth cluster and Θ_l is the mean angle of the lth cluster, which is uniformly distributed over the interval 0 to 2π. The ray angle ω_{kl} within a cluster is modeled as a zero-mean Laplacian pdf with a standard deviation σ:

$$f(\omega_{kl}) = \frac{1}{\sqrt{2}\sigma} \exp\left(-|\sqrt{2}\omega_{kl}/\sigma|\right) \tag{11.43}$$

In order for (11.43) to represent a legitimate pdf, it must satisfy the integral relationship in (6.3). This will be the case if $\sigma \ll \pi$. The distribution parameters

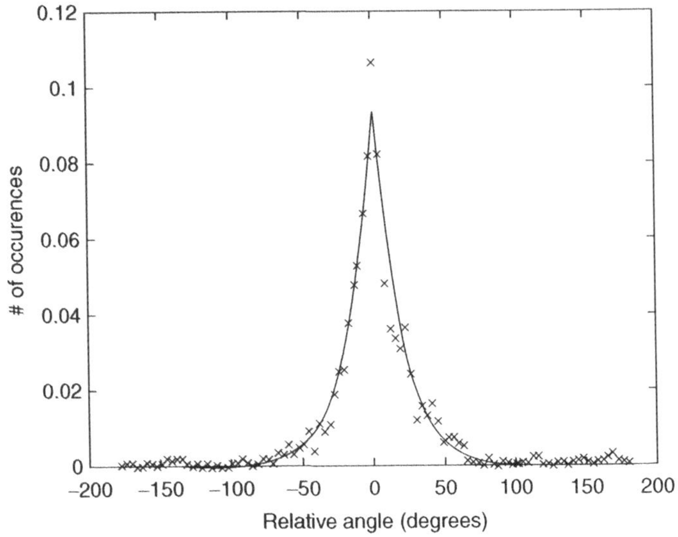

FIGURE 11.7 Histogram of relative ray arrivals with respect to the cluster mean for the Clyde Building. Superimposed is the best-fit Laplacian distribution ($\sigma = 25.5°$) [156].

of the cluster means Θ_l is found by identifying each of the clusters in a given data set. The mean angle of arrival for each cluster is calculated. The cluster mean is subtracted from the absolute angle of each ray in the cluster to give a relative arrival angle with respect to the cluster mean. The relative arrivals are collected over the ensemble of all data sets, and a histogram can be generated. The histogram is fit to the closest Laplacian distribution by use of a least mean square algorithm, which gives the estimated value for σ. An example of the measured data and best-fit Laplacian distribution for a reinforced concrete and cinder block building is shown in Figure 11.7 [156].

Because the angle of arrival will continue to be important in diversity applications for overcoming multipath interference, more research in this area is justified. This is particularly the case since results up to now are either experimental or a best-fit model to experimental data.

11.5 REVERBERATION CHAMBER SIMULATION

In Sections 9.2 and 9.3, the effects of an unintended direct-path signal (unstirred energy) on the performance of a reverberation chamber for radiated immunity testing were analyzed. In that case, the direct-path signal was undesired and resulted in degradation of chamber performance. However, it is possible to make use of the controlled combination of the direct-path signal and the stirred field to simulate a realistic multipath environment for testing wireless communication devices [158].

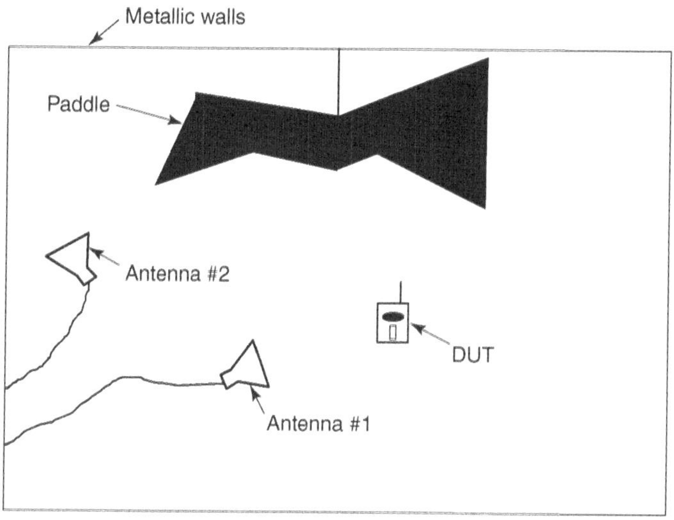

FIGURE 11.8 Reverberation chamber configuration for both a one-antenna and a two antenna approach. Antenna #1 points toward the center of the chamber [158].

This multipath environment is relevant for both indoor and outdoor wireless propagation.

For example, as shown in (9.34), the magnitude of the θ component of the electric field $|E_\theta|$ has a Rice PDF:

$$f(|E_\theta|) = \frac{|E_\theta|}{\sigma^2} \mathrm{I}_0\left(\frac{|E_{s\theta}||E_{d\theta}|}{2\sigma^2}\right) \exp\left(-\frac{|E_{s\theta}|^2 + |E_{d\theta}|^2}{2\sigma^2}\right) \mathrm{U}(|E_\theta|), \tag{11.44}$$

where $|E_{s\theta}|$ is the magnitude of the stirred field, $|E_{d\theta}|$ is the magnitude of the direct field, σ^2 is the variance of the real and imaginary parts of the stirred field, as shown in (9.33), I_0 is the modified Bessel function of zero order [25], and U is the unit step function. The geometry is shown in Figure 11.8 for the case where Antenna #1 (θ polarized) is transmitting and Antenna #2 is removed. Figure 9.8 shows scatter plots of the scattering matrix S_{21} for one case where the direct path signal is negligible and three cases where the presence of the direct path has caused the cluster of data to move off the origin.

The Rice K-factor is conventionally defined as [141,159,160]:

$$K = \frac{|E_{d\theta}|^2}{2\sigma^2} \tag{11.45}$$

If the direct path is negligible, $K = 0$ and the PDF is Rayleigh, as shown in (9.36). When there is no multipath stirred field, $K = \infty$, and the field is deterministic. In the next two sections, we introduce two methods for obtaining any K-factor for simulation application.

11.5.1 A Controllable *K*-Factor Using One Transmitting Antenna

The test configuration shown in Figure 11.8 (with Antenna #2 removed) is discussed in this section. One antenna points toward a device under test (DUT) placed in the center of the chamber. As before, we assume that the only unstirred component is the direct coupling term from the antenna (all wall reflections are assumed to interact with the stirrer). We again assume that the transmitting antenna is θ polarized. The transmitting antenna has a directivity pattern $D(\theta, \phi)$ which will just be written as D. Then, the square of the direct field can be written [158]:

$$|E_{d\theta}|^2 = \frac{\eta}{4\pi r^2} P_t D, \tag{11.46}$$

where r is the distance between the transmitting antenna and the DUT, and P_t is the transmitted power. To evaluate (11.45), we also need the variance σ^2 of the real and imaginary parts of the stirred field. The variance is related to the frequency and chamber characteristics as [158]

$$\sigma^2 = \frac{\eta \lambda Q P_t}{12\pi V} \tag{11.47}$$

If we substitute (11.46) and (11.47) into (11.45), we obtain the following for K:

$$K = \frac{3}{2} \frac{V}{\lambda Q} \frac{D}{r^2} \tag{11.48}$$

Because K in (11.48) depends on a number of quantities, it is possible to obtain a large range of values for the K-factor. Since K is proportional to D, a directional antenna can be rotated with respect to the DUT, thereby changing the K-factor. If D is small, K is small (approaching a Rayleigh environment). If r is large, K is small (approaching a Rayleigh environment). If r is small, K is large. Hence, if the separation between the antenna and the DUT is varied, then the K-factor can be adjusted to some desired value. Since K is inversely proportional to chamber Q, the K-factor can be changed to a desired value by varying Q. The chamber Q can be varied by loading the chamber with lossy materials. Increased loading decreases the chamber Q, as shown in Figure 9.6.

11.5.2 A Controllable *K*-Factor Using Two Transmitting Antennas

The test configuration shown in Figure 11.8 (with both Antennas #1 and #2 transmitting) is discussed in this section. Antenna #1 is pointing toward the center of the chamber where the DUT is placed, and Antenna #2 is pointed away from the center of the chamber. Once again, we assume that the only unstirred component of the electric field is the direct coupling term from Antenna #1. As in the previous section, we assume that Antenna #1 is θ polarized. Hence, the square of the direct field is similar to that in (11.46):

$$|E_{d\theta}|^2 = \frac{\eta}{4\pi r^2} P_{t1} D_1, \tag{11.49}$$

where P_{t1} is the power transmitted by Antenna #1 and D_1 is the directivity of Antenna #1. The variance of the real and imaginary parts of the stirred field is similar to that in (11.47):

$$\sigma^2 = \frac{\eta\lambda Q(P_{t1} + P_{t2})}{12\pi V}, \tag{11.50}$$

where P_{t2} is the power transmitted by Antenna #2. The K-factor is obtained by substituting (11.49) and (11.50) into (11.45):

$$K = \frac{3}{2}\frac{V}{\lambda Q}\frac{D_1}{r^2}\frac{P_{t1}}{P_{t1} + P_{t2}} \tag{11.51}$$

This result is independent of the directivity of the Antenna #2, which is pointed away from the DUT.

The potential advantage of using two transmitting antennas is that K can be varied over a large range by varying only the power ratio P_{t1}/P_{t2}. If $P_{t1}/P_{t2} \gg 1$, then (11.51) reduces to (11.48), the result for a single transmitting antenna. If $P_{t1}/P_{t2} \ll 1$, then (11.51) reduces to:

$$K = \frac{3}{2}\frac{V}{\lambda Q}\frac{D_1}{r^2}\frac{P_{t1}}{P_{t2}} \tag{11.52}$$

If P_{t1}/P_{t2} is reduced to a very small value, $K \to 0$ and the PDF approaches Rayleigh.

11.5.3 Effective *K*-Factor

When the K-factor is measured for different chamber and transmitting antenna characteristics, the DUT is replaced with a probe or receiving antenna. Figure 11.9 shows the experimental setup for measurement of the K-factor of the chamber. When testing a wireless device, one of the horn antennas in Figure 11.9 is replaced with a DUT (cell phone or other wireless device). Figure 11.9 is the experimental setup used in the next section and consists of two antennas: both transmitting and receiving horn antennas. The expressions for the K-factor in the previous two sections were applied to a component of the electric field. This gave the same results as that for a DUT that had omnidirectional properties for pattern and was polarization matched to the transmitting antenna. If the DUT (or receiving antenna) does not have these properties, then the DUT (or receiving antenna) will see an effective K-factor. The expressions for the K-factor in the previous two sections can be modified to take these effects into account by introducing D_{DUT} (the directivity of the DUT) and $\vec{\rho}_t$ and $\vec{\rho}_{\text{DUT}}$ (the polarization unit vectors of the transmitting antenna and the DUT, respectively).

A factor $2(\vec{\rho}_t \bullet \vec{\rho}_{DUT})^2$ results from the fact the DUT is polarization matched to the direct path (when $\vec{\rho}_t \bullet \vec{\rho} = 1$), but the DUT has a $\frac{1}{2}$ polarization mismatch to the stirred field as shown in (7.103). A factor of $\frac{1}{3}$ comes from the general theory for any DUT in a stirred field because all three rectangular components of the stirred field are statistically equal. With these modifications, the K-factor for the one antenna factor

FIGURE 11.9 Chamber configuration for testing. In measuring the K-factor of a chamber, one horn antenna is used as a source and the other horn antenna is used as a probe. When testing a wireless device, one of the horn antennas is replaced with a device under test (for example, a cell phone or other wireless device). The absorber in the chamber is used to control the chamber Q [158].

in (11.48) becomes:

$$K = \frac{V}{\lambda Q}\frac{1}{r^2}D_t D_{\mathrm{DUT}}(\vec{\rho}_t \bullet \rho_{\mathrm{DUT}})^2, \tag{11.53}$$

and the K-factor for the two-antenna method becomes:

$$K = \frac{V}{\lambda Q}\frac{1}{r^2}\frac{P_{t1}}{P_{t1}+P_{t2}}D_t D_{\mathrm{DUT}}(\vec{\rho}_t \bullet \rho_{\mathrm{DUT}})^2 \tag{11.54}$$

If the DUT (or receiving antenna) is omnidirectional and polarization matched to the transmitting antenna, then (11.53) and (11.54) reduce to (11.48) and (11.51). The polarization properties of the transmitting antenna and the DUT can be used as an additional means of controlling the K-factor.

11.5.4 Experimental Results

In order to verify the functional dependence for the K-factor in (11.53) for one transmitting antenna, measurements were performed in the NIST reverberation chamber [158]. The chamber dimensions are $2.8 \times 3.1 \times 4.6$ m, and the measurement setup is shown in Figure 11.9. Two horn antennas were placed inside the chamber and connected to a vector network analyzer. The scattering parameter S_{21} between the two antennas was measured. This is a common approach used to determine the statistical behavior of a reverberation chamber [66].

The distance between the two horn antennas, the azimuth of the receiving antenna, and the relative polarization of the receiving antenna can be adjusted to control the direct-path component and, in turn, to change the K-factor. Only the relative positions of the two antennas are important because the stirred-field statistics are spatially uniform for a well stirred chamber. Statistics for S_{21} were obtained by measuring at 1601 stirrer positions at each of 201 frequencies from 1 to 6 GHz [158]. Twice the variance of the real or imaginary part of S_{21} measured in the reverberation chamber can be written as [158]:

$$2\sigma_R^2 = \langle |S_{21} - \langle S_{21} \rangle|^2 \rangle \tag{11.55}$$

The magnitude of the mean value of S_{21} measured in the reverberation chamber can be written as [158]:

$$d_R = |\langle S_{21} \rangle| \tag{11.56}$$

This is essentially the magnitude of the direct-path signal. In analogy to (11.45), the K-factor can be written:

$$K = \frac{d_R^2}{2\sigma_R^2} = \frac{|\langle S_{21} \rangle^2|}{\langle |S_{21} - \langle S_{21} \rangle|^2 \rangle} \tag{11.57}$$

This is seen visually by referring to the scattering plots in Figure 11.10: σ_R is the radius of the clutter of data and d_R is the distance of the centroid of the clutter from the origin.

The value d_R should be the same as the direct component d_A measured in an anechoic chamber for an identical antenna configuration, where $d_A = |S_{21AC}|$ and S_{21AC} is the scattering parameter measured in an anechoic chamber. An ideal anechoic chamber would have no wall reflections and S_{21AC} would be only the direct component. This is verified in Figure 11.11, which shows d_A^2 as measured in the NIST anechoic chamber (thick smooth curve) and d_R^2 as measured in the NIST reverberation chamber for four different loading configurations (zero, one, two, and four pieces of 60 cm absorber) [158]. Some of the absorber is visible in Figure 11.9. The trends of the curves are similar, but the data from the reverberation chamber are substantially noisier than the data from the anechoic chamber. The noise in these data can be explained by the physical design of the NIST reverberation chamber. The assumption in obtaining (11.56) is that all wall reflections in the reverberation chamber interact with the stirrer (paddle); i.e., the only unstirred component is the direct coupling term from the transmitting antenna.

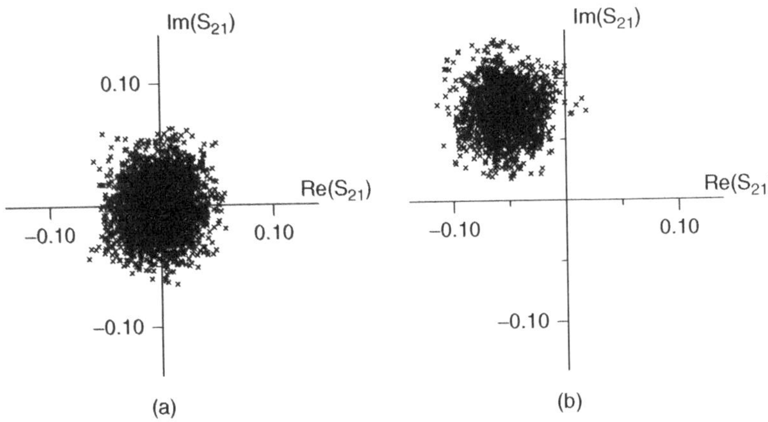

FIGURE 11.10 Scatter plots of measured S_{21} for two antennas in the NIST reverberation chamber at a frequency of 2 GHz: (a) little direct coupling and (b) strong direct coupling [158].

These results indicate that the NIST reverberation chamber is not optimized for this type of measurement and that there are reflected components that are not altered (stirred) by the paddle, which are referred to as unstirred multipath (UMP) components. The UMP components are most likely due to the large volume of the

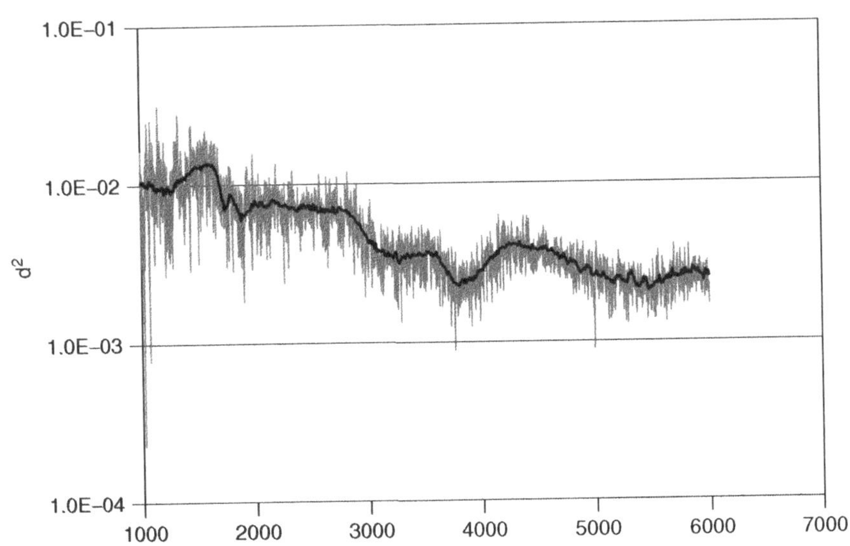

FIGURE 11.11 Values of d_R^2 for each different absorber configuration in the NIST reverberation chamber. The set of indistinguishable curves consists of data taken with zero, one, two, and four pieces of absorber. The thick black curve represents the data taken in the anechoic chamber. All data were taken at 1 m separation [158].

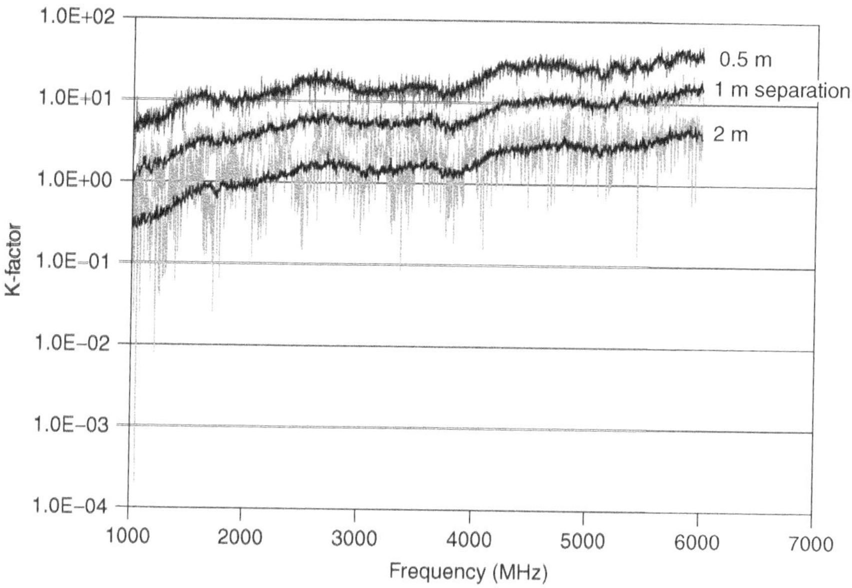

FIGURE 11.12 *K*-factor for three different antenna separations. The thick black curve running over each data set represents the *K*-factor obtained by use of d_A [158].

chamber that does not interact with the paddle. The NIST chamber was one of the first reverberation chambers built over 20 years ago and has only one paddle at the top of the chamber; thus many wall reflections near the bottom part of the chamber will not be affected by the paddle. Newer chambers use two or more paddles in the chamber, such that more wall reflections interact with the paddle. Harima [161] showed smaller variations (±2 dB from 1 to 18 GHz) in a chamber with three paddles located on the ceiling and two walls. (However, extra paddles are not necessarily required for EM immunity and emissions tests, for which the NIST chamber was initially intended.)

Figure 11.12 shows the effect of antenna separation on *K*-factor as determined from (11.57) for frequencies from one to six GHz. As expected from (11.48), the *K*-factor can be decreased by increasing antenna separation. Also shown in the figure are results (the thick smooth curve) based on determining the direct coupling term from anechoic chamber measurements of d_A and using it in place of d_R in (11.57). Once again, the smoother results obtained with d_A are because of the UMP components in the reverberation chamber.

Figure 11.13 shows the effect of loading the chamber (decreasing the Q) on the *K*-factor. The antennas were copolarized and positioned 1 m apart. Placing two or six pieces of 60 cm absorber in the corners of the reverberation chamber lowered the Q and increased the *K*-factor for the entire frequency range of one to six GHz. The thick black curve represents the *K*-factor obtained by use of d_A from the anechoic chamber in place of d_R. This technique for increasing K can be taken only so far because increasing

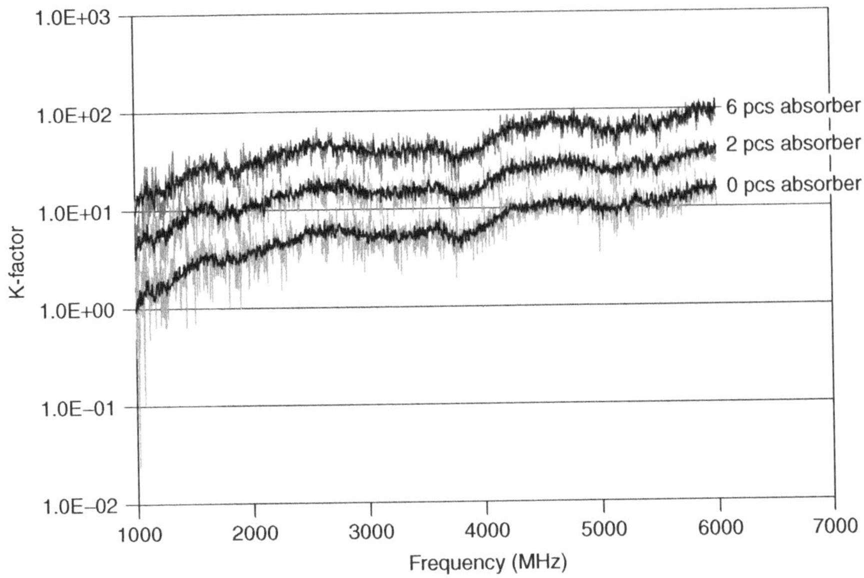

FIGURE 11.13 *K*-factor for different numbers of absorber pieces. The thick black curve represents the *K*-factor obtained using d_A. All data sets were taken with the antennas at 1 m separation [158].

losses (reducing Q) results in poorer reverberation chamber performance due to poorer stirring [112].

The K-factor can also be changed by adjusting the relative orientation of the transmitting or receiving antenna. Results obtained by changing the relative azimuth of one of the antennas are shown in Figure 11.14. The change in K due to use of this technique varies depending on the pattern of each individual antenna and varies over frequency. Although K was decreased by increasing azimuthal angle, it was difficult to decrease K much below one because of unstirred multipath.

The effect of changing the relative polarization was also studied. As shown in Figure 11.15, a relative polarization of 45° decreases K by a factor of two at all frequencies, as expected. Also shown in Figure 11.15 is the measured K when the antennas were cross-polarized, but still facing each other, as well as an estimate based on measurements of the same configuration in an anechoic chamber. The minimum measured K in Figure 11.15 is significantly lower than that in Figure 11.14. Equation (11.54) indicates that ideal cross-polarized antennas would give a zero value of K, but this is not quite achieved because of some nonzero cross-polarization coupling between the two antennas.

A final technique used to manipulate the K-factor is to include a second transmitting antenna as shown in Figure 11.8. The mathematical result is given in (11.51). To do this, a radio-frequency signal splitter was used with one arm connected to the direct-illumination antenna, and the other arm connected to the other antenna, which was directed at the paddle. From (11.51) with $P_{t1} = P_{t2}$, the K-factor was reduced

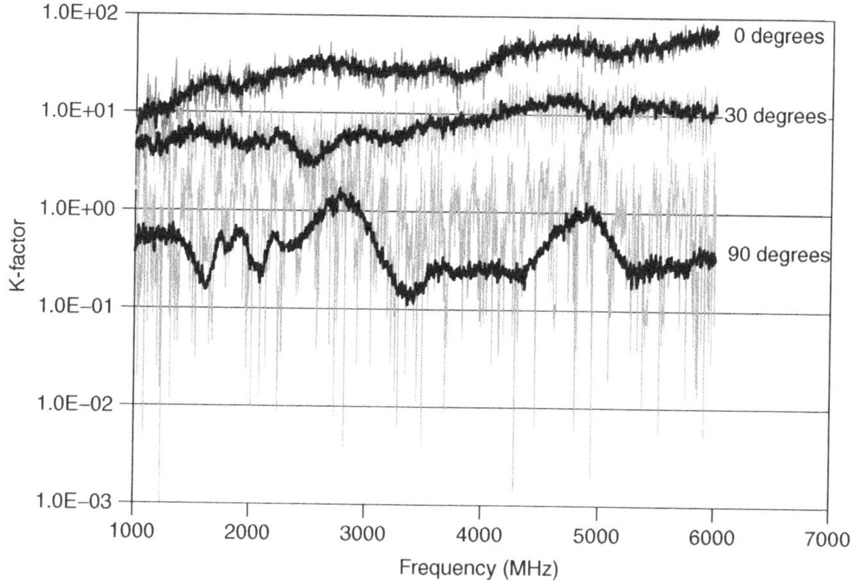

FIGURE 11.14 Experimental results obtained from varying the relative azimuth of the antennas. The thick black curve over each data set represents the K-factor obtained using d_A. Each data set was taken at 1 m antenna separation with four pieces of absorber in the chamber [158].

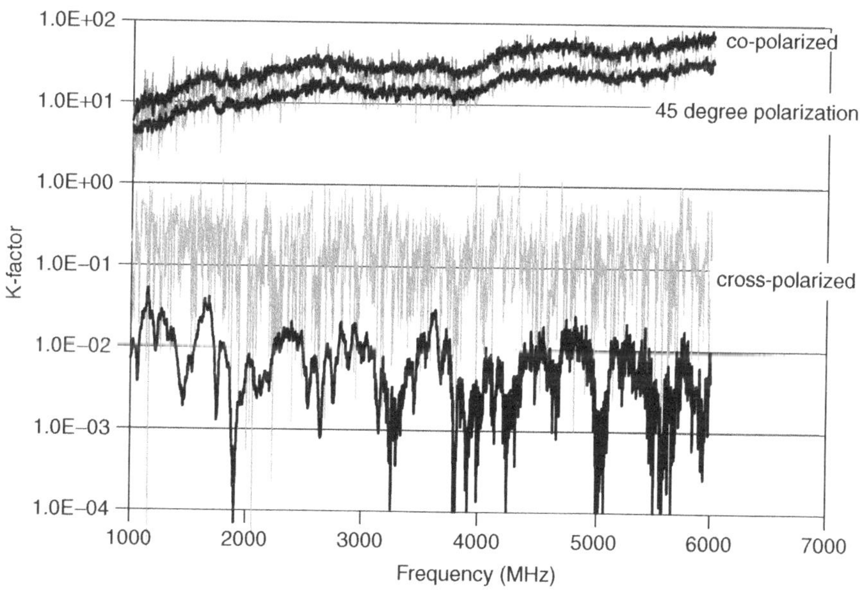

FIGURE 11.15 Experimental results obtained from varying the polarization of the antennas. The thick black curve over each data set represents the K-factor obtained using d_A. All data sets were taken at 1 m antenna separation with four pieces of absorber in the chamber [158].

by a factor of two. Experimental results [158] showed that unstirred multipath components can cause the measured reduction to vary from two. Uncertainties in reverberation chamber measurements are discussed in [115] and [119].

PROBLEMS

11-1 Consider (11.1) with X_σ set equal to zero. Show that this reduced case of (11.1) yields an electric (or magnetic) field strength that decays as a power law: $|E| = \frac{\sqrt{K_p}}{d^{n/2}}$, where K_p is a constant independent of d. Determine the expression for K_p in terms of $PL(d_0)$ and d_0.

11-2 Show that the path loss in (11.4) yields an electric (or magnetic) field strength that decays as follows: $|E| = \frac{\sqrt{K_e}}{d} 10^{-\alpha d/20}$. If $\alpha = 0$, $n = 2$, and $K_e = K_p$, show that the expressions for $|E|$ in Problems 11-1 and 11-2 are identical.

11-3 Consider (11.2) for the case where the transmitting and receiving antennas are located on different floors. Calculate the path loss PL for the case where $FAF = 20$ dB, $PL(d_0) = 10$ dB, $n_{SF} = 2$, $d_0 = 1$ m, and $d = 50$ m.

11-4 In (11.4), what is the spherical spreading loss at $d = 50$ m referred to $d_0 = 1$ m?

11-5 In (11.4), what is the attenuation loss at $d = 50$ m for frequencies of 850 MHz, 1.9 GHz, 4.0 GHz, and 5.8 GHz?

11-6 Consider a variation of (11.5), where the exponentially decaying pulse is terminated at t_L: $\langle P_r(t)\rangle = P_0 \exp(-t/\tau)[U(t) - U(t_L)]$. Derive the mean delay time from (11.6) and the RMS delay spread from (11.7). In both cases this involves replacing the infinite upper limit of the integrals with t_L.

11-7 For a reverberation model, the decay time when the main loss is leakage through electrically large apertures is independent of frequency, as shown in (11.9). If wall loss is dominant, what is the frequency dependence of the decay time $\tau_1 = Q_1/\omega$?

11-8 Following up on Problem 11-6, if antenna extraction is the main loss, what is the frequency dependence of the decay time $\tau_4 = Q_4/\omega_4$?

11-9 From (11.22), what is the value of the absorption coefficient α for the case $|\Gamma_{TE}|^2 = |\Gamma_{tm}|^2 = 0.8$?

11-10 Consider (11.48) for the K factor in a reverberation chamber. Obtain the expression for K when Q is determined by wall loss, $Q \approx Q_1$, where Q_1 is given by (7.123). Why is the result independent of V?

11-11 For the result in Problem 11-10, calculate the value of K for $D = 10$, $\lambda = 0.3$ m, $r = 1$ m, $A = 24$ m^2, $\mu_r = 1$, and $\sigma_W = 5.7 \times 10^7$.

APPENDIX A

Vector Analysis

Rectangular (x, y, z), cylindrical (ρ, ϕ, z), and spherical (r, θ, ϕ) coordinates are normally oriented as shown in Figure A.1. Coordinate transformations are then given by:

$$\begin{aligned} x &= \rho \cos\phi = r \sin\theta \cos\phi, \\ y &= r \sin\phi = r \sin\theta \sin\phi, \\ z &= r \cos\theta, \\ \rho &= \sqrt{x^2 + y^2} = r \sin\theta, \\ \phi &= \tan^{-1}\frac{y}{x}, \\ r &= \sqrt{x^2 + y^2 + z^2} = \sqrt{\rho^2 + z^2}, \\ \theta &= \tan^{-1}\frac{\sqrt{x^2 + y^2}}{z} = \tan^{-1}\frac{\rho}{z} \end{aligned} \tag{A1}$$

The unit vectors for the three coordinate systems are denoted $(\hat{x}, \hat{y}, \hat{z})$, $(\hat{\rho}, \hat{\phi}, \hat{z})$, and $(\hat{r}, \hat{\theta}, \hat{\phi})$. In rectangular coordinates, we can write a general vector $\vec{A}$ as:

$$\vec{A} = \hat{x}A_x + \hat{y}A_y + \hat{z}A_z \tag{A2}$$

Vector addition is defined by:

$$\vec{A} + \vec{B} = \hat{x}(A_x + B_x) + \hat{y}(A_y + B_y) + \hat{z}(A_z + B_z) \tag{A3}$$

Scalar multiplication (dot product) is defined by:

$$\vec{A} \bullet \vec{B} = A_x B_x + A_y B_y + A_z B_z \tag{A4}$$

Vector multiplication (cross product) is defined by:

$$\vec{A} \times \vec{B} = \begin{vmatrix} \hat{x} & \hat{y} & \hat{z} \\ A_x & A_y & A_z \\ B_x & B_y & B_z \end{vmatrix} \tag{A5}$$

Electromagnetic Fields in Cavities: Deterministic and Statistical Theories, by David A. Hill

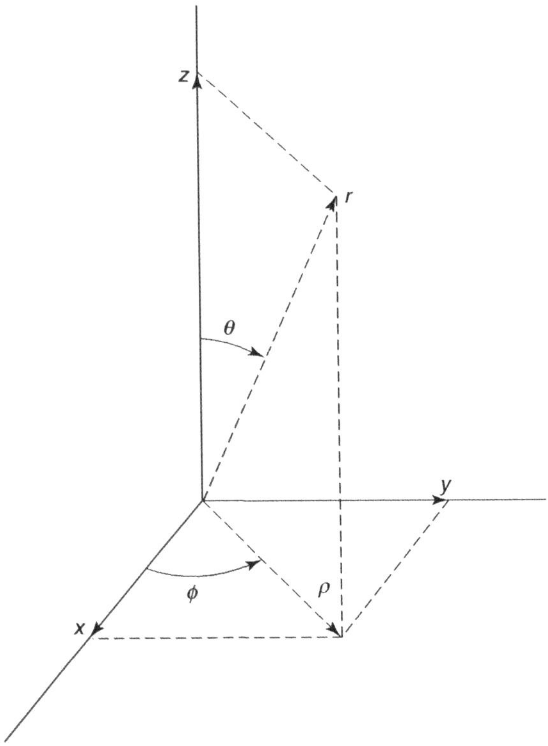

FIGURE A.1 Rectangular (x, y, z), cylindrical (ρ, ϕ, z), and spherical (r, θ, ϕ) coordinates.

The right side of (A5) is a determinant to be expanded in the standard manner. In cylindrical and spherical coordinates, the related forms are analogous to (A2)–(A5).

The important differential operators are the gradient (∇w), divergence ($\nabla \bullet \vec{A}$), curl ($\nabla \times \vec{A}$), and Laplacian ($\nabla^2 w$). In rectangular coordinates, the vector operator del (∇) is:

$$\nabla = \hat{x}\frac{\partial}{\partial x} + \hat{y}\frac{\partial}{\partial y} + \hat{z}\frac{\partial}{\partial z}, \tag{A6}$$

and the differential operations are written [3]:

$$\nabla w = \hat{x}\frac{\partial w}{\partial x} + \hat{y}\frac{\partial w}{\partial y} + \hat{z}\frac{\partial w}{\partial z}, \tag{A7}$$

$$\nabla \bullet \vec{A} = \frac{\partial A_x}{\partial x} + \frac{\partial A_y}{\partial y} + \frac{\partial A_z}{\partial z}, \tag{A8}$$

$$\nabla \times \vec{A} = \begin{vmatrix} \hat{x} & \hat{y} & \hat{z} \\ \dfrac{\partial}{\partial x} & \dfrac{\partial}{\partial y} & \dfrac{\partial}{\partial z} \\ A_x & A_y & A_z \end{vmatrix}, \tag{A9}$$

$$\nabla^2 w = \frac{\partial^2 w}{\partial x^2} + \frac{\partial^2 w}{\partial y^2} + \frac{\partial^2 w}{\partial z^2} \tag{A10}$$

In cylindrical coordinates, the differential operations are written:

$$\nabla w = \hat{\rho}\frac{\partial w}{\partial \rho} + \hat{\phi}\frac{1}{\rho}\frac{\partial w}{\partial \phi} + \frac{\partial w}{\partial z}, \tag{A11}$$

$$\nabla \bullet \vec{A} = \frac{1}{\rho}\frac{\partial}{\partial \rho}(\rho A_\rho) + \frac{1}{\rho}\frac{\partial A_\phi}{\partial \phi} + \frac{\partial A_z}{\partial z}, \tag{A12}$$

$$\nabla \times \vec{A} = \hat{\rho}\left(\frac{1}{\rho}\frac{\partial A_z}{\partial \phi} - \frac{\partial A_\phi}{\partial z}\right) + \hat{\phi}\left(\frac{\partial A_\rho}{\partial z} - \frac{\partial A_z}{\partial \rho}\right) + \hat{z}\left[\frac{1}{\rho}\frac{\partial}{\partial \rho}(\rho A_\phi) - \frac{1}{\rho}\frac{\partial A_\rho}{\partial \phi}\right], \tag{A13}$$

$$\nabla^2 w = \frac{1}{\rho}\frac{\partial}{\partial \rho}\left(\rho\frac{\partial w}{\partial \rho}\right) + \frac{1}{\rho^2}\frac{\partial^2 w}{\partial \phi^2} + \frac{\partial^2 w}{\partial z^2} \tag{A14}$$

In spherical coordinates, the differential operations are written:

$$\nabla w = \hat{r}\frac{\partial w}{\partial r} + \hat{\theta}\frac{1}{r}\frac{\partial w}{\partial \theta} + \hat{\phi}\frac{1}{r\sin\theta}\frac{\partial w}{\partial \phi}, \tag{A15}$$

$$\nabla \bullet \vec{A} = \frac{1}{r^2}\frac{\partial}{\partial r}(r^2 A_r) + \frac{1}{r\sin\theta}\frac{\partial}{\partial \theta}(A_\theta \sin\theta) + \frac{1}{r\sin\theta}\frac{\partial A_\theta}{\partial \phi}, \tag{A16}$$

$$\begin{aligned}\nabla \times \vec{A} = {} & \hat{r}\frac{1}{r\sin\theta}\left[\frac{\partial}{\partial \theta}(A_\phi \sin\theta) - \frac{\partial A_\theta}{\partial \phi}\right] + \hat{\theta}\frac{1}{r}\left[\frac{1}{\sin\theta}\frac{\partial A_r}{\partial \phi} - \frac{\partial}{\partial r}(rA_\phi)\right] \\ & + \hat{\phi}\frac{1}{r}\left[\frac{\partial}{\partial r}(rA_\theta) - \frac{\partial A_r}{\partial \theta}\right],\end{aligned} \tag{A17}$$

$$\nabla^2 w = \frac{1}{r^2}\frac{\partial}{\partial r}\left(r^2\frac{\partial w}{\partial r}\right) + \frac{1}{r^2\sin\theta}\frac{\partial}{\partial \theta}\left(\sin\theta\frac{\partial w}{\partial \theta}\right) + \frac{1}{r^2\sin^2\theta}\frac{\partial^2 w}{\partial \phi^2} \tag{A18}$$

Vector identities (independent of the coordinate system) exist for the following dot products, cross products, and differentiation [2], [3], [162]:

$$\vec{A} \bullet (\vec{B} \times \vec{C}) = \vec{B} \bullet (\vec{C} \times \vec{A}) = \vec{C} \bullet (\vec{A} \times \vec{B}), \tag{A19}$$

$$\vec{A} \times (\vec{B} \times \vec{C}) = (\vec{A} \bullet \vec{C})\vec{B} - (\vec{A} \bullet \vec{B})\vec{C}, \tag{A20}$$

$$\nabla(ab) = a\nabla b + b\nabla a, \tag{A21}$$

$$\nabla \bullet (a\vec{B}) = a\nabla \bullet \vec{B} + \vec{B} \bullet \nabla a, \tag{A22}$$

$$\nabla \times (a\vec{B}) = a\nabla \times \vec{B} - \vec{B} \times \nabla a, \tag{A23}$$

$$\nabla \bullet (\vec{A} \times \vec{B}) = \vec{B} \bullet \nabla \times \vec{A} - \vec{A} \bullet \nabla \times \vec{B}, \tag{A24}$$

$$\nabla(\vec{A} \bullet \vec{B}) = (\vec{A} \bullet \nabla)\vec{B} + (\vec{B} \bullet \nabla)\vec{A} + \vec{A} \times (\nabla \times \vec{B}) + \vec{B} \times (\nabla \times \vec{A}), \quad \text{(A25)}$$

$$\nabla \times (\vec{A} \times \vec{B}) = \vec{A}\nabla \bullet \vec{B} - \vec{B}\nabla \bullet \vec{A} - (\vec{A} \bullet \nabla)\vec{B} + (\vec{B} \bullet \nabla)\vec{A}, \quad \text{(A26)}$$

$$\nabla \bullet (\nabla a) = \nabla^2 a, \quad \text{(A27)}$$

$$\nabla \bullet (\nabla \vec{A}) = \nabla^2 \vec{A}, \quad \text{(A28)}$$

$$\nabla \times (\nabla \times \vec{A}) = \nabla(\nabla \bullet \vec{A}) - \nabla^2 \vec{A}, \quad \text{(A29)}$$

$$\nabla \times (\nabla a) = 0, \quad \text{(A30)}$$

$$\nabla \bullet (\nabla \times \vec{A}) = 0 \quad \text{(A31)}$$

Dyadic identities also exist for the following dot products, cross products, and differentiation [2]:

$$\vec{A} \bullet (\vec{B} \times \overleftrightarrow{C}) = -\vec{B} \bullet (\vec{A} \times \overleftrightarrow{C}) = (\vec{A} \times \vec{B}) \bullet \overleftrightarrow{C}, \quad \text{(A32)}$$

$$\vec{A} \times (\vec{B} \times \overleftrightarrow{C}) = \vec{B} \bullet (\vec{A} \times \overleftrightarrow{C}) - (\vec{A} \bullet \vec{B})\overleftrightarrow{C}, \quad \text{(A33)}$$

$$\nabla(a\vec{B}) = a\nabla\vec{B} + (\nabla a)\vec{B}, \quad \text{(A34)}$$

$$\nabla \bullet (a\overleftrightarrow{B}) = a\nabla \bullet \overleftrightarrow{B} + (\nabla a) \bullet \overleftrightarrow{B}, \quad \text{(A35)}$$

$$\nabla \times (a\overleftrightarrow{B}) = a\nabla \times \overleftrightarrow{B} + (\nabla a) \times \overleftrightarrow{B}, \quad \text{(A36)}$$

$$\nabla \times (\nabla \times \overleftrightarrow{A}) = \nabla(\nabla \bullet \overleftrightarrow{A}) - \nabla^2 \overleftrightarrow{A}, \quad \text{(A37)}$$

$$\nabla \bullet (\nabla \times \overleftrightarrow{A}) = 0. \quad \text{(A38)}$$

The following integral theorems are also useful [2].

Divergence Theorem:

$$\iiint \nabla \bullet \vec{A} \, dV = \oiint (\hat{n} \bullet \vec{A}) \, dS \quad \text{(A39)}$$

Curl Theorem:

$$\iiint \nabla \times \vec{A} \, dV = \oiint (\hat{n} \times \vec{A}) \, dS \quad \text{(A40)}$$

Gradient Theorem:

$$\iiint \nabla a \, dV = \oiint \hat{n} a \, dS \quad \text{(A41)}$$

Stokes' Theorem:

$$\iint \hat{n} \bullet \nabla \times \vec{A}\,\mathrm{d}S = \oint \vec{A} \bullet \mathrm{d}\vec{l} \tag{A42}$$

Cross-Gradient Theorem:

$$\iint \hat{n} \times \nabla a\,\mathrm{d}S = \oint a\,\mathrm{d}\vec{l} \tag{A43}$$

Cross-Del-Cross Theorem:

$$\iint (\hat{n} \times \nabla) \times \vec{A}\,\mathrm{d}S = -\oint \vec{A} \times \mathrm{d}\vec{l} \tag{A44}$$

APPENDIX B

Associated Legendre Functions

The associated Legendre equation is [3]:

$$\frac{1}{\sin\theta}\frac{\mathrm{d}}{\mathrm{d}\theta}\left(\sin\theta\frac{\mathrm{d}y}{\mathrm{d}\theta}\right)+\left[\nu(\nu+1)-\frac{m^2}{\sin^2\theta}\right]y=0 \tag{B1}$$

For spherical coordinates and spherical cavities, we have the case where ν is an integer n and θ has the range, $0 \le \theta \le \pi$. In this case, the two independent solutions of (B1) are the associated Legendre functions [25] of the first kind $\mathrm{P}_n^m(\cos\theta)$ and the second kind $Q_n^m(\cos\theta)$. Since Q_n^m is singular at $\cos\theta = \pm 1$, it is not useful for describing fields in spherical cavities. Hence, from here on we will consider only P_n^m.

Equation (B1) can be put into another useful form by making the substitution, $u = \cos\theta$. The equivalent result is:

$$(1-u^2)\frac{\mathrm{d}^2y}{\mathrm{d}u^2}-2u\frac{\mathrm{d}y}{\mathrm{d}u}+\left[n(n+1)-\frac{m^2}{1-u^2}\right]y=0 \tag{B2}$$

Consider first the case, $m = 0$, where (B2) reduces to the ordinary Legendre equation:

$$(1-u^2)\frac{\mathrm{d}^2y}{\mathrm{d}u^2}-2u\frac{\mathrm{d}y}{\mathrm{d}u}+n(n+1)y=0 \tag{B3}$$

The solutions to (B3) that are finite over the range, $-1 \le u \le 1$, are the Legendre polynomials $\mathrm{P}_n(u)$, which can be written as a finite sum [3]:

$$\mathrm{P}_n(u)=\sum_{l=0}^{L}\frac{(-1)^l(2n-2l)!}{2^n l!(n-l)!(n-2l)!}u^{n-2l}, \tag{B4}$$

where $L = n/2$ or $(n-1)/2$, whichever is an integer. An alternative expression for the Legendre polynomials is given by Rodrigues' formula:

$$\mathrm{P}_n(u)=\frac{1}{2^n n!}\frac{\mathrm{d}^n}{\mathrm{d}u^n}(u^2-1)^n \tag{B5}$$

Electromagnetic Fields in Cavities: Deterministic and Statistical Theories, by David A. Hill

The five-lowest order Legendre polynomials are:

$$\begin{aligned}
P_0(u) &= 1, \\
P_1(u) &= u, \\
P_2(u) &= \frac{1}{2}(3u^2-1), \\
P_3(u) &= \frac{1}{2}(5u^3-3u), \\
P_4(u) &= \frac{1}{8}(35u^4-30u^2+3)
\end{aligned} \tag{B6}$$

Equation (B6) can also be written in terms of θ [3]:

$$\begin{aligned}
P_0(\cos\theta) &= 1, \\
P_1(\cos\theta) &= \cos\theta, \\
P_2(\cos\theta) &= \frac{1}{4}(3\cos 2\theta+1), \\
P_3(\cos\theta) &= \frac{1}{8}(5\cos 3\theta+3\cos\theta), \\
P_4(\cos\theta) &= \frac{1}{64}(35\cos 4\theta+20\cos 2\theta+9)
\end{aligned} \tag{B7}$$

Solutions to the associated Legendre equation (B2) can be obtained by differentiating the Legendre polynomials:

$$P_n^m(u) = (-1)^m(1-u^2)^{m/2}\frac{d^m P_n(u)}{du^m} \tag{B8}$$

For $m > n$, $P_n^m(u) = 0$. Also, $P_n^0(u) = P_n(u)$. Lower-order associated Legendre functions through $n = 3$ are:

$$\begin{aligned}
P_1^1(u) &= -(1-u^2)^{1/2}, \\
P_2^1(u) &= -3(1-u^2)^{1/2}u, \\
P_2^2(u) &= 3(1-u^2), \\
P_3^1(u) &= \frac{3}{2}(1-u^2)^{1/2}(1-5u^2), \\
P_3^2(u) &= 15(1-u^2)u \\
P_3^3(u) &= -15(1-u^2)^{3/2}
\end{aligned} \tag{B9}$$

A useful way to calculate a large number of associated Legendre functions is via recurrence relations. A recurrence formula in n is [3]:

$$(m-n-1)P_{n+1}^m(u)+(2n+1)uP_n^m(u)-(m+n)P_{n-1}^m(u) = 0 \tag{B10}$$

A recurrence formula in m is:

$$P_n^{m+1}(u) + \frac{2mu}{(1-u^2)^{1/2}} P_n^m(u) + (m+n)(n-m+1)P_n^{m-1}(u) = 0 \qquad \text{(B11)}$$

Some formulas also exist for derivatives with respect to the argument:

$$\begin{aligned}
P_n^{m\prime}(u) &= \frac{1}{1-u^2}\left[-nuP_n^m(u) + (n+m)P_{n-1}^m(u)\right] \\
&= \frac{1}{1-u^2}\left[(n+1)uP_n^m(u) - (n-m+1)P_{n+1}^m(u)\right] \\
&= \frac{mu}{1-u^2}P_n^m(u) + \frac{(n+m)(n-m+1)}{(1-u^2)^{1/2}}P_n^{m-1}(u) \\
&= -\frac{mu}{1-u^2}P_n^m(u) - \frac{1}{(1-u^2)^{1/2}}P_n^{m+1}
\end{aligned} \qquad \text{(B12)}$$

The recurrence formulas in (B10) and (B11) and the derivative formulas in (B12) also apply to the associated Legendre functions of the second kind.

APPENDIX C

Spherical Bessel Functions

As indicated in Section 4.1, the radial functions R required in spherical geometries satisfy the following differential equation:

$$\frac{\mathrm{d}}{\mathrm{d}r}\left(r^2\frac{\mathrm{d}R}{\mathrm{d}r}\right)+\left[(kr)^2-n(n+1)\right]R=0 \tag{C1}$$

For spherical cavities, we require only the spherical Bessel function of the first kind $\mathrm{j}_n(kr)$ [25], [163], where n is an integer, because it is finite at the origin. The spherical Hankel functions are useful in radiation problems [163], but they are singular at the origin. The spherical Bessel function of the first kind is related to the cylindrical Bessel function of order $n+1/2$ [25]:

$$\mathrm{j}_n(kr)=\sqrt{\frac{\pi}{2kr}}\mathrm{J}_{n+1/2}(kr) \tag{C2}$$

However, in one way the spherical Bessel functions are simpler than the cylindrical Bessel functions because they can be written as a finite number of terms. For example, the first five spherical Bessel functions of the first kind (with the argument kr replaced by x) are:

$$\begin{aligned}
\mathrm{j}_0(x)&=x^{-1}\sin x,\\
\mathrm{j}_1(x)&=x^{-1}[\ \cos x+x^{-1}\sin x],\\
\mathrm{j}_2(x)&=x^{-1}[-3x^{-1}\cos x+(-1+3x^{-2})\sin x],\\
\mathrm{j}_3(x)&=x^{-1}[(1-15x^{-2})\cos x+(-6+15x^{-3})\sin x],\\
\mathrm{j}_4(x)&=x^{-1}[(10x^{-1}-105x^{-3})\cos x+(1-45x^{-2}+105x^{-4})\sin x]
\end{aligned} \tag{C3}$$

The following limiting value of $\mathrm{j}_n(x)$ as x approaches zero is consistent with (C3) [25]:

$$\mathrm{j}_n(x)\underset{x\to 0}{\to}\frac{x^n}{1\cdot 3\cdot 5\ldots(2n+1)} \tag{C4}$$

The results in (C3) can be obtained from Rayleigh's formula:

$$\mathrm{j}_n(x) = x^n \left(-\frac{1}{x}\frac{\mathrm{d}}{\mathrm{d}x} \right)^n \frac{\sin x}{x}, \tag{C5}$$

which is valid for any non-negative integer value of n.

For calculating large numbers of values of $\mathrm{j}_n(x)$, the following recurrence relation can be useful:

$$\mathrm{j}_{n-1}(x) + \mathrm{j}_{n+1}(x) = (2n+1)x^{-1}\,\mathrm{j}_n(x) \tag{C6}$$

The following formulas for derivatives with respect to the argument are also available:

$$\begin{aligned} j'_n(x) &= \frac{1}{(2n+1)}\left[n\mathrm{j}_{n-1}(x) - (n+1)\mathrm{j}_{n+1}(x)\right] \\ &= \mathrm{j}_{n-1}(x) - \frac{n+1}{x}\mathrm{j}_n(x) \\ &= \frac{n}{x}\mathrm{j}_n(x) - \mathrm{j}_{n+1}(x) \end{aligned} \tag{C7}$$

The relations in (C6) and (C7) also apply to the spherical Hankel and Neumann functions [25].

APPENDIX D

The Role of Chaos in Cavity Fields

A qualitative description of chaos is that "a chaotic system is a *deterministic* system that exhibits *random* behavior" [164]. The literature on chaos is very large (for example, see [165–168] and the references in these books), but references on chaos that are specific to electromagnetics [169] are more limited. The reason for this is that Maxwell's equations in linear media are linear equations that traditionally would not be expected to generate chaotic behavior. A dipole antenna with a nonlinear load [170, 171] and transmission through a nonlinear material [172] are nonlinear examples in electromagnetics where chaos has been observed and analyzed.

However, ray chaos [173–175], characterized by the exponential divergence of trajectories of initially nearby rays, can occur in linear propagation environments due to the nonlinear eikonal equation that determines ray trajectories. The (nonlinear) eikonal equation can be derived by expanding an asymptotic solution of the Helmholtz equation (or Maxwell's equations for the vector electromagnetic case) in inverse powers of wavenumber k [176–178]:

$$(\nabla\varphi)^2 = n^2, \tag{D1}$$

where φ is the ray phase and n is the (possibly inhomogeneous) refractive index. Equation (D1) was used in [175] to track rays reflected from a periodic grid coated by an inhomogeneous dielectric, and ray chaos (exponential divergence of closely spaced incident rays) was demonstrated. A fit to exponential divergence was also shown, and this can be used to determine the (positive) Lyapounov exponent [174], one of the main indicators of chaos. Along with (D1), the transport equation(s) [176–178] can be used to determine coefficient(s) of the propagating ray factor, but we will not pursue that portion of the ray solution.

Even when n is homogenous, reflecting boundaries can cause ray chaos. Examples of exterior scattering geometries that can lead to ray chaos are aircrafts with ducts [173, 179] and multiple cylinders [180].

For our purposes, we are more interested in (interior) cavity geometries that can produce ray chaos [174]. Assume that n is homogeneous (for example, free space)

Electromagnetic Fields in Cavities: Deterministic and Statistical Theories, by David A. Hill

so that the ray behavior is determined by the geometry of the cavity walls (assumed to be perfectly reflecting). In ray chaos, the ray trajectories approach almost every point in the cavity arbitrarily closely, with uniformly distributed arrival angles [174]. Integrable geometries, typically related to the analytic integrability of ray path evolution, imply regular (non-chaotic) ray paths. There are several definitions for integrable systems (see [165–167] for details). Coordinate-separable geometries are always integrable. For example, the rectangular cavity (Chapter 2), circular cylindrical cavity (Chapter 3), and spherical cavity (Chapter 4) are all coordinate separable, and can be analyzed by separation of variables. Hence, their analyses are appropriate for deterministic theory, Part I of this book. Nonseparability does not necessarily imply nonintegrability. Some polygonal cavities are nonseparable, but still integrable [174].

Strictly speaking, ray chaos is applicable only in the zero-wavelength (infinite-frequency) limit. However, for small, but nonzero, wavelengths in complex cavities [181], some properties of chaos (such as sensitivity to initial conditions or to cavity geometry perturbations) appear. This has been called the realm of "wave chaos" [182–184]. In this case, "the full-wave properties of ray-chaotic systems turn out to be *naturally described* in statistical terms" [169]. The most commonly used statistical model is a superposition of a large number of plane waves with uniformly distributed arrival directions, polarizations, and phases [185]. This random plane wave (RPW) model accounts very well for the properties of the wave functions of ray-chaotic cavities [186]. The early work with the RPW treated single modes of two-dimensional cavities so that source-free solutions of the scalar Helmholtz equation were obtained. However, the extension of the RPW model to vector Maxwell's equations in three dimensions follows naturally, as shown in Chapter 7. An arbitrary source is also included via conservation of power, as shown in Chapter 7.

Since the idea of mechanical stirring of reverberation chambers by wall motion rather than the typical paddle wheel has been proposed, the two-dimensional analysis of a wall with a sinusoidal motion [187] is worth mentioning. Wall motion introduces a nonlinearity to the boundary value problem, and the onset of chaos as a function of wall displacement was determined by the increase of the Lyapounov exponent.

APPENDIX E

Short Electric Dipole Response

Consider a short electric dipole of effective length L oriented in the z direction, as shown in Figure E.1. The components $S_{r\alpha}$ and $S_{r\beta}$ of the dipole receiving function are given by [69]:

$$S_{r\alpha} = \frac{L \sin \alpha}{2\, R_r} \quad \text{and} \quad S_{r\beta} = 0, \tag{E1}$$

where R_r is the radiation resistance. In (E1), $S_{r\alpha}$ is derived by dividing the induced voltage by twice the radiation resistance for a matched load. Because β components of the electric field are orthogonal to the z-directed dipole, $S_{r\beta} = 0$.

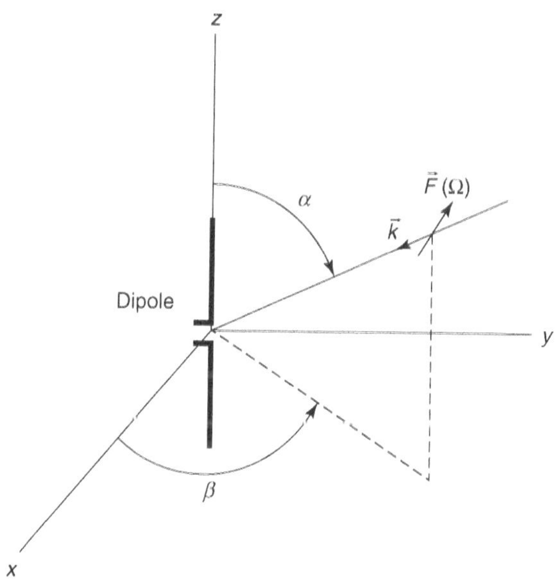

FIGURE E.1 Short dipole antenna illuminated by a plane-wave component of the electric field.

If (E1) is substituted into (7.101) the angular integration can be carried out to obtain:

$$\langle P_r \rangle = \frac{E_0^2 L^2}{12\, R_r} \tag{E2}$$

The radiation resistance of a short electric dipole is [3]:

$$R_r = \frac{2\pi\eta L^2}{3\lambda^2} \tag{E3}$$

Substitution of (E3) into (E2) yields the desired final result:

$$\langle P_r \rangle = \frac{1}{2}\frac{E_0^2}{\eta}\frac{\lambda^2}{4\pi} \tag{E4}$$

Equation (E4) is identical to (7.103), which was derived for general antennas. The polarization mismatch factor of $\frac{1}{2}$ is particularly clear for the electric dipole antenna because $S_{r\beta} = 0$.

APPENDIX F

Small Loop Antenna Response

Another electrically small antenna of practical interest is the small loop, as shown in Figure F.1. For a small loop of area A centered on the z axis in the xy plane, the components of the receiving function are given by [69]:

$$S_{r\alpha} = 0 \quad \text{and} \quad S_{r\beta} = \frac{-i\omega\mu A \sin\alpha}{2\eta R_r} \tag{F1}$$

The result for $S_{r\beta}$ is obtained by: (1) determining the magnetic flux penetrating the loop; (2) multiplying by $-i\omega$ to determine the induced voltage; and (3) dividing by

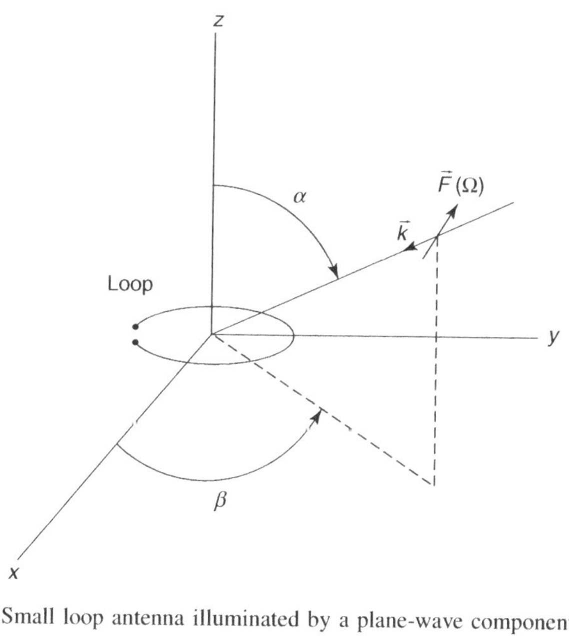

FIGURE F.1 Small loop antenna illuminated by a plane-wave component of the electric field.

Electromagnetic Fields in Cavities: Deterministic and Statistical Theories, by David A. Hill

$2R_r$ to determine the induced current induced in a matched load. Because β components of the magnetic field are orthogonal to the z axis of the loop, $S_{r\alpha} = 0$. The other way to see this is that α components of the electric field are orthogonal to the loop conductor in the xy plane.

If (F1) is substituted into (7.101), the angular integration can be carried out to obtain:

$$\langle P_r \rangle = \frac{E_0^2 \omega^2 \mu^2 A^2}{12\eta^2 R_r} \tag{F2}$$

The radiation resistance of a small loop is [3]:

$$R_r = \frac{2\pi\eta}{3}\left(\frac{kA}{\lambda}\right)^2 \tag{F3}$$

Substitution of (F3) into (F2) yields the desired final result:

$$\langle P_r \rangle = \frac{1}{2}\frac{E_0^2}{\eta}\frac{\lambda^2}{4\pi} \tag{F4}$$

which is identical to (7.103) for general antennas and (E4) for a short electric dipole. The polarization mismatch factor of $\frac{1}{2}$ is also clear for a small loop because $S_{r\alpha} = 0$.

APPENDIX G

Ray Theory for Chamber Analysis

The mathematical link between mode theory and ray theory for a perfectly conducting, rectangular cavity is the three-dimensional Poisson sum formula [188,189]. This formula allows the dyadic Green's function to be converted from a triple sum of modes to a triple sum of rays. The mathematical details are fairly involved and will not be covered here. However, the physical interpretation is clearly pictured in terms of multiple images, as shown in Figure G.1. For simplicity, the source is a z-directed electric dipole, and the multiple images represent multiple ray bounces in the $y = y'$ plane. Similar diagrams could be generated for other sources and locations.

The computation of the field at a point in the cavity is tedious because of the triple sum of image contributions. In fact, the sum is not convergent for some frequencies and field locations. This has to be the case because the equivalent mode sum has infinities at

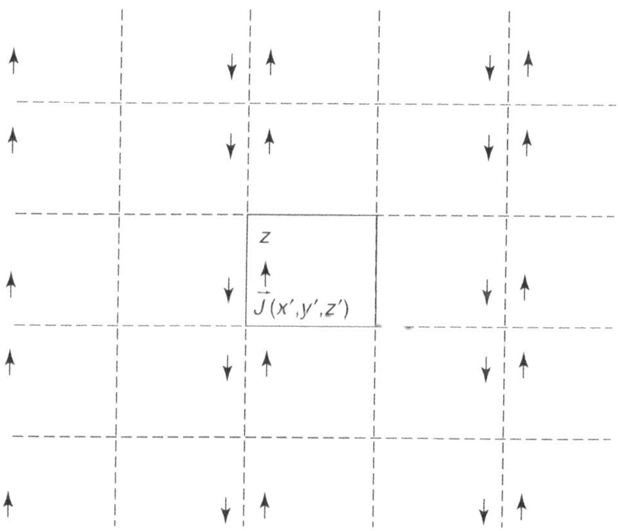

FIGURE G.1 Multiple images for a z-directed dipole source in a rectangular cavity.

Electromagnetic Fields in Cavities: Deterministic and Statistical Theories, by David A. Hill

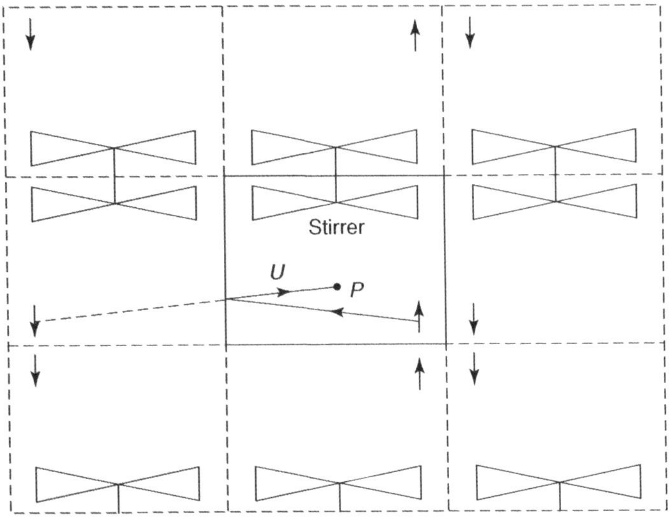

FIGURE G.2 Images of source and stirrer in a rectangular cavity. The (single-bounce) ray U is not affected by the stirrer (hence contributes to unstirred energy) [18].

the resonant frequencies of each cavity mode. The mode representation is made finite for imperfectly conducting walls by introducing a finite Q (hence the resonant frequencies become complex). The ray sum can be made finite for imperfectly conducting walls by introducing a reflection coefficient (which has magnitude less than one) at each wall bounce. This has been done for studying the field buildup in a rectangular cavity when the source is a turned-on sinusoid [99].

Multiple image theory can be extended to include the effect of a mechanical stirrer. Each image cell then contains an image of the mechanical stirrer with location and orientation as shown in Figure G.2. The solution of the large boundary-value problem would be extremely difficult, even with the ray-tracing approximation. However, the multiple-image diagram in Figure G.2 can be used to provide some insight into stirrer design. The goals of stirring are to randomize the field and to eliminate any deterministic component. Another way to state these goals is to minimize the ratio of unstirred to stirred energy. Unstirred energy arrives at the observation point without interacting with the stirrer. An example is (single-bounce) ray U in Figure G.2. An improved stirring strategy then would be to design the stirrer (or stirrers) to eliminate as many direct rays as possible. The conclusion that follows is that the stirrer(s)' dimensions must be comparable to chamber size rather than just comparable to a wavelength. This conclusion is consistent with recent chamber measurements [66].

APPENDIX H

Absorption by a Homogeneous Sphere

Section 7.6 discusses reverberation chamber losses and the resultant quality factor Q. Absorption cross section due to lossy objects located within the chamber is in general given by an average over incidence angle and polarization, as shown in (7.125). For a homogeneous sphere, the absorption cross section is independent of incidence angle and polarization, so the averages are not necessary. Hence, we select a spherical absorber as a simple example that has an analytical solution [41].

An incident plane-wave field with known frequency and intensity propagates toward the sphere. The problem is to determine the field penetrating into the sphere so that the absorption loss can be determined. The classical solution of this problem is due to Mie [190], based on the formulation of the vector wave equation:

$$\nabla^2 \vec{E} + k^2 m^2 \vec{E} = 0, \tag{H1}$$

in a source-free region satisfying appropriate boundary conditions. $\vec{E}$ is the unknown electric field inside the sphere, k is the free-space wavenumber, and m is the refractive index defined as:

$$m^2 = \varepsilon_r + i\sigma/(\omega\varepsilon_0), \tag{H2}$$

where ε_r is the relative dielectric constant (permittivity normalized to the free-space value ε_0) and σ is the conductivity of the material. In free space, $\varepsilon_r = 1$, $\sigma = 0$, and $m = 1$. We use m for refractive index in order to be consistent with common notation [88], but this should not be confused with the impedance mismatch factor in (7.106).

To solve the vector wave equation, we have to obtain solutions to the corresponding scalar wave equation:

$$\nabla^2 u + k^2 m^2 u + 0, \tag{H3}$$

in spherical coordinates. Since (H3) is a second-order partial differential equation, we have two independent solutions, u_1 and u_2. Let us represent the incident wave outside

Electromagnetic Fields in Cavities: Deterministic and Statistical Theories, by David A. Hill

the sphere where, $m = 1$, as:

$$
\begin{aligned}
u_1 &= \cos\phi \sum_{n=1}^{\infty} i^n \frac{2n+1}{n(n+1)} \mathrm{P}_n^1(\cos\theta)\mathrm{j}_n(kr), \\
u_2 &= \sin\phi \sum_{n=1}^{\infty} i^n \frac{2n+1}{n(n+1)} \mathrm{P}_n^1(\cos\theta)\mathrm{j}_n(kr),
\end{aligned} \tag{H4}
$$

where (ϕ, θ, r) are the spherical coordinates with $r = 0$ at the sphere center, P_n^1 is the associated Legendre function, and j_n is the spherical Bessel function.

The scattered wave outside the sphere can then be expressed as:

$$
\begin{aligned}
u_1' &= \cos\phi \sum_{n=1}^{\infty} (-a_n) i^n \frac{2n+1}{n(n+1)} \mathrm{P}_n^1(\cos\theta) h_n^{(2)}(kr), \\
u_2' &= \sin\phi \sum_{n=1}^{\infty} (-b_n) i^n \frac{2n+1}{n(n+1)} \mathrm{P}_n^1(\cos\theta) h_n^{(2)}(kr),
\end{aligned} \tag{H5}
$$

where $\mathrm{h}_n^{(2)}$ is the spherical Hankel function [25], and a_n and b_n are the unknown coefficients to be determined. The wave that penetrates into the sphere can be written:

$$
\begin{aligned}
u_1'' &= \cos\phi \sum_{n=1}^{\infty} (mc_n) i^n \frac{2n+1}{n(n+1)} \mathrm{P}_n^1(\cos\theta)\mathrm{j}_n(mkr), \\
u_2'' &= \sin\phi \sum_{n=1}^{\infty} (md_n) i^n \frac{2n+1}{n(n+1)} \mathrm{P}_n^1(\cos\theta)\mathrm{j}_n(mkr),
\end{aligned} \tag{H6}
$$

where m is complex and c_n and d_n are other unknown coefficients to be determined.

The boundary condition requires that at the sphere surface $r = a$,

$$
u_1 + u_1' = u_1'' \quad \text{and} \quad u_2 + u_2' = u_2'' \tag{H7}
$$

After some algebraic simplification, (H4)-(H7) yield:

$$
a_n = A/B, b_n = C/D, c_n = -i/B, \quad \text{and} \quad d_n = -i/D, \tag{H8}
$$

where:

$$
\begin{aligned}
A &= \Psi_n'(y)\Psi_n(x) - m\Psi_n(y)\Psi_n'(x), \\
B &= \Psi_n'(y)\xi_n(x) - m\Psi_n(y)\xi_n'(x), \\
C &= m\Psi_n'(y)\Psi_n(x) - \Psi_n(y)\Psi_n'(x), \\
D &= m\Psi_n'(y)\xi_n(x) - \Psi_n(y)\xi'(x),
\end{aligned} \tag{H9}
$$

with $x = ka$ and $y = mka = mx$. In (H9), Ψ_n and $\xi_n = \Psi_n - i\chi_n$ are the Riccati-Bessel functions [88], and Ψ_n' and ξ_n' are the derivatives with respect to their arguments.

The spherical Bessel functions are discussed in Appendix C. The Riccati-Bessel functions for low integer orders can be written:

$$\begin{aligned}
\Psi_0(z) &= \sin z,\\
\Psi_1(z) &= z^{-1}\sin z - \cos z,\\
\Psi_2(z) &= (3z^{-2}-1)\sin z - 3z^{-1}\cos z,\\
\Psi_3(z) &= (15z^{-3}-6z^{-1})\sin z - (15z^{-2}-1),\\
\Psi_4(z) &= (105z^{-4}-45z^{-2}+1)\sin z - (105z^{-3}-10z^{-1})\cos z;
\end{aligned} \tag{H10}$$

$$\begin{aligned}
\chi_0(z) &= \cos z,\\
\chi_1(z) &= \sin z + z^{-1}\cos z,\\
\chi_2(z) &= 3z^{-1}\sin z + (3z^{-2}-1)\cos z,\\
\chi_3(z) &= (15z^{-2}-1)\sin z + (15z^{-3}-6z^{-1})\cos z,\\
\chi_4(z) &= (105z^{-3}-10z^{-1})\sin z + (105z^{-4}-45z^{-2}+1)\cos z.
\end{aligned} \tag{H11}$$

The derivatives of the Riccati-Bessel functions required in (H9) are then

$$\begin{aligned}
\Psi_0'(z) &= \cos z,\\
\Psi_1'(z) &= (-z^{-2}+1)\sin z + z^{-1}\cos z,\\
\Psi_2'(z) &= (-6z^{-3}+3z^{-1})\sin z + (6z^{-2}-1)\cos z,\\
\Psi_3'(z) &= (-45z^{-4}+21z^{-2}-1)\sin z + (45z^{-3}-6z)\cos z,\\
\Psi_4'(z) &= (-420z^{-5}+195z^{-3}-10z^{-1})\sin z + (420z^{-4}-55z^{-2}+1)\cos z;
\end{aligned} \tag{H12}$$

$$\begin{aligned}
\chi_0'(z) &= -\sin z,\\
\chi_1'(z) &= -z^{-1}\sin z + (-z^{-2}+1)\cos z,\\
\chi_2'(z) &= (-6z^{-2}+1)\sin z + (-6z^{-3}+3z^{-1})\cos z,\\
\chi_3'(z) &= (-45z^{-3}+6z^{-1})\sin z + (-45z^{-4}+21z^{-2}-1)\cos z,\\
\chi_4'(z) &= (420z^{-4}+55z^{-2}-1)\sin z + (-420z^{-5}+195z^{-3}-10z^{-1})\cos z
\end{aligned} \tag{H13}$$

In (H10)-(H13), $z = x$ (real) or $z = y$ (complex). Once the frequency f (thus ω), sphere radius a, and sphere material constants ε_r and σ are specified, the coefficients a_n, b_n, c_n, and $d_n$can be computed from (H8) and (H9), and the field distributions inside and outside the sphere can be determined from (H4)-(H6). The number of terms required in (H4)-(H6) depends on ka.

Two efficiency factors, the total factor η_t and the scattered factor η_s, are useful in determining the absorption. They are determined by:

$$\eta_t = 2(ka)^{-2}\sum_{n=1}^{\infty}(2n+1)\mathrm{Re}(a_n+b_n), \tag{H14}$$

$$\eta_s = 2(ka)^{-2}\sum_{n=1}^{\infty}(2n+1)\left(|a_n|^2+|b_n|^2\right). \tag{H15}$$

The efficiency factor η_a for absorption is then:

$$\eta_a = \eta_t - \eta_s \tag{H16}$$

The absorption cross section of the sphere is obtained from:

$$\sigma_a = (\pi a^2)\eta_a, \tag{H17}$$

which in turn is used to compute the power loss due to absorption. Because of the symmetry of the sphere, the averaging over incidence angle and polarization has no effect: $\langle \sigma_a \rangle = \sigma_a$.

When ka becomes very large, this theory is not convenient because the summations converge slowly. In this case, a geometrical optics approximation [88, Sec. 14.23] can be used to compute $\langle \sigma_a \rangle$. The computer program in [41] uses this approximation to compute $\langle \sigma_a \rangle$ and Q_2 when ka becomes large.

APPENDIX I

Transmission Cross Section of a Small Circular Aperture

Consider a small circular aperture of radius a ($ka \ll 1$) in a planar sheet, as shown in Figure 8.3. The transmitted fields can be written as the fields of a tangential magnetic dipole p_m and a normal electric dipole moment p_e that can be written as the product of an aperture polarizability times the appropriate incident field [85,104]:

$$p_m = \alpha_m H_{tan}^{sc} \quad \text{and} \quad p_e = \varepsilon_0 \alpha_e E_n^{sc}, \tag{I1}$$

where H_{tan}^{sc} is the tangential magnetic field at the center of the short-circuited aperture and E_n^{sc} is the normal electric field at the center of the short-circuited aperture. The magnetic and electric polarizabitities, α_m and α_e, are given by [85,104]:

$$\alpha_m = 4a^3/3 \quad \text{and} \quad \alpha_e = 2a^3/3 \tag{I2}$$

The dipole moments radiate in the presence of the ground plane (so their images are included), and the total transmitted power (radiated into one half-space) is [3]:

$$P_t = \frac{4\pi\eta_0}{3\lambda^2}\left(k^2|p_m|^2 + |p_e|^2\right) \tag{I3}$$

We consider the cases of parallel and perpendicular polarizations separately. For parallel polarization, the short-circuited fields are:

$$H_{tan}^{sc} = 2H_i \quad \text{and} \quad E_n^{sc} = 2E_i \sin\theta^i, \tag{I4}$$

where the incident fields can be related to the incident power density S_i by:

$$S_i = \eta_0 H_i^2 \quad \text{and} \quad S_i = E_i^2/\eta_0 \tag{I5}$$

Electromagnetic Fields in Cavities: Deterministic and Statistical Theories, by David A. Hill

From (I1)-(I5), we can write the transmission cross section for parallel polarization as

$$\sigma_{tpar} = P_t/S_i = \frac{64}{27\pi}k^4a^6\left(1+\frac{1}{4}\sin^2\theta^i\right), \tag{I6}$$

which is the result needed in Section 8.1.

For perpendicular polarization, the short-circuited fields are:

$$H_{tan}^{sc} = 2H_i\cos\theta^i \quad \text{and} \quad E_n^{sc} = 0 \tag{I7}$$

From (I1)-(I3), (I5), and (I7), we can write the transmission cross section for perpendicular polarization as:

$$\sigma_{tperp} = \frac{64}{27\pi}k^4a^6\cos^2\theta^i, \tag{I8}$$

which is the other result needed in Section 8.1.

APPENDIX J

Scaling

For applications involving large objects, such as aircraft, laboratory measurements are more conveniently done on smaller scale models. For the time-harmonic form of Maxwell's equations, scaling of frequency and length in nondispersive, lossless media is well known. For the example of frequency-independent antennas [191], "the entire electrical performance is frequency-independent if all length dimensions are scaled in inverse proportion to frequency."

To consider the more general case of lossy media [41], we begin with the time-harmonic, source-free form of Maxwell's equations:

$$\begin{aligned}\nabla \times \vec{H}(\vec{r}, \omega) &= [-i\omega\varepsilon(\vec{r}) + \sigma(\vec{r})]\vec{E}(\vec{r}, \omega),\\ \nabla \times \vec{E}(\vec{r}, \omega) &= i\omega\mu\vec{H}(\vec{r}, \omega),\end{aligned} \tag{J1}$$

where the magnetic permeability μ, the permittivity ε, and the conductivity σ are assumed to be independent of frequency, but can be functions of position $\vec{r}$. Suppose that we wish to scale (multiply) lengths by a real factor $1/s$ (that can be greater or less than one):

$$\vec{r}\,' = \vec{r}/s \quad \text{or} \quad \vec{r} = s\vec{r}\,' \tag{J2}$$

If $s > 1$, then the new primed lengths are less than the original lengths.

To examine the scaling possibilities of (J1), we rewrite the del (∇) operator as follows:

$$\nabla = \hat{x}\frac{\partial}{\partial x} + \hat{y}\frac{\partial}{\partial y} + \hat{z}\frac{\partial}{\partial z} = \frac{1}{s}\left(\hat{x}\frac{\partial}{\partial x'} + \hat{y}\frac{\partial}{\partial y'} + \hat{z}\frac{\partial}{\partial z'}\right) = \frac{1}{s}\nabla' \tag{J3}$$

where $\hat{x}$, $\hat{y}$, and $\hat{z}$ are unit vectors that remain unchanged in the primed coordinate system. If we substitute (J3) into (J1) and multiply by s, then we have:

$$\begin{aligned}\nabla' \times \vec{H}(s\vec{r}\,') &= [-i\omega s\varepsilon(s\vec{r}\,') + \sigma(s\vec{r}\,')]\vec{E}(s\vec{r}\,'),\\ \nabla' \times \vec{E}(s\vec{r}\,') &= i\omega s\mu(s\vec{r}\,')\vec{H}(s\vec{r}\,')\end{aligned} \tag{J4}$$

Electromagnetic Fields in Cavities: Deterministic and Statistical Theories, by David A. Hill

The object now is to scale quantities on the right side of (J4) to bring it to the same form as given in (J1). There are two possibilities: (1) we could scale ϵ and μ by s or (2) we could scale ω and σ by s. The first possibility is generally of no value for scaled experiments. The second possibility is the standard length/frequency scaling for lossless media, where σ is either 0 or ∞. For those cases, scaling of σ has no effect on the results.

We choose the second possibility and the following specific scaling:

$$\vec{H}'(\vec{r}\,') = \vec{H}(s\vec{r}\,'), \vec{E}'(\vec{r}\,') = \vec{E}(s\vec{r}\,'), \vec{r}\,' = \frac{1}{s}\vec{r}, \omega' = s\omega,$$

and

$$\sigma'(\vec{r}\,') = s\sigma(sr'), \varepsilon'(r') = \varepsilon(s\vec{r}\,'), \mu'(\vec{r}\,') = \mu(s\vec{r}\,') \tag{J5}$$

If we substitute (J5) into (J1), we obtain:

$$\begin{aligned} \nabla' \times \vec{H}'(\vec{r}\,', \omega') &= [-i\omega'\varepsilon'(\vec{r}\,') + \sigma'(r')]\vec{E}'(\vec{r}\,', \omega'), \\ \nabla' \times \vec{E}'(\vec{r}\,', \omega') &= i\omega'\mu'(\vec{r}\,')\vec{H}(\vec{r}\,', \omega') \end{aligned} \tag{J6}$$

Equations (J6) are identical to Maxwell's equations (J1) except that all quantities are primed. So they are equivalent under the scaling transformations in (J3) and (J5). To summarize, we scale all distances by $1/s$, frequency by s, and conductivity by s. If we wish to perform a reduced-size ($s > 1$) scale-model experiment, then we increase frequency by a factor s as expected, *but we also need to increase conductivity by a factor s*. This obviously presents a materials problem, but some ways around this problem are discussed in the remainder of this Appendix.

The required conductivity scaling can be explained in an equivalent manner. The first (J1) equation can also be written

$$\nabla \times \vec{H}(\vec{r}, \omega) = -i\omega\varepsilon_c(\vec{r})\vec{E}(\vec{r}, \omega), \tag{J7}$$

where

$$\varepsilon_c(\vec{r}) = \varepsilon_r(\vec{r}) + i\sigma(\vec{r})/\omega, \tag{J8}$$

and ε_r is the real part of the complex permittivity ε_c. Since our frequency scaling requires multiplying ω by s, we must also multiply σ by s to keep the imaginary part of the complex permittivity ε_c from changing.

For general cavity applications, if the conductivity is not scaled according to (J5), the field distributions will change and the resonant frequencies and Qs will also change in an unpredictable manner. However, if the walls are highly conducting, the resonant frequencies will not depend on the wall conductivity and will scale as s times the resonant frequencies of the original cavity.

Consider now the composite Q of a cavity with highly conducting walls. We examine the individual Qs separately, as in Section 7.6. The expression for Q_1 is given by (7.123). For the scaling in (J5), the new primed quantities are:

$$Q_1' = \frac{3V'}{2\mu_r S'\delta'} = Q_1, \tag{J9}$$

where

$$V' = \frac{V}{s^3}, S' = \frac{S}{s^2}, \delta' = \frac{\delta}{s}, \omega' = s\omega, \quad \text{and} \quad \sigma' = s\sigma \tag{J10}$$

We assume that the magnetic permeability is unchanged. If it is not possible to scale wall conductivity, then we have:

$$\sigma' = \sigma, \delta' = \delta s^{-1/2}, \quad \text{and} \quad Q_1' = Q_1 s^{-1/2} \tag{J11}$$

Thus Q_1' of the scaled cavity will decrease if the frequency is scaled up ($s > 1$), lengths are scaled down, and the wall conductivity is not scaled.

Consider now Q_2. If the loading objects have high conductivity (as for metal), then Q_2' will change by the factor $s^{-1/2}$ as in (J11) because the loading objects have the same loss dependence on frequency and conductivity as the walls. A different situation arises when the cavity losses are due primarily to objects of low conductivity and low permittivity (nonmetal objects). In this case the Born approximation [192] states that the field distribution is not strongly affected by the loading objects. Thus the resonant frequencies are not significantly changed, and the cavity loss is proportional to the conductivity σ of the loading objects. Then Q_2 is inversely proportional to the conductivity or the loading objects:

$$Q_2 \propto \omega/\sigma \tag{J12}$$

If we scale frequency, length, and conductivity, then the new Q_2' is:

$$Q_2' = Q_2 \propto \omega'/\sigma', \quad \text{where} \quad \omega' = s\omega \quad \text{and} \quad \sigma' = s\sigma \tag{J13}$$

If we are not able to scale conductivity, then:

$$Q_2' = sQ_2, \quad \text{where} \quad \omega' = s\omega \quad \text{and} \quad \sigma' = \sigma \tag{J14}$$

Here the change in Q' is in the opposite direction as that in (J11) where the wall conductivity is unscaled. The actual situation for low conductivity objects (such as people and nonmetal furniture) is that the unscaled conductivity is equal to frequency times the imaginary part of the permittivity:

$$\sigma = \omega \,\text{Im}(\varepsilon_c) \tag{J15}$$

If $\text{Im}(\varepsilon_c)$ does not change with frequency, then the proper conductivity scaling automatically occurs with no change in loading material.

Consider now Q_3 and Q_4. For aperture leakage losses and antenna reception in a fixed load impedance, length and frequency scaling are sufficient to maintain Q_3 and Q_4 with no change upon scaling.

Thus we have three situations with regard to cavity scaling where frequency is scaled up ($s > 1$), length is inversely scaled, and conductivity is unscaled. For dominant wall losses, Q' drops by a factor $s^{-1/2}$, as given by (J11). For dominant loading losses due to low conductivity, Q' increases by a factor s, as given by (J14). (This change will be less or even zero for dielectrics with a nearly constant loss tangent.) For dominant aperture leakage or antenna reception losses, Q' is unchanged.

In summary, the resonant frequencies will scale with s if the field distributions change little. The cavity Q' can be higher or lower than the original Q if the conductivity is unscaled. However, the magnitude and direction of change can be predicted if the dominant loss mechanism is known.

REFERENCES

CITED REFERENCES

[1] D. Kajfez, *Q Factor*. Oxford, MS: Vector Fields, 1994.

[2] C.-T. Tai, *Dyadic Green Functions in Electromagnetic Theory*. New York: IEEE Press, 1997.

[3] R.F. Harrington, *Time-Harmonic Electromagnetic Fields,* Second Edition. New York: Wiley-IEEE Press, 2001.

[4] G.S. Smith, *An Introduction to Classical Electromagnetic Radiation*. Cambridge, UK: Cambridge University Press, 1997.

[5] I.V. Lindell, A.H. Sihvola, S.A. Tretyakov, and A.J. Viitanen, *Electromagnetic Waves in Chiral and Bi-Isotropic Media*. Boston: Artech House, 1994.

[6] F.E. Borgnis and C.H.Pappas, "Electromagnetic waveguides and resonators," *Encyclopedia of Physics, Volume XVI, Electromagnetic Fields and Waves* (ed., S. Flugge). Berlin: Springer-Verlag, 1958.

[7] E. Argence and T. Kahan, *Theory of Waveguides and Cavity Resonators*. New York: Hart Publishing Co., 1968.

[8] H. Weyl, "Über die randwertaufgabe der randwertaufgabe der Strahlungstheorie und asymptotische Spektralgesetze," *J. Reine U. Angew. Math.*, vol. 143, pp. 177–202, 1913.

[9] B.H. Liu, D.C. Chang, and M.T. Ma,"Eigenmodes and the composite quality factor of a reverberating chamber," U.S. Nat. Bur. Stand. Tech. Note 1066, 1983.

[10] T.B.A. Senior and J.L. Volakis, *Approximate Boundary Conditions in Electromagnetics*. London: IEE Press, 1995.

[11] D.A. Hill, "A reflection coefficient derivation for the *Q* of a reverberation chamber," *IEEE Trans. Electromagn. Compat.*, vol. 38, pp. 591–592, 1996.

[12] K. Kurokawa, "The expansions of electromagnetic fields in cavities," *IRE Trans. Microwave Theory Tech.*, vol. 6, pp. 178–187, 1958.

[13] R.E. Collin, *Field Theory of Guided Waves*, Second Edition. Piscataway, NJ: IEEE Press, 1991.

[14] R.A. Waldron, "Perturbation theory of resonant cavities," *Proc. IEE*, vol. 107C, pp. 272–274, 1960.

Electromagnetic Fields in Cavities: Deterministic and Statistical Theories, by David A. Hill

[15] J. Van Bladel, *Electromagnetics Fields*, Second Edition. New York: Wiley-IEEE Press, 2007.

[16] H.C. Van de Hulst, *Light Scattering by Small Particles*. New York: Dover, 1981.

[17] "Standard test methods for complex permittivity (dielectric constant) of solid electrical insulating materials at microwave frequencies and temperatures to 1650 °C," *American Society for Testing and Materials*, D 2520, 1995.

[18] D.A. Hill,"Electromagnetic theory of reverberation chambers," U.S. Nat. Inst. Stand. Technol. Tech. Note 1506, 1998.

[19] M.L. Crawford and G.H. Koepke,"Design, evaluation, and use of a reverberation chamber for performing electromagnetic susceptibility/vulnerability measurements," U.S. Nat. Bur. Stand. Tech. Note 1092, 1986.

[20] A.D. Yaghjian, "Electric dyadic Green's functions in the source region," *Proc. IEEE*, vol. 68, pp. 248–263, 1980.

[21] J. van Bladel, *Singular Electromagnetic Fields and Sources*. Oxford: Clarendon, 1991.

[22] C.-T. Tai,"Singular terms in the eigen-function expansion of dyadic Green's function of the electric type," EMP Interaction Note 65, 1980.

[23] J.J. Green and T. Kohane, "Testing of ferrite materials for microwave applications," *Semiconductor Products and Solid State Technology*, vol. 7, pp. 46–54, 1964.

[24] C.E.Patton and T. Kohane, "Ultrasensitive technique for microwave susceptibility determination down to 10^{-5}," *Review of Scientific Instruments*, vol. 43, pp. 76–79, 1972.

[25] M. Abramowitz and I.A. Stegun, *Handbook of Mathematical Functions*. U.S. National Bureau of Standards, Applied Mathematics Series 55, 1964.

[26] I.S Gradshteyn I.M. Ryzhik, *Tables of Integrals, Series, and Products*. New York: Academic Press, 1965.

[27] R.F. Soohoo and P. Christensen, "Theory and method for magnetic resonance measurements," *J. Appl. Phys.*, vol. 40, pp. 1565–1566, 1969.

[28] R.A. Waldron, *Theory of Guided Electromagnetic Waves*. London: Van Nostrand Reinhold, 1970.

[29] E. Jahnke and F. Emde, *Tables of Functions*. New York: Dover Publications, 1945.

[30] J.R. Wait, *Geo-Electromagnetism*. New York: Academic Press, 1982.

[31] M.L. Burrows, *ELF Communication Antennas*. Stevenage, UK: Peter Peregrinus Ltd., 1978.

[32] P.V. Bliokh, A.P. Nicholaenko, and Iu.R. Fillipov, *Schumann Resonances in the Earth-Ionosphere Cavity*. Stevanage, UK: Peter Perigrinus Ltd., 1980.

[33] J.R. Wait, *Electromagnetic Waves in Stratified Media*. New York: IEEE Press, Third Edition, 1995.

[34] J.D. Jackson, *Classical Electrodynamics*. New York: John Wiley & Sons, 1999.

[35] J. Galejs, *Terrestrial Propagation of Long Electromagnetic Waves*. Oxford: Pergamon Press, 1972.

[36] R.H. Price, H.T. Davis, and E.P. Wenaas, "Determination of the statistical distribution of electromagnetic-field amplitudes in complex cavities," *Phys. Rev. E*, vol. 48, pp. 4716–4729, 1993.

[37] T.H. Lehman,"A statistical theory of electromagnetic fields in complex cavities," EMP Interaction Note 494," 1993.

[38] D.A. Hill, M.T. Ma, A.R. Ondrejka, B.F. Riddle, M.L. Crawford, and R.T. Johnk, "Aperture excitation of electrically large, lossy cavities," *IEEE Trans. Electromagn. Compat.*, vol. 36, pp. 169–178, 1994.

[39] R. Holland and R. St. John, *Statistical Electromagnetics*. Philadelphia: Taylor & Francis, 1999.

[40] K.S.H. Lee, editor, *EMP Interaction: Principles, Techniques, and Reference Data*. Washington: Hemisphere Pub. Corp., 1986.

[41] D.A. Hill, J.W. Adams, M.T. Ma, A.R. Ondrejka, B.F. Riddle, M.L. Crawford, and R.T. Johnk,"Aperture excitation of electrically large, lossy cavities," U.S. Nat. Inst. Stand. Technol. Tech. Note 1361, 1993.

[42] R. Vaughn and J. Bach Anderson, *Channels, Propagation and Antennas for Mobile Communications*. London: IEE Press, 2003.

[43] S. Loredo, L. Valle,R.P. Torres, "Accuracy analysis of GO/UTD radio-channel modeling in indoor scenarios at 1.8 and 2.5 GHz," *IEEE Ant. Propagat. Mag.*, vol. 43, pp. 37–51, 2001.

[44] J.M Keenan and A.J. Motley, "Radio coverage in buildings," *British Telecom Technology Journal*, vol. 8, pp. 19–24, 1990.

[45] COST231: "Digital mobile radio towards future generations," Final Report, European Commission, 1991.

[46] D.M.J. Devasirvatham, C. Banerjee, R.R. Murray, and D.A. Rappaport, "Four-frequency radiowave propagation measurements of the indoor environment in a large metropolitan commercial building," *Globecom'91*, pp. 1281–1286, 1991.

[47] W. Spencer, M. Rice, B. Jeffs, and M. Jensen, "A statistical model for angle of arrival in indoor multipath propagation," *Proc. VTC'97*, pp. 1415–1419, 1997.

[48] D.A. Hill, "Electronic mode stirring," *IEEE Trans. Electromagn. Compat.*, vol. 36, pp. 294–299, 1994.

[49] M.L. Crawford, T.A. Loughry, M.O. Hatfield, and G.J. Freyer,"Band-limited, white Gaussian noise excitation for reverberation chambers and applications to radiated susceptibility testing," U.S. Nat. Inst. Stand. Technol. Tech. Note 1375, 1996.

[50] A. Taflove and S.C. Hagness, *Computational Electrodynamics: The Finite-Difference Time-Domain Method, 3rd ed.* Norwood, MA: Artech House, 2005.

[51] P.M. Morse and K.U. Ingard, *Theoretical Acoustics*. New York: McGraw-Hill Book Co., 1968.

[52] R.K. Cook, R.V. Waterhouse, R.D. Berendt, S. Edelman, and M.C. Thompson, "Measurement of correlation coefficients in reverberant sound fields," *J. Acoust. Soc. Amer.*, vol. 27, pp. 1072–1077, 1955.

[53] A. Ishimaru, *Wave Propagation and Scattering in Random Media*. New York: Academic Press, 1978.

[54] S. Chandrasekar, *Radiative Transfer*. New York: Dover, 1960.

[55] L. Mandel and E. Wolf, *Optical Coherence and Quantum Optics*. Cambridge, UK: Cambridge University Press, 1995.

[56] A.J. Mackay, "Application of the generalized radiance function for prediction of the mean RCS of bent chaotic ducts with apertures not normal to the duct axis," *IEE Proc.-Radar Sonar Navig.*, vol. 149, pp. 9–15, 2002.

[57] A. Papoulis, *Probability, Random Variables, and Stochastic Processes*. New York: McGraw-Hill Book Co., 1965.

[58] P. Beckmann, *Probability in Communication Engineering*. New York: Harcourt, Brace & World, Inc., 1967.

[59] S.M. Ross, *Introduction to Probability and Statistics for Engineers and Scientists, Second Edition*. San Diego, CA: Academic Press, 2000.

[60] P. Olofsson, *Probability, Statistics, and Stochastic Processes*. Hoboken, NJ: Wiley-Interscience, 2005.

[61] S.M. Rytov, Yu.A. Kravtsov, and V.I. Tatarskii, *Principles of Statistical Radiophysics*, Vols. 1–4. Berlin: Springer-Verlag, 1989.

[62] S. Gasiorowicz, *Quantum Physics*, 3rd *ed.* Hoboken, NJ: John Wiley & Sons, 2003.

[63] B.V. Gnedenko, *The Theory of Probability*. New York: Chelsea Publ. Co., 1962.

[64] J. Baker-Jarvis and M. Racine, "Solving differential equations by a maximum entropy-minimum norm method with applications to Fokker-Planck equations," *J. Math. Phys.*, vol. 30, pp. 1459–1463, 1989.

[65] J.N. Kapur and H.K. Kesavan, *Entropy Optimization Principles with Applications*. Boston: Academic Press, 1992.

[66] J. Ladbury, G. Koepke, and D. Camell,"Evaluation of the NASA Langley Research Center Mode-Stirred Chamber Facility," U.S. Nat. Inst. Stand. Technol. Tech. Note 1508, 1999.

[67] H.A. Mendes, "A new approach to electromagnetic field-strength measurements in shielded enclosures," *Wescon*, Los Angeles, CA, 1968.

[68] P. Corona, G. Latmiral, and E. Paolini, "Performance and analysis of a reverberating enclosure with variable geometry," *IEEE Trans. Electromagn. Compat.*, vol. 22, pp. 2–5, 1980.

[69] D.A. Hill, "Plane wave integral representation for fields in reverberation chambers," *IEEE Trans. Electromagn. Compat.*, vol. 40, pp. 209–217, 1998.

[70] J.A. Stratton, *Electromagnetic Theory*. New York: McGraw-Hill, 1941.

[71] R.C. Wittmann and D.N. Black, "Quiet-zone evaluation using a spherical synthetic-aperture radar," *IEEE Antennas Propagat. Soc. Int. Symp.*, Montreal, Canada, July 1997, pp. 148–151.

[72] J.G. Kostas and B. Boverie, "Statistical model for a mode-stirred chamber," *IEEE Trans. Electromagn. Compat.*, vol. 33, pp. 366–370, 1991.

[73] D.A. Hill and J.M. Ladbury, "Spatial-correlation functions of fields and energy density in a reverberation chamber," *IEEE Trans. Electromagn. Compat.*, vol. 44, pp. 95–101, 2002.

[74] D.A. Hill, "Linear dipole response in a reverberation chamber," *IEEE Trans. Electromagn. Compat.*, vol. 41, pp. 365–368, 1999.

[75] D.A. Hill, "Spatial correlation function for fields in reverberation chambers," *IEEE Trans. Electromagn. Compat.*, vol. 37, p. 138, 1995.

[76] E. Wolf, "New theory of radiative energy transfer in free electromagnetic fields," *Phys. Rev. D*, vol. 13, pp. 869–886, 1976.

[77] R.K. Cook, R.V. Waterhouse, R.D. Berendt, S. Edelman, and M.C. Thompson, "Measurement of correlation coefficients in reverberant sound fields," *J. Acoust. Soc. Amer.*, vol. 27, pp. 1072–1077, 1955.

[78] B. Eckhardt, U. Dörr, U. Kuhl, and H.-J. Stöckmann, "Correlations of electromagnetic fields in chaotic cavities," *Europhys. Lett.*, vol. 46, pp. 134–140, 1999.

[79] A.K. Mittra and T.R. Trost, "Statistical simulations and measurements inside a microwave reverberation chamber," *Proc. Int. Symp. Electromagn. Compat.*, Austin, TX. Aug. 1997, pp. 48–53.

[80] A. Mittra,"Some critical parameters for the statistical characterization of power density within a microwave reverberation chamber," Ph.D. dissertation, Dept. Elect. Engr., Texas Tech. Univ., Lubbock, TX, 1996.

[81] D.M. Kerns,*Plane-Wave Scattering Theory of Antennas and Antenna-Antenna Interactions*. U.S. Nat. Bur. Stand. Monograph 162; 1981.

[82] P.K. Park and C.T. Tai, "Receiving antennas," Ch. 6 in *Antenna Handbook* (ed. Y.T. Lo and S.W. Lee). New York: Van Nostrand Reinhold Co., 1988.

[83] C.T. Tai, "On the definition of effective aperture of antennas," *IEEE Trans. Antennas Propagat.*, vol. 9, pp. 224–225, 1961.

[84] D.A. Hill, D.G. Camell, K.H. Cavcey, and G.H. Koepke, "Radiated emissions and immunity of microstrip transmission line: theory and reverberation chamber measurements," *IEEE Trans. Electromagn. Compat.*, vol. 38, pp. 165–172, 1996.

[85] D.A. Hill, M.L. Crawford, M. Kanda,D.I. Wu, "Aperture coupling to a coaxial air line: theory and experiment," *IEEE Trans. Electromagn. Compat.*, vol. 35, pp. 69–74, 1993.

[86] D.A. Hill, D.G. Camell, K.H. Cavcey, and G.H. Koepke,"Radiated emissions and immunity of microstrip transmission lines: theory and measurements," U.S. Nat. Inst. Stand. Technol. Tech. Note 1377, 1995.

[87] J.M. Dunn, "Local, high-frequency analysis of the fields in a mode-stirred chamber," *IEEE Trans. Electromagn. Compat.*, vol. 32, pp. 53–58, 1990.

[88] H.C. Van de Hulst, *Light Scattering by Small Particles*. New York: Dover, 1981.

[89] C.M. Butler, Y. Rahmat-Samii, and R. Mittra, "Electromagnetic penetration through apertures in conducting surfaces," *IEEE Trans. Antennas Propagat.*, vol. 26, pp. 82–93, 1978.

[90] T.A. Loughry,"Frequency stirring: an alternate approach to mechanical mode-stirring for the conduct of electromagnetic susceptibility testing," Phillips Laboratory, Kirtland Air Force Base, NM Technical Report 91-1036, 1991.

[91] R.E. Richardson, "Mode-stirred calibration factor, relaxation time, and scaling laws," *IEEE Trans. Instrum. Meas.*, vol. 34, pp. 573–580, 1985.

[92] A.T. De Hoop and D. Quak,"Maxwell fields and Kirchhoff circuits in electromagnetic interference," Technical University of Delft, Netherlands, Report Et/EM 1995-34, 1995.

[93] D.A. Hill, "Reciprocity in reverberation chamber measurements," *IEEE Trans. Electromagn. Compat.*, vol. 45, pp. 117–119, 2003.

[94] G.D. Monteath, *Applications of the Electromagnetic Reciprocity Principle*. Oxford, UK: Pergamon, 1973.

[95] G.L. James, *Geometrical Theory of Diffraction for Electromagnetic Waves*. Stevenage, UK: Peter Perigrinus, 1976.

[96] J.D. Kraus, *Antennas*. New York: McGraw-Hill, 1950.

[97] D.A. Hill, "Boundary fields in reverberation chambers," *IEEE Trans. Electromagn. Compat.*, vol. 47, pp. 281–290, 2005.

[98] J.M. Ladbury and D.A. Hill, "Enhanced backscatter in a reverberation chamber," *IEEE Int. Symp. Electromagn. Compat.*, Honolulu, Hawaii, July 2007.

[99] D.H. Kwon, R.J. Burkholder, and P.H. Pathak, "Ray analysis of electromagnetic field buildup and quality factor of electrically large shielded enclosures," *IEEE Trans. Electromagn. Compat.*, vol. 40, pp. 19–26, 1998.

[100] P.-E. Wolf and G. Maret, "Weak localization and coherent backscattering of photons in disordered media," *Phys. Rev. Letters*, vol. 55, pp. 2696–2699, 1985.

[101] A. Ishimaru, J.S. Chen, P. Phu, and K. Yoshitomi, "Numerical, analytical, and experimental studies of scattering from very rough surfaces and backscattering enhancement," *Waves in Random Media*, vol. 1, pp. S91–S107, 1991.

[102] P. Phu, A. Ishimaru, and Y. Kuga, "Controlled millimeter-wave experiments and numerical simulations on the enhanced backscattering from one-dimensional very rough surfaces," *Rad. Sci.*, vol. 28, pp. 533–548, 1993.

[103] H.A. Bethe, "Theory of diffraction by small holes," *Phys. Rev.*, vol. 66, pp. 163–182, 1944.

[104] J. Meixner and W. Andrejewski, "Strenge Theorie der beugung ebener elektromagnetischer Wellen an der vollkommen leitenden eben Schirm," *Annalen der Physik.*, vol. 7, pp. 157–168, 1950.

[105] H. Levine and J. Schwinger, "On the theory of electromagnetic wave diffraction by an aperture in an infinite plane conducting screen," *Comm. Pure Appl. Math.*, vol. 3, pp. 355–391, 1950.

[106] K.S.H. Lee and F.-C. Yang, "Trends and bounds in RF coupling to a wire inside a slotted cavity," *IEEE Trans. Electromagn. Compat.*, vol. 34, pp. 154–160, 1992.

[107] J.A. Saxton and J.A. Lane, "Electrical properties of sea water," *Wireless Engineer*, vol. 29, pp. 269–275, 1952.

[108] J.M. Ladbury, T. Lehman, and G.H. Koepke, "Coupling to devices in electrically large cavities, or why classical EMC evaluation techniques are becoming obsolete," *IEEE Int. Symp. Electromagn. Compat.*, pp. 648–655, Aug. 2002.

[109] D.I. Wu and D.C. Chang, "The effect of an electrically large stirrer in a mode-stirred chamber," *IEEE Trans. Electromagn. Compat.*, vol. 31, pp. 164–169, 1989.

[110] P.M. Morse and H. Feshbach, *Methods of Theoretical Physics*. New York, McGraw-Hill, 1953.

[111] P. Corona, G. Ferrara, and M. Migliaccio, "Reverberating chamber electromagnetic field in presence of an unstirred component," *IEEE Trans. Electromagn. Compat.*, vol. 42, pp. 111–115, 2000.

[112] C.L. Holloway, D.A. Hill, J.M. Ladbury, and G. Koepke, "Requirements for an effective reverberation chamber: unloaded or loaded," *IEEE Trans. Electromagn. Compat.*, vol. 48, pp. 187–194, 2006.

[113] *ASTM-ES7 and ASTM-D4935 Standard for Measuring the Shielding Effectiveness in the Far Field*, vol. 10.02, ASTM, Philadelphis, PA, 1995.

[114] P.F. Wilson and M.T. Ma,"A study of techniques for measuring the electromagnetic shielding effectiveness of materials," U.S. Nat. Bur. Stand. Tech. Note 1095, 1986.

[115] C.L. Holloway, D.A. Hill, J. Ladbury, G. Koepke, and R. Garzia, "Shielding effectiveness measurements of materials using nested reverberation chambers," *IEEE Trans. Electromagn. Compat.*, vol. 45, pp. 350–356, 2003.

[116] M.O. Hatfield, "Shielding effectiveness measurements using mode-stirred chambers: a comparison of two approaches," *IEEE Trans. Electromagn. Compat.*, vol. 30, pp. 229–238, 1988.

[117] T.A. Loughry and S.H. Burbazani, "The effects of intrinsic test fixture isolation on material shielding effectiveness measurements using nested mode-stirred chambers," *IEEE Trans. Electromagn. Compat.*, vol. 37, pp. 449–452, 1995.

[118] *IEEE Standard Dictionary of Electrical and Electronics Terms*. ANSI/IEEE Std 100-1984, New York: IEEE, 1984.

[119] D.A. Hill and M. Kanda,"Measurement uncertainty of radiated emissions," U.S. Nat. Inst. Stand. Technol. Tech. Note 1508, 1997.

[120] C.L. Holloway, J. Ladbury, J. Coder, G. Koepke, and D.A. Hill, "Measuring the shielding effectiveness of small enclosures/cavities with a reverberation chamber," *IEEE Internat. Symp. Electromagn. Compat., Honolulu, Hawaii, July* 2007.

[121] K. Rosengren and P.-S. Kildal, "Radiation efficiency, correlation, diversity gain, and capacity of a six monopole antenna array for a MIMO system: Theory, simulation and measurement in reverberation chamber," *Proc. Inst. Elect. Eng. Microwave, Antennas, Propag.*, vol. 152, pp. 7–16, 2005.

[122] U. Carlberg, P.-S. Kildal, A. Wolfgang, O. Sotoudeh, and C. Orienius, "Calculated and measured absorption cross sections of lossy objects in reverberation chambers," *IEEE Trans. Electromagn. Compat.*, vol. 46, pp. 146–154, 2004.

[123] P. Hallbjorner, U. Carlberg, K. Madsen, and J. Andersson, "Extracting electrical material parameters of electrically large dielectric objects from reverberation chamber measurements of absorption cross section," *IEEE Trans. Electromagn. Compat.*, vol. 47, pp. 291–303, 2005.

[124] H. Hashemi, "The indoor radio propagation channel," *Proc. IEEE*, vol. 81, pp. 943–968, 1993.

[125] R.A. Valenzuela, "A ray tracing approach to predicting indoor wireless transmission," *IEEE Vehicular Technology Conference*, pp. 214–218, 1993.

[126] D. Molkdar, "Review on radio propagation into and within buildings," *IEE Proc.-H*, vol. 38, pp. 197–210, 1959.

[127] L.P. Rice, "Radio transmission into buildings at 35 and 150 mc," *Bell Syst. Tech. J.*, vol. 38, pp. 197–210, 1959.

[128] P.I. Wells, "The attenuation of UHF radio signals by houses," *IEEE Trans. Vehicular Techn.*, vol. 26, pp. 358–362, 1977.

[129] D.C. Cox, R.R. Murray, and A.W. Norris, "Measurements of 800 MHz radio transmission into buildings with metallic walls," *Bell Syst. Tech. J.*, vol. 32, pp. 230–238, 1983.

[130] D.C. Cox, R.R. Murray, and A.W. Norris, "800-MHz attenuation measured in and around suburban houses," *AT&T Bell Lab. Tech. J.*, vol. 63, pp. 921–954,1984.

[131] T.S. Rappaport, *Wireless Communications: Principles and Practice*. Upper Saddle River, NJ: Prentice-Hall, 1996.

[132] J.B. Anderson, T.S. Rappaport, and S. Yoshida,"Propagation measurements and models for wireless communications channels," *IEEE Communications Magazine*, November 1994.

[133] S.Y. Seidel and T.S. Rappaport, "914 MHz path loss prediction models for wireless communications in multifloored buildings," *IEEE Trans. Antennas Propagat.*, vol. 40, pp. 207–217, 1992.

[134] C.L. Holloway, M.G. Cotton, and P. McKenna, "A model for predicting the power delay profile characteristics inside a room," *IEEE Trans. Vehicular Techn.*, vol. 48, pp. 1110–1120, 1999.

[135] D.A. Hill, M.L. Crawford, R.T. Johnk, A.R. Ondrejkea, and D.G. Camell,"Measurement of shielding effectiveness and cavity characteristics of airplanes," U.S. Nat. Inst. Stand. Technol. Interagency Report 5023, 1994.

[136] K.A. Remley, G. Koepke, C. Grosvenor, R.T. Johnk, J. Ladbury, D. Camell, and J. Coder,"NIST tests of the wireless environment on a production floor," Natl. Inst. of Stand. Technol. Tech. Note 1550, 2008.

[137] T.S. Rappaport, "Characterization of UHF multipath radio channels in factory buildings," *IEEE Trans. Antennas Propagat.*, vol. 37, pp. 1058–1069, 1989.

[138] J. Proakis, *Digital Communications*. New York: McGraw-Hill, 1983, Ch. 7.

[139] A.A.M. Saleh and R.A. Valenzuela, "A statistical model for indoor multipath propagation," *IEEE J. Selected Areas Commun.*, vol. 5, pp. 138–146, 1987.

[140] D.M.J. Devasirvatham, "Time delay spread and signal level measurements of 850 MHz radio waves in building environments," *IEEE Trans. Antennas Propagat.*, vol. 34, pp. 1300–1308, 1986.

[141] W. Jakes, Jr., *Microwave Mobile Communications*. New York: Wiley-Interscience, 1974.

[142] R.W. Young, "Sabine reverberation equation and power calculations," *J. Acoustic. Soc. Amer.*, vol. 31, pp. 912–921, 1959.

[143] L.M. Brekhovskikh, *Waves in Layered Media*. New York: Academic Press, 1960, Ch. 1.

[144] C.A. Balanis, *Advanced Engineering Electromagnetics*. New York: Wiley, 1989, Ch. 5.

[145] C.F. Eyring, "Reverberation time in dead rooms," *J. Acoustic. Soc. Amer.*, vol. 1, pp. 217–241, 1930.

[146] R.R. DeLyser, C.L. Holloway, R.J. Johnk, A.R. Ondrejka, and M. Kanda, "Figure of merit for low frequency anechoic chambers based on absorber reflection coefficients," *IEEE Trans. Electromag. Compat.*, vol. 38, pp. 576–584, 1996.

[147] E.K. Dunens and R.F. Lambert, "Impulsive sound-level response statistics in a reverberant enclosure," *J. Acoust. Soc. Amer.*, vol. 61, pp. 1524–1532, 1977.

[148] R.H. Espeland, E.J. Violette, and K.C. Allen,"Millimeter wave wideband diagnostic probe measurements at 30.3 GHz on an 11.8 km link," NTIA Tech. Memo. TM-83-95, U.S. Dept. Commerce, Boulder, CO, 1983.

[149] P.B. Papazian, Y. Lo, E.E. Pol, M.P. Roadifer, T.G. Hoople, and R.J. Achatz,"Wideband propagation measurements for wireless indoor communication," NTIA Rep. 93-292, U.S. Dept. Commerce, Boulder, CO, 1993.

[150] W.B. Westphal and A. Sils,"Dielectric constant and loss data," Tech. Rep. AFML-TR-72-39, MIT, Cambridge, 1972.

[151] A. Papoulis, *The Fourier Integral and Its Applications*. New York, McGraw-Hill, 1962.

[152] Ph. De Doncker and R. Meys, "Statistical response of antennas under uncorrelated plane wave spectrum illumination," *Electromagnetics*, vol. 24, 409–423, 2004.

[153] T. Lo and J. Litva, "Angles of arrival of indoor multipath," *Electronics Let.*, vol. 28, pp. 1687–1689, 1992.

[154] S. Guerin, "Indoor wideband and narrowband propagation measurements around 60.5 GHz in an empty and furnished room," *IEEE Vehicular Technol. Conf.*, pp. 160–164, 1996.

[155] J.-G. Wang, A.S. Mohan, and T.A. Aubrey, "Angles-of-arrival of multipath signals in indoor environments," *IEEE Vehicular Technol. Conf.*, pp. 155–159, 1996.

[156] Q. Spencer, M. Rice, B. Jeffs, and M. Jensen, "A statistical model for angle of arrival in indoor multipath propagation," *IEEE Vehicular Technol. Conf.*, pp. 1415–1419, 1997.

[157] Q. Spencer, M. Rice, B. Jeffs, and M. Jensen, "Indoor wideband time/angle of arrival multipath propagation results," *IEEE Vehicular Technol.Conf.*, pp. 1410–1414, 1997.

[158] C.L. Holloway, D.A. Hill, J.M. Ladbury, P.F. Wilson, G. Koepke, and J. Coder, "On the use of reverberation chambers to simulate a Rician radio environment for the testing of wireless devices," *IEEE Trans. Antennas Propagat.*, vol. 54, pp. 3167–3177, 2006.

[159] R. Steele, *Mobile Radio Communications.* New York: IEEE Press, 1974.

[160] G.D. Durgin, *Space-Time Wireless Channels.* Upper Saddle River, N.J.: Prentice Hall, 2003.

[161] K. Harima, "Determination of EMI antenna factor using reverberation chamber," *Proc. 2005 IEEE Int. Symp. Electromagn. Compat.*, Chicago, IL, pp. 93–96, 2005.

[162] D.L. Sengupta and V.V. Liepa, *Applied Electromagnetics and Electromagnetic Compatibility.* Hoboken, NJ: Wiley, 2006.

[163] J.E. Hansen, editor, *Spherical Near-Field Antenna Measurements.* London: Peter Perigrinus Ltd., 1988.

[164] S. Parker and L.O. Chua, "Chaos: A tutorial for engineers," *Proc. IEEE*, vol. 75, pp. 982–1008, 1987.

[165] A.J. Lichtenberg and M.A. Lieberman, *Regular and Stochastic Motion.* New York: Springer-Verlag, 1983.

[166] M.C. Gutzwiller, *Chaos in Classical and Quantum Mechanics.* New York: Springer-Verlag, 1990.

[167] E. Ott, *Chaos in Dynamical Systems.* Cambridge: Cambridge University Press, 1993.

[168] L.E. Reichl, *The Transition to Chaos.* New York: Springer-Verlag, 2004.

[169] I.M. Pinto, "Electromagnetic chaos: A tutorial," *Proc. 8th Int. Conf. Electromagnetics in Advanced Applications (ICEAA '03)*, Torino, Italy, pp. 511–514, 2003.

[170] T. Matsumoto, L.O. Chua,S. Tanaka, "Simplest chaotic nonautonomous circuit," *Phys. Rev. A*, vol. 30, pp. 1155–1157, 1984.

[171] T. Matsumoto, L.O. Chus, and M. Komuro, "The double scroll," *IEEE Trans. Circuits Syst.*, vol. 32, pp. 797–818, 1985.

[172] K. Ikeda, H. Daido, and O. Akimoto, "Optical turbulence: Chaotic behaviour of transmitted light from a ring cavity," *Phys. Rev. Let.*, vol. 45, pp. 709–712, 1980.

[173] A.J. Mackay, "Application of chaos theory to ray tracing in ducts," *Proc. IEE Radar, Sonar, Nav.*, vol. 164, pp. 298–304, 1999.

[174] V. Galdi, I.M. Pinto, and L.B. Felsen, "Wave propagation in ray-chaotic enclosures: Paradigns, oddities and examples," *IEEE Antennas Propagat. Mag.*, vol. 47, pp. 62–81, 2005.

[175] G. Castaldi, V. Fiumara, V. Galdi, V. Pierro, I.M. Pinto, and L.B. Felsen, "Ray-chaotic footprints in deterministic wave dynamics: A test model with coupled Floquet-type and ducted-type mode characteristics," *IEEE Trans. Antennas Propagat.*, vol. 53, pp. 753–765, 2005.

[176] R.G. Kouyoumjian, "Asymptotic high-frequency methods," *Proc. IEEE*, vol. 53, pp. 864–876, 1965.

[177] V.M. Babič and V.S. Buldyrev, *Short-Wavelength Diffraction Theory.* Berlin: Springer-Verlag, 1991.

[178] V.A. Borovikov and B.Ye. Kinber, *Geometrical Theory of Diffraction*. London: IEE, 1994.

[179] A.J. Mackay, "An application of chaos theory to the high frequency RCS prediction of engine ducts," in *Ultra-Wideband Short-Pulse Electromagnetics 5*, P.D. Smith and S.R. Cloude, Eds. New York: Kluwer/Academic, 2002, pp. 723–730.

[180] T. Kottos, U. Smilansky, J. Fortuny, and G. Nesti, "Chaotic scattering of microwaves," *Radio Sci.*, vol. 34, pp. 747–758, 1999.

[181] G. Orjubin, E. Richalot, O. Picon, and O. Legrand, "Chaoticity of a reverberation chamber assessed from the analysis of modal distributions obtained by FEM," *IEEE Trans. Electromagn. Compat.*, vol. 49, pp. 762–771, 2007.

[182] P. Šeba, "Wave chaos in singular quantum billiard," *Phys. Rev. Let.*, vol. 64, pp. 1855–1858, 1990.

[183] S. Hemmady, X. Zheng, E. Ott, T.M. Antonsen, and S.M. Anlage, "Universal impedance fluctuations in wave chaotic systems," *Phys. Rev. Let.*, vol. 94, pp. 014102-1–014102-4, 2005.

[184] S. Hemmady, X. Zheng, T.M. Antonsen, Jr., E. Ott, and S.M. Anlage, "Universal statistics of the scattering coefficient of chaotic microwave cavities," *Phys. Rev. E*, vol. 71, pp. 056215-1–056215-9, 2005.

[185] M.V. Berry, "Regular and irregular semiclassical wavefunctions," *J. Phys. A: Math. Gen.*, vol. 10, pp. 2083–2091, 1977.

[186] S.W. McDonald and A.N. Kaufman, "Wave chaos in the stadium: Statistical properties of the short-wave solution of the Helmholtz equation," *Phys. Rev. A*, vol. 37, pp. 3067–3086, 1988.

[187] L. Cappatta, M. Feo, V. Fiumara, V. Pierro, and I.M. Pinto, "Electromagnetic chaos in mode-stirred reverberation enclosures," *IEEE Trans. Electromagn. Compat.*, vol. 40, pp. 185–192, 1998.

[188] D.I. Wu and D.C. Chang,"An investigation of a ray-mode representation of the Green's function in a rectangular cavity," U.S. Nat. Bur. Stand. Tech. Note 1312, 1987.

[189] M.A.K. Hamid and W.A. Johnson, "Ray-optical solution for the dyadic Green's function in a rectangular cavity," *Electron. Let.*, vol. 6, pp. 317–319, 1970.

[190] G. Mie, "Beiträge zur optic trüber medien, speziell kolloidaler metallösungen," *Ann. Physik*, vol. 25, p. 377–445, 1908.

[191] V.H. Rumsey, *Frequency-Independent Antennas*. New York: Academic Press, 1966.

[192] D.A. Hill, "Electromagnetic scattering by buried objects of low contrast," *IEEE Trans. Geosci. Rem. Sens.*, vol. 6, pp. 195–203, 1988.

RELATED REFERENCES

General Cavities and Applications

G. Goubau, *Electromagnetic Waveguides and Cavities*. New York: Pergamon Press, 1961.

D.S. Jones, *The Theory of Electromagnetism*. New York: MacMillan, 1964, Ch. 4.

C.G. Montgomery, R.H. Dicke, and E.M. Purcell (eds.), *Principles of Microwave Circuits*. London: Peter Perigrinus, 1987, Ch.7 by R. Beringer.

D.M. Pozar, *Microwave Engineering*. Reading, MA: Addison Wesley, 1990, Ch. 7.

S. Ramo and J.R. Whinnery, *Fields and Waves in Modern Radio*. New York: Wiley, 1953, Ch. 10.

S.A. Schelkunoff, *Electromagnetic Waves*. Princeton: Van Nostrand, 1960, Chs. VIII and X.

J.C. Slater, *Microwave Electronics*. New York: Van Nostrand, 1954, Chs. IV–VII.

W.R. Smythe, *Static and Dynamic Electricity*. New York: McGraw-Hill, 1968, pp. 526–545.

Dyadic Green's Functions

K.M. Chen, "A simple physical picture of tensor Green's function in source region," *Proc. IEEE*, vol. 65, pp. 1202–1204, 1977.

R.E. Collin, "On the incompleteness of E and H modes in waveguides," *Can. J. Phys.*, vol. 51, pp. 1135–1140, 1973.

R.E. Collin, "Dyadic Green's function expansions in spherical coordinates," *Electromagnetics*, vol. 6, pp. 183–207, 1986.

J.G. Fikioris, "Electromagnetic field inside a current-carrying region," *J. Math. Phys.*, vol. 6, pp. 1617–1620, 1965.

W.A. Johnson, A.Q. Howard, and D.G. Dudley, "On the irrotational component of electric Green's dyadic," *Rad. Sci.*, vol. 14, pp. 961–967, 1979.

M. Kisliuk, "The dyadic Green's functions for cylindrical waveguides and cavities," *IEEE Trans. Microwave Theory Tech.*, vol. 28, pp. 894–898, 1980.

D.E. Livesay and K.M.Chen, "Electromagnetic fields induced inside arbitrarily shaped biological bodies," *IEEE Trans. Microwave Theory Tech.*, vol. 22, pp. 1273–1280, 1974.

P.H. Pathak, "On the eigenfunction expansion of electromagnetic dyadic Green's functions," *IEEE Trans. Antenn. Propagat.*, vol. 31, pp. 837–846, 1983.

L.W. Pearson, "On the spectral expansion of the electric and magnetic dyadic Green's functions in cylindrical coordinates," *Rad. Sci.*, vol. 18, pp. 166–174, 1983.

Y. Rahmat-Samii, "On the question of computation of the dyadic Green's function at the source region in waveguides and cavities," *IEEE Trans. Microwave Theory Tech.*, vol. 23, pp. 762–765, 1975.

C.T. Tai, "On the eigenfunction expansion of dyadic Green's functions," *Proc. IEEE*, vol. 61, pp. 480–481, 1973.

C.T. Tai and P. Rozenfeld, "Different representations of dyadic Green's functions for a rectangular cavity," *IEEE Trans. Microwave Theory Tech.*, vol. 24, pp. 597–601, 1976.

C.T. Tai, "Equivalent layers of surface charge, current sheet, and polarization in the eigenfunction expansion of Green's Functions in electromagnetic theory," *IEEE Trans. Antenn. Propagat.*, vol. 29, pp. 733–739, 1981.

J.J.H. Wang, "Analysis of a three-dimensional arbitrarily shaped dielectric or biological body inside a rectangular waveguide," *IEEE Trans. Microwave Theory Tech.*, vol. 26, pp. 457–462, 1978.

J.J.H. Wang, "A unified and consistent view on the singularities of the electric dyadic Green's function in the source region," *IEEE Trans. Antenn. Propagat.*, vol. 30, pp. 463–468, 1982.

D.I. Wu and D.C. Chang, "A hybrid representation of the Green's function in an overmoded rectangular cavity," *IEEE Trans. Microwave Theory Tech.*, pp. 1334–1342, 1988.

A.D. Yaghjian, "A delta-distribution derivation of the electric field in the source region," *Electromagnetics*, vol. 2, pp. 161–167, 1982.

Reverberation Chambers

Electromagnetic Compatibility (EMC): Part 4: Testing and Measurement Techniques: Section 21: Reverberation Chambers, International Electrotechnical Commission Standard JWG REV SC77B-CISPR/A, IEC 61000-4-21, Geneva, Switzerland, 2003.

L.R. Arnaut, "Effect of local stir and spatial averaging on the measurement and testing in mode-tuned and mode-stirred reverberation chambers," *IEEE Trans. Electromagn. Compat.*, vol. 43, pp. 305–325, 2001.

L.R. Arnaut, "Compound exponential distributions for undermoded reverberation chambers," *IEEE Trans. Electromagn. Compat.*, vol. 44, pp. 442–457, 2002.

L.R. Arnaut, "Limit distributions for imperfect electromagnetic reverberation," *IEEE Trans. Electromagn. Compat.*, vol. 45, pp. 357–377, 2003.

L.R. Arnaut, "On the maximum rate of fluctuation in mode-stirred reverberation," *IEEE Trans. Electromagn. Compat.*, vol. 47, pp. 781–804, 2005.

L.R. Arnaut, "Effect of size, orientation, and eccentricity of mode stirrers on their performance in reverberation chambers," *IEEE Trans. Electromagn. Compat.*, vol. 48, pp. 600–602, 2006.

L.R. Arnaut, "Time-domain measurement and analysis of mechanical step transitions in mode-tuned reverberation: characterization of instantaneous field," *IEEE Trans. Electromagn. Compat.*, vol. 49, pp. 772–784, 2007.

L.R. Arnaut and D.A. Knight, "Observation of coherent precursors in pulsed mode-stirred reverberation fields," *Phys. Rev. Lett.*, vol. 98, 053903, 2007.

L.R. Arnaut and P.D. West, "Electromagnetic reverberation near a perfectly conducting boundary," *IEEE Trans. Electromagn. Compat.*, vol. 48, pp. 359–371, 2006.

C.F. Bunting, "Statistical characterization and the simulation of a reverberation chamber using finite-element techniques," *IEEE Trans. Electromagn. Compat.*, vol. 44, pp. 214–221, 2002.

G. Ferrara, M. Migliaccio, and A. Sorrentino, "Characterization of GSM non-line-of-sight propagation channels generated in a reverberating chamber by using bit error rates," *IEEE Trans. Electromagn. Compat.*, vol. 49, pp. 467–473, 2007.

G. Gradoni, F. Moglie, A.P. Pastore, and V.M. Primiani, "Numerical and experimental analysis of the field to enclosure coupling in reverberation chamber and comparison with anechoic chamber," *IEEE Trans. Electromagn. Compat.*, vol. 48, pp. 203–211, 2006.

P. Hallbjörner, "Estimating the number of independent samples in reverberation chamber measurements from sample differences," *IEEE Trans. Electromagn. Compat.*, vol. 48, pp. 354–358, 2006.

M. Höijer, "Maximum power available to stress onto the critical component in the equipment under test when performing a radiated susceptibility test in the reverberation chamber," *IEEE Trans. Electromagn. Compat.*, vol. 48, pp. 372–384, 2006.

P.-S. Kildal and K. Rosengren, "Electromagnetic analysis of effective and apparent diversity gain of two parallel dipoles," *IEEE Antennas and Wireless Propagation Letters*, vol. 2, pp. 9–13, 2003.

P.-S. Kildal, K. Rosengren, J. Byun, and J. Lee, "Definition of effective diversity gain and how to measure it in a reverberation chamber," *Microwave and Optical Technology Letters*, vol. 34, pp. 56–59, 2002.

H.G. Krauthäuser, "On the measurement of total radiated power in uncalibrated reverberation chambers," *IEEE Trans. Electromagn. Compat.*, vol. 49, pp. 270–279.

C. Lemoine, P. Besnier, and M. Drissi, "Investigation of reverberation chamber measurements through high-power goodness-of-fit tests," *IEEE Trans. Electromagn. Compat.*, vol. 49, pp. 745–755, 2007.

G. Lerosey and J. de Rosny, "Scattering cross section measurement in reverberation chamber," *IEEE Trans. Electromagn. Compat.*, vol. 49, pp. 280–284, 2007.

M. Lienard and P. Degauque, "Simulation of dual array multipath cannels using mode-stirred reverberation chambers," *Electronics Letters*, vol. 40 pp. 578–579, 2004.

F. Moglie and A.P. Pastore, "FDTD analysis of plane wave superposition to simulate susceptibility tests in reverberation chambers," *IEEE Trans. Electromagn. Compat.*, vol. 48, pp. 195–202, 2006.

G. Orjubin, "Maximum field inside a reverberation chamber modeled by the generalized extreme value distribution," *IEEE Trans. Electromagn. Compat.*, vol. 49, pp. 104–113, 2007.

G. Orjubin, "On the FEM modal approach for a reverberation chamber analysis," *IEEE Trans. Electromagn. Compat.*, vol. 49, pp. 76–85, 2007.

M. Otterskog and K. Madsen, "On creating a nonisotropic propagation environment inside a scattered field chamber," *Microwave and Optical Technology Letters*, vol. 43, pp. 192–195, 2004.

V.M. Primiani, F. Moglie, and A.P. Pastore, "A metrology application of reverberation chambers: the current probe calibration," *IEEE Trans. Electromagn. Compat.*, vol. 49, pp. 114–122, 2007.

K. Rosengren and P.-S. Kildal, "Study of distributions of modes and plane waves in reverberation chambers for characterization of antennas in multipath environment," *Microwave and Optical Technology Letters*, vol. 30, pp. 386–391, 2001.

K. Rosengren, P.-S. Kildal, C. Carlsson, and J. Carlsson, "Characterization of antennas for mobile and wireless terminals in reverberation chambers: improved accuracy by platform stirring," *Microwave and Optical Technology Letters*, vol. 30, pp. 391–397, 2001.

E. Voges and T. Eisenburger, "Electrical mode stirring in reverberating chambers by reactively loaded antenna," *IEEE Trans. Electromagn. Compat.*, vol. 49, pp. 756–761, 2007.

N. Wellander, O. Lundén, and M. Bäckstrom, "Experimental investigation and mathematical modeling of design parameters for efficient stirrers and mode-stirred reverberation chambers," *IEEE Trans. Electromagn. Compat.*, vol. 49, pp. 94–103, 2007.

Z. Yuan, J. He, S. Chen, R. Zeng, and T. Li, "Evaluation of transmit antenna position in reverberation chamber," *IEEE Trans. Electromagn. Compat.*, vol. 49, pp. 86–93, 2007

Aperture Penetration

C.L. Andrews, "Diffraction pattern in a circular aperture measured in the microwave region," *J. Appl. Phys.*, vol. 22, pp. 761–767, 1950.

F. Bekefi, "Diffraction of electromagnetic waves by a aperture in a large screen," *J. Appl. Phys.*, vol. 24, pp. 1123–1130, 1953.

C.F. Bunting and S.-H. Yu, "Field penetration in a rectangular box using numerical techniques: an effort to obtain statistical shielding effectiveness," *IEEE Trans. Electromagn. Compat.*, vol. 46, pp. 160–168, 2004.

C.M. Butler and K.R. Umashandar, "Electromagnetic excitation of a wire through an aperture-perforated, conducting screen," *IEEE Trans. Antennas propagat.*, vol. 25, pp. 456–462, 1976.

W.P. Jr., Carpes L. Pichon, and A. Razek, "Analysis of the coupling of an incident wave with a wire inside a cavity using an FEM in frequency and time domains," *IEEE Trans. Electromagn. Compat.*, vol. 44, pp. 470–475, 2002.

C.C. Chen, "Transmission of microwave through perforated flat plates of finite thickness," *IEEE Trans. Microwave Theory Tech.*, vol. 21, pp. 1–6, 1973.

S.B. Cohn, "The electric polarizability of apertures of arbitrary shape," *Proc. IRE*, vol. 40, 1069–1071, 1952.

F. De Meulenaere and J. Van Bladel, "Polarizability of some small apertures," *IEEE Trans. Anennas Propagat., IEEE Trans. Anntennas Propagat.*, vol. 25, 198–205, 1977.

R.A. Hurd and B.K. Sachdeva, "Scattering by a dielectric loaded slit in a conducting plane," *Radio Sci.*, vol. 10, pp. 565–572, 1975.

S.N. Karp and A. Russek, "Diffraction by a wide slit," *J. Appl. Phys.*, vol. 27, pp. 886–894, 1956.

S.C. Kashyap, M.A. Hamid, and N.J. Mostowy, "Diffraction pattern of a slit in a thick conducting screen," *J. App. Phys.*, pp. 894–895, 1971.

S.C. Kashyap and M.A.K. Hamik, "Diffraction characteristics of a slit in a thick conducting screen," *IEEE Trans. Antennas Propagat.*, vol. 19, pp. 499–507, 1971.

G.F. Koch and K.S. Kolbig, "The transmission coefficient of elliptical and rectangular apertures for electromagnetic waves," *IEEE Trans. Antennas Propagat.*, vol. 16, 1968.

K.C. Lang, "Babinet's principle for a perfectly conducting screen with aperture covered by resistive sheet," *IEEE Trans. Antennas Propagat.*, vol. 21, pp. 738–740, 1973.

C. Lertsirmit, D.R. Jackson, and D.R. Wilton, "An efficient hybrid method for calculating the EMC coupling to a device on a printed circuit board inside a cavity," *Electromagnetics*, vol. 25, pp. 637–654, 2005.

C. Lertsirmit, D.R. Jackson, and D.R. Wilton, "Time domain coupling to a device on a printed circuit board inside a cavity," *Rad. Sci.*, vol. 40, RS6S14, 2005.

J.L. Lin, W.L. Curtis, and M.C. Vincent, "On the field distribution of an aperture," *IEEE Trans. Antennas Propagat.*, vol. 22, pp. 467–471, 1974.

N.A. McDonald, "Electric and magnetic coupling through small apertures in shield walls of any thickness," *IEEE Trans. Microwave Theory Tech.*, vol. 20, pp. 689–695, 1972.

R.F. Millar, "Radiation and reception properties of a wide slot in a parallel-plate transmission line, Parts I and II," *Can. J. Phys.*, vol. 37, pp. 144–169, 1959.

F.L. Neerhoff and G. Mur, "Diffraction of a plane electromagnetic wave by a slit in a thick screen placed between two different media," *Appl. Sci. Res.*, vol. 28, pp. 73–88, 1973.

V. Rajamani, C.F. Bunting, M.D. Deshpande, and Z.A. Khan, "Validation of modal/MoM in shielding effectiveness studies of rectangular enclosures with apertures," *IEEE Trans. Electromagn. Compat.*, vol. 48, pp. 348–353, 2006.

Y. Rahmat-Samii and R. Mittra, "Electromagnetic coupling through small apertures in a conducting screen," *IEEE Trans. Antennas Propagat.*, vol. 25, pp. 180–187, 1977.

H.H. Snyder, "On certain wave transmission coefficients for elliptical and rectangular apertures," *IEEE Trans. Antennas Propagat.*, vol. 17, pp. 107–109, 1969.

T. Teichmann and E.P. Wigner, "Electromagnetic field expansions in loss-free cavities excited through holes," *J. Appl. Phys.*, vol. 24, pp. 262–267, 1953.

J. Van Bladel, "Small holes in a waveguide wall," *Proc. Inst. Elec. Eng.*, vol. 118, pp. 43–50, 1971.

Indoor Wireless Propagation

S.E. Alexander, "Radio propagation within buildings at 900 MHz," *Electronics Let.*, vol. 18, pp. 913–914, 1982.

R.J.C. Bultitude, "Measurement, characterization and modeling of indoor 800/900 MHz radio channels for digital communications," *IEEE Commun. Mag.*, vol. 25, pp. 5–12, 1987.

R.J.C. Bultitude, S.A. Mahmoud, and W.A. Sullivan, "A comparison of indoor radio propagation characteristics at 910 MHz and 1.75 GHz," *IEEE J. Select. Areas in Comm.*, vol. 7, pp. 20–30, 1989.

COST231: "Digital mobile radio towards future generation systems," Final Report, European Commission, 1999.

D.C. Cox, R.R. Murray, and W.W. Norris, "Antenna height dependence of 800 MHz attenuation measured in houses," *IEEE Trans. Vehicular Tech.*, vol 34, pp. 108–115, 1985.

D.C. Cox, R.R. Murray, H.W. Arnold, A.W. Norris, and M.F. Wazowics, "Cross-polarization coupling measured for 800 MHz radio transmission in and around houses and large buildings, *IEEE Trans. Antennas Propagat.*, vol 34, pp. 83–87, 1986.

D.M.J. Devasirvatham, "Time delay spread measurements of wideband radio signals within a building," *Electronics Let.*, vol. 20, pp. 950–951, 1984.

D.M.J. Devasirvatham, "A comparison of time delay spreadand signal level measurements within two dissimilar office buildings," *IEEE Trans. Antennas Propagat.*, vol. 35, 319–324, 1987.

D.M.J. Desasirvatham, R.R. Murray, and C. Banerjee, "Time delay spread measurements at 850 MHz and 1.7 GHz inside a metropolitan office building," *Electronics Let.*, vol. 25, pp. 194–196, 1989.

R. Ganesh and K. Pahlavan, "On the arrival of paths in fading multipath indoor radio channels," *Electronics Let.*, vol. 25, pp. 763–765, 1989.

D.A. Hawbaker and T.S. Rappaport, "Indoor wideband radiowave propagation measurements at 1.3 GHz and 4.0 GHz," *Electronics Let.*, vol. 26, pp., 1990.

H.H. Hoffman and D.C. Cox, "Attenuation of 900 MHz radio waves propagating into a metal building," *IEEE Trans. Antennas Propagat.*, vol. 30, pp 808–811, 1982.

J. Horikoshi, K. Tanaka, and T. Morinaga, "1.2 GHz band wave propagation measurements in concrete buildings for indoor radio communications," *IEEE Trans. Vehicular Tech.*, vol. 35, pp. 146–152, 1986.

S.J. Howard and K. Pahlavan, "Doppler spread measurements of the indoor radio channel," *Electronics Let.*, vol. 26, pp. 107–109, 1990.

S.J. Howard and K. Pahlavan, "Measurement and analysis of the indoor radio channel in the frequency domain," *IEEE Trans. Instrumentation Meas.*, vol. 39, pp. 751–755, 1990.

S.J. Howard and K. Pahlavan, "Autoregressive modeling of wide-band indoor radio propagation," *IEEE Trans. Commun.*, vol. 40, pp. 1540–1552, 1992.

K. Pahlavan, R. Ganesh, and T. Hotaling, "Multipath propagation measurements on manufacturing floors at 910 MHz," *Electronics Let.*, vol. 25, pp. 225–227, 1989.

K. Pahlavan and S.J. Howard, "Statistical AR models for the frequency selective indoor radio channel," *Electronics Let.*, vol. 26, pp. 1133–1135, 1990.

D.A. Palmer and A.J. Motley, "Controlled radio coverage within buildings," *British Telecomm. Technol. J.*, vol. 4, pp. 55–57, 1986.

T.S. Rappaport and C.D. McGillem, "Characterizing the UHF factory radio channel," *Electronics Let.*, vol. 23, pp. 1015–1016, 1987.

T.S. Rappaport, and C.D. McGillem, "UHF fading in factories," *IEEE J. Selected Areas Communications*, vol. 7, pp. 40–48, 1989.

T.S. Rappaport, S.Y. Siedel, and K. Takamizawa, "Statistical channel impulse response models for factory and open plan building radio communication system design," *IEEE Trans. Communications*, vol. 39, pp. 794–807, 1991.

T.A. Russel, C.W. Bostian, and T.S. Rappaport, "A deterministic approach to predicting microwave diffraction by buildings for microcellular systems," *IEEE Trans. Antennas Propagat.*, vol. 41, pp. 1640–1649, 1993.

A.M. Saleh and R. Valenzuela, "A statistical model for indoor multipath propagation," *IEEE J. Selected Areas Communications*, vol. 5, pp. 128–137, 1987.

S.Y. Seidel and T.S. Rappaport, "Site-specific propagation prediction for wireless in-building personal communication system design," *IEEE Trans. Vehicular Tech.*, vol. 43, 1994.

T.A. Sexton and K. Pahlavan, "Channel modeling and adaptive equalization of indoor radio channels," *IEEE J. Selected Areas Commun.*, vol. 7, pp. 114–120, 1989.

P.F.M. Smulders and A.G. Wagemans, "Wideband indoor radio propagation measurement at 58 GHz," *Electron. Letters*, vol. 28, pp. 1270–1272, 1992.

S.R. Todd, M.S. El-Tanany, and S.A. Mahmoud, "Space and frequency division measurements of 1.7 GHz indoor radio channel using a four-branch receiver," *IEEE Trans. Vehic. Techn.*, vol. 41, pp. 312–320, 1992.

A.M.D. Turkmani and A.F. Toledo, "Radio transmission at 1800 MHz into and within multistory buildings," *IEE Proc.-Part I*, vol. 138, pp. 577–584, 1991.

INDEX

Printed and bound by CPI Group (UK) Ltd, Croydon, CR0 4YY

16/09/2026

14963412-0002